SYSTEMS APPROACH TO MANAGEMENT OF DISASTERS

. . . The weakest thing in the world
Overcomes the strongest thing in the world
What doesn't exist finds room where there's none
Wordless instruction
Effortless help
Few in the world can match this.

Lao Tzu (6th century BC)
Tao Te Ching (verse 43)

. . . To understand all this, it is necessary that
we should learn to look at things from a new
point of view.

D.T. Suzuki (1870–1966)
Living by Zen

... The truly noble and resolved spirit raises
itself, and becomes more conspicuous in
times of disaster and ill-fortune.

Plutarch (46 AD – 120 AD)
Lives, Eumenes

SYSTEMS APPROACH TO MANAGEMENT OF DISASTERS

Methods and Applications

Slobodan P. Simonović

A JOHN WILEY & SONS, INC., PUBLICATION

Published by John Wiley & Sons, Inc., Hoboken, New Jersey.
Published simultaneously in Canada.

Library of Congress Cataloging-in-Publication Data

Simonović, Slobodan P.
Systems approach to management of disasters: methods and applications / Slobodan P. Simonović.
p. cm.
Includes bibliographical references and index.
ISBN 978-0-470-52809-9 (cloth)
1. Emergency management. 2. System analysis. I. Title
HD49.S58 2011
363.34′52–dc22 2010016635

Printed in the United States of America

10 9 8 7 6 5 4 3 2 1

To Tanja, Dijana, and Damjan

Contents

List of Figures and Tables

Figures

Tables

About the Author

Slobodan P. Simonović was born and raised in Belgrade, Yugoslavia. He obtained his undergraduate degree in civil engineering (water resources division) from the University of Belgrade in 1974. By joining the interdisciplinary master's degree program at the University of Belgrade, he was able to direct his education further into the application of formal systems theory to water resources systems management. His MSc included training in formal systems theory from the Department of Electrical Engineering and in water resources engineering from the Department of Civil Engineering. He graduated in 1976.

From 1974 until 1978, he worked as a researcher for the Jaroslav Cerni Institute for Water Resources Development in Belgrade. In 1978, he continued his graduate education at the University of California in Davis, where he obtained his PhD in engineering in 1981. Until 1986, he worked as consulting engineer for the large international consulting company Energoproject in Belgrade. In 1986, he moved to Canada and joined the Department of Civil Engineering at the University of Manitoba in Winnipeg. He was professor in the department until 1996, when he became the Director of the Natural Resources Institute, an interdisciplinary graduate program in natural resources management at the University of Manitoba.

In 2000, Dr Simonović moved to London, Ontario, where he became professor at the Department of Civil and Environmental Engineering and an engineering research chair of the Institute for Catastrophic Loss Reduction at the University of Western Ontario.

Dr Simonović teaches courses in civil engineering and water resources systems. He still actively works for national and international professional organizations (the Canadian Society of Civil Engineers, Canadian Water Resources Association, International Association of Hydrological Sciences, and International Hydrologic Program of UNESCO). He has received a number of awards for excellence in teaching, research and outreach, and has been invited to present special courses for practicing water resources engineers in many countries. He currently serves as associate editor of four water resources journals, and participates actively in the organization of national and international meetings. He has more than 350 professional publications.

Dr Simonović's primary research interest focuses on the application of a systems approach to, and the development of decision support tools for, the management of complex water and environmental systems. Most of his work is related to the integration of risk, reliability, uncertainty, simulation, and optimization in hydrology and water resources management. He has undertaken applied research projects that integrate mathematical modeling, database management, geographical information systems, and intelligent interface development into decision support tools for water resources decision makers. Most of his research is conducted through the Facility for Intelligent Decision Support (FIDS) at the University of Western Ontario.

His subject expertise focuses on systems modeling, risk and reliability, water resources and environmental systems analysis, computer-based decision support systems development, and water resources education and training. Particular topical areas of expertise are reservoirs, flood control, hydropower energy, operational hydrology, and climate change.

Foreword

Alleluia! This is not a conventional way in which to begin a Foreword, but in this case, it is justified. Slobodan Simonović has produced a book that is a bold and vital step toward revising, even revolutionizing, our approach to disaster management. This helps in the movement toward the establishment of a recognized professional domain with appropriate theory, methods, and a community of practice. It is long overdue. At the same time, it has to be said that Slobodan's effort is not the first, and nor will it be the last. It gives a big push in the right direction. It should be widely read and widely used by students and professionals.

The problem of disasters and their management has proved to be highly intractable. The frequency and magnitude of disasters are growing worldwide. Scientific knowledge of the natural forces in the earth's crust and in the atmosphere that initiate disaster events has expanded enormously in the last 50 years. So has our capacity to predict, to make accurate forecasts, and to issue warnings. Also our knowledge of where extreme events are more likely to occur, and of the available measures for mitigation (or what the climate change community refers to as "adaptation"), has grown significantly. We have the scientific understanding, the technology, and the wealth to be able to mitigate disasters much more effectively. But instead things continue to get worse at an apparently accelerating rate.

The international community of scientists and policymakers has known this for a long time. The decade of the 1990s was designated by the United Nations as the International Decade for Natural Disaster Reduction. Since 2000, a new UN coordinating body—the International Strategy for Disaster Reduction—has been leading the global effort to contain disaster losses. And yet the losses continue to climb at an accelerating rate.

There are many ideas and theories of why this continues to happen. Common to them is a recognition that we have failed to effectively use and apply internationally, at the national level, and in many communities the knowledge we have. At the same time, it is recognized that we do not adequately understand the causes of this persistent failure. Slobodan Simonović's book is based on the idea that part of the explanation at least is that we have not properly integrated the knowledge and methods available to us. Our disaster management is still too highly fragmented into the stovepipes and silos of different disciplines, areas of expertise, and administration. It remains bedeviled by reliance on short-term emergency response and myopic short-term thinking. Simonović argues cogently for the use of a systems approach, and much of what he writes is a demonstration and a set of guides and instructions on how to develop and apply such an approach, incorporating optimization, simulation, and multiple-objective analysis. Each chapter is accompanied by a set of exercises intended for student use.

Slobodan has substantial experience in the application of systems analysis in the water resources management field. More recently he has been heavily involved in disaster risk management in Canada, especially the flood risks in Winnipeg, Manitoba, and so has turned his attention to the development of a systems approach to disaster management. Although the book may have special relevance to Canadians, its message and its contents are universal. Human response to disasters has long been dominated by humanitarian relief and rehabilitation and emergency preparedness. The trend toward addressing the multiple and complex causes of rising disaster losses has been underway for some time. The book is a further reinforcement of this trend. Slobodan Simonović is modest about his contribution. He recognizes that there is need to further strengthen and broaden the approach presented here. He specifically recognizes the need to incorporate the increased risks associated with climate change and to deepen our understanding of the dimensions of population growth and migration and community-level resilience. He also recognizes the need for better treatment of uncertainty. In sum, we are observing the slow and steady emergence of a new profession—that of disaster risk management. Slobodan Simonović's book is a vital contribution to the foundations and practice of this field.

Ian Burton, PhD, FRSC
Emeritus Professor, University of Toronto
Scientist Emeritus, Meteorological Service of Canada

This must be the first book ever published on systems analysis approach to disaster management. It is timely and precious. The author must be congratulated, appreciated, and encouraged for his further lead in this increasingly important subject.

The author has long experiences of applying systems analyses technology in the field of water resources management. In this book, his vast knowledge on systems analyses has been reintegrated into disaster management, specifically on systems dynamics simulation, linear programming, and multiobjective analyses. Concrete examples such as simulation of evacuation procedures, optimal placement of evacuation assets, and selection of flood protection alternatives, illustrated from the author's practical experiences, are quite valuable. The evacuation simulation in Chapter 5 drew my special attention as it shows how to analyze the effects of "warning methods and mode of evacuation order dissemination," a very important practical question in emergency response.

The author humbly says that they are still rather simple, but all the examples would certainly serve for mind broadening and refreshing of practitioners who are facing complicated problems in everyday disaster management. This book eloquently demonstrates how remarkable systems thinking might be in sustainable management of a complex system with "equity, efficiency, and integrity."

The author has long and diverse experiences of witnessing flood disasters and consulting with municipal flood disaster management agencies in countries such as Serbia, Canada, China, and Egypt. In particular, his experience of "Flood of the

Century" of the Red River in 1997 in Manitoba made him extensively engage in the International Joint Commission (IJC). The causes and effects of this great flood are vividly described in Chapter 1. This experience, I believe, has become the basis of his compiling this book.

At UNESCO-ICHARM (International Center for Water Hazard and Risk Management), Tsukuba, Japan, we have a graduate program on water-related disaster management. Students are practical engineers in the water field. I am glad to see Chapters 2 and 3 of this book introducing concepts of integrated approach and systems approach, which best suit those who start learning disaster management, and for students interested in systems tools, the CD-ROM that goes with the book illustrates the principles with numerical examples.

I hope this book will be used by many students and practitioners in disaster management to step ahead for better integrated disaster management in societies at risk.

Kuniyoshi Takeuchi, Ph.D.,
Director, International Center for Water Hazard and Risk
Management under the auspices of UNESCO (ICHARM)
Tsukuba, Japan

Preface

I am one of the lucky few who have the opportunity to work all their professional life in an area that they enjoy. The most enjoyable activity for me is to integrate the knowledge from different fields into an approach for solving complex problems. My work has brought me into contact with many great people, responsible professionals, talented engineers, capable managers, and dedicated politicians. In my capacity as an academic I have also had an opportunity to work with the abundant young talent that continues to feed the workforce. I learned a lot from all these people. I learned many things about the profession, I learned a lot about different cultures, and most importantly, I learned about life. Thank you.

My interest in natural disasters as one would expect grew from my main area of expertise—water resources systems management. From early days of my professional carrier I was involved with floods and flood management, first from the engineering point of view and then later from the management point of view. Flood problems along Morava, Sava, and Danube rivers in my country of origin—Serbia—were among the first professional challenges I had to deal with, after graduation. In 1997, I was teaching at the University of Manitoba and living in Winnipeg. That was the year of the "Flood of the Century." The governments of Canada and the United States have agreed that steps must be taken to reduce the impact of future flooding on the Red River. In June 1997, they asked the International Joint Commission (IJC) to analyze the causes and effects of the Red River flood of that year. The IJC appointed the International Red River Basin Task Force to examine a range of alternatives to prevent or reduce future flood damage. I was appointed to the task force and the following experience changed my life.

My work has taken me all over the world. I have had an opportunity to see the water problems in the developed and developing world, in small villages and large urban centers. Projects I have been involved with range in scale from the local to the international. I have discussed the flooding issues with farmers of the Sihu area in China as well as the Minister for Irrigation and Water Resources of Egypt. I hope that my professional expertise continues to contribute to the solution of some of these problems. It definitely inspires me to continue to work with greater effort and more dedication.

For more than 30 years of personal research, consulting, teaching, involvement in policy, implementation of projects, and presentation of experiences through the pages of many professional journals, I have worked hard to raise the awareness of the importance of interdisciplinary approach to the solution of complex problems. The main thrust of my work is the use of systems approach in dealing with complexity.

I have accumulated tremendous experience over the years. In that time I realized that there is an opportunity to contribute to the area of disaster management by transferring some of the knowledge and experience from the implementation of the systems thinking and systems tools to various steps of the disaster management cycle. Writing this book offered me a moment of reflection, and it elaborates on lessons learned from the past to develop ideas for the future.

The main goal of this book is to introduce the systems approach to the disasters management community as an alternative approach that can provide support for interdisciplinary activities involved in the management of disasters. The systems approach draws on the fields of operations research and economics to create skills in solving complex management problems. The field of operations research evolved from its origins during the Second World War, and the area known as mathematical programming found wide application as a means to simulate and optimize complex design and operational problems in many fields (of natural, social and health sciences and engineering). A primary emphasis of systems analysis in disaster management as I see it is on providing an improved basis for decision making. A large number of analytical, computer-based tools, from simulation and optimization to multiobjective analysis, are available for formulating, analyzing, and solving disaster management problems.

Large and more frequent disasters in last few decades have brought a remarkable transformation of attitude by the disaster management community toward integration of economic, social and environmental concerns related to disasters, and of action to deal with them.

The early period of hazards research was characterized by taking knowledge from various fields of science and engineering that is applicable to natural and related technological hazards and using it in disaster management. The most significant contribution in the last 10 years is a fundamental shift in the character of how the citizens, communities, governments, and businesses conduct themselves in relation to the natural environment they occupy. Pressures from a growing population and the associated needs for food production and rapid urbanization contribute to an exponential increase in human and material losses from natural and technological disasters.

Disaster management being divided among disciplinary boundaries has faced an uphill battle with the regulatory approaches that are used in many countries around the world. They have not been conducive to the integrative character of the systems approach that is inherent in simulation and optimization management models. Fortunately, recent trends in regulation include consideration of the entire region under threat, explicit consideration of all costs and benefits, elaboration of a large number of alternatives to reduce the damages, and the greater participation of all stakeholders in decision making. Systems approaches based on simulation, optimization, and multiobjective analyses have great potential for providing appropriate support for effective disaster management in this emerging context.

In 1987, with the publication of the Brundtland Commission's report *Our Common Future*, decision making in many fields began to be influenced by a sustainability paradigm. It can safely be assumed that sustainability is now the major unifying concept promoted, accepted, and discussed by governments throughout most of the world. The original report introduced the concept of sustainable development as "the ability to meet the needs of the present without compromising the needs of future generations". This concept as applied to contemporary hazards mitigation aims at implementing approaches that could result in disaster-resilient communities.

Applying the principles of sustainability to disaster decision making requires major changes in the objectives on which decisions are based, and an understanding of the complicated interrelationships between existing ecological, economic, and social factors. The broadest objectives for achieving sustainability are equity, economic efficiency, and environmental integrity. In addition, sustainable decision making regarding natural hazards faces the challenge of time; that is, it must identify and account for long-term consequences.

To make disaster management decisions designed to produce sustainable disaster-resilient, communities also calls for a change in procedural policies and implementation. If the choice is to select projects with this outcome, it will require major changes in both substantive and procedural policies. Sustainability is an integrating process. It encompasses technology, ecology, and the social and political infrastructure of society. It is not a state that may ever be reached completely. It is, however, one for which the disaster management community and decision makers strive.

The evolution of disaster management is occurring in the context of rapid technological change. In the same period that brought us the systems approach, environmental awareness, and sustainability, we were exposed to the dynamic development of computer hardware and software systems. The power of the large mainframe computers of the early 1970s is now exceeded many times over by the average laptop computer. The computer has moved out of data processing, through the user's office and into knowledge processing. Whether it takes the form of a laptop personal computer or a desktop multiprocessing workstation is not important. The important point is that the computer acts as a partner for more effective decision making.

Systems can be defined as a collection of various structural and nonstructural elements that are connected and organized in such a way as to achieve some specific objective through the control and distribution of material resources, energy, and information. The systems approach is a paradigm concerned with systems and interrelationships among their components. Today, more than ever, we face the need for appropriate tools that can assist in dealing with difficulties introduced by the increase in the complexity of disaster management problems, consideration of environmental impacts and the introduction of the principles of sustainability. The systems approach is one such tool. It uses rigorous methods to help determine the preferred plans and designs for complex, often large-scale systems. It combines knowledge of the available analytic tools, an understanding of when each is appropriate, and a skill in applying them to practical problems. It is both mathematical and intuitive, as is all disaster management cycle of hazard mitigation, preparation, emergency/event/crisis management, and recovery.

Despite many efforts, systems thinking is in a less secure position in the social sciences than it was 30 years ago. Many theorists still write it off as another version of functionalism, discredited in their eyes because of its inability to deal with the subtlety and dynamics of organizational processes and, in particular, power and conflict. Practitioners continue to see the approach as too theoretical to be helpful with their everyday concerns. Progress there might have been but the full potential of systems ideas still remains to be realized.

The aim of this book is directly related to the current state of systems thinking as an approach within the social and specially disaster management sciences. Its purpose is to offer systems thinking as a coherent approach to inquiry and disaster problem management so that it can again occupy a role at the leading edge of development in the applied disciplines. With this book I would like to contribute to the change of disaster management practice and respond to a clear need to redefine the education of disaster management professionals and increase their abilities to (a) work in an interdisciplinary environment; (b) develop a new framework for hazard mitigation, preparation, emergency/event/crisis management, and recovery that will take into consideration current complex socioeconomic conditions; and (c) provide the context for disaster management in conditions of uncertainty.

The main objectives of this book are to introduce the systems approach as the theoretical background for modern disaster management, and to focus on three main sets of tools: simulation, optimization, and multiobjective analysis. At the same time, this book will allow me to reflect on the past 30 years of practicing and teaching water resources systems management. The process of reflection unlocks theory from practice in one field, brings to the surface insights gained from experience, and offers a framework for uncovering many hidden aspects of applying a theoretical approach in the search for a solution of practical problems in other areas. Insights gained from reflection can then be used to elaborate and present a theoretical approach in a different way, which I hope will prove more understandable to the students of the discipline and more acceptable to the practicing disaster managers. Therefore, my sincere hope is that this book will be able to serve multiple communities: as a text for teaching systems analysis and as a guide for the application of a systems approach to disaster management.

The text presented in this book is supported by a number of computer programs that can be used in applying the theory presented here to the solution of real-world problems.

THE ORGANIZATION OF THE BOOK

This book is organized into 4 parts and 8 chapters. Part I provides an introductory discussion and sets the scene. In Chapter 1, there is a brief overview of my personal experience, which provided my motivation for writing this book. I then define the main terms used in integrated disaster management in Chapter 2.

Part II is devoted to the introduction of the systems theory, mathematical formalization, and classification of methods. The material presented in this section should be of practical relevance during the process of formulating a disaster management problem as a systems problem and selecting an appropriate tool for the solution of the problem. In Chapter 3, I focus on systems thinking as a philosophical background of the systems approach and then I formally introduce the systems approach, define systems terms, and look at how they are applied in disaster management. This chapter ends with a set of system formulation examples from the disaster management domain. Chapter 4 introduces systems tools and techniques and provides their main characteristics.

Part III is technical in nature and it is aimed at disaster management practitioners. Chapter 5 concerns the simulation approach. It provides a detailed description of system dynamics simulation. Development of system dynamics simulation models is illustrated with two examples: a simple epidemic model and a more complex epidemic model with recovery. The chapter ends with real application of system dynamics to flood evacuation simulation.

Optimization is addressed in Chapter 6, with a focus on one of the most widely used technique—linear programming. In addition to the introduction of linear programming and simplex method for its solution, I am presenting in this chapter two special types of linear programming problems that have great application potential in disaster management—transportation problems and network problems. Appropriate algorithms for the solution of special problems are presented, including transportation simplex method, shortest path method, minimum spanning three method, and the maximum flow method. The chapter ends with the presentation of the linear programming application to optimal placement of casualty evacuation assets.

Chapter 7 focuses on multiobjective analysis. A very practical approach is taken to the material in this chapter. Because it approaches multiobjective analysis from an application point of view, it deals with a number of important issues in addition to the selection of an appropriate technique. Two deterministic multiobjective analysis techniques are presented for single and group decision making. The first one, the weighting method, is a technique for generating nondominated solutions. The second one, the compromise programming, is a technique for ranking a discrete set of solutions and identifying one that provides the best compromise between the set of criteria used in evaluation. The chapter ends with an example application in the selection of flood management alternatives.

This book ends with the presentation of my vision for the future of disaster management. In Part IV, Chapter 8 presents this view. This section also provides additional references for readers with a deeper interest in some of the concepts discussed.

SOFTWARE CD-ROM

The application of methodologies introduced in this book is supported through a set of computer programs contained on the accompanying CD-ROM. The state-of-the-art simulation software Vensim PLE (Personal Learning Edition) is enclosed for the implementation of system dynamics simulation. This program was developed by Ventana Systems, which has kindly given permission for its use in this context.

The CD-ROM includes two more original computer programs developed in the user-friendly Windows environment, for the illustration and implementation of the methods outlined in this book. They are LINPRO, a linear programming optimization tool; and COMPRO, for the implementation of the multiobjective analysis tool of compromise programming.

Each program is presented in the same way on the CD-ROM, with (a) a Readme file with installation instructions; (b) a folder containing the main program files; and (c) a folder containing all the examples presented and discussed in the text.

Vensim PLE is accompanied by a short tutorial developed by Craig W. Kirkwood of Arizona State University. I am grateful to the author for permission to provide it here. The other two programs have very extensive help manuals, which are integrated into the Windows environment. These provide detailed instructions on program use, data preparation, data import, and interpretation of the results. This software component of the book is not intended as a commercial product. It has been developed to illustrate the application of the methodological approaches presented in this book, and to allow the solution of real disaster management problems. However, the responsibility for its appropriate use is in the hands of the user.

USE OF THIS BOOK

This book and the accompanying CD-ROM have four main purposes:

1. They provide material for an undergraduate course in disaster management. A course might be based on Chapters 1 through 4, and possibly parts of Chapters 5, 6, and 7.
2. They also provide support for a graduate course in disaster management, with an emphasis on the analytical aspects of application of the systems approach to management of disasters. Such a course might draw on Chapters 1 through 4, and details in Chapters 5, and/or 6, and/or 7. Both undergraduate and graduate courses could use the computer programs provided on the CD-ROM.
3. Disaster management practitioners should find the focus on the application of the methodologies presented to be particularly helpful, and could use the programs for the solution of real disaster management problems. There is discussion of a number of specific applications in Chapters 5, 6, and 7 that may be of assistance.
4. Specific parts of this book can be used as a tool for specialized short courses for practitioners. For example, material from Chapter 5 and parts of Chapter 4 could support a short course on "System dynamics simulation and integrated disaster management." A course on "System analysis for emergency management optimization" could be based on Chapters 3, 4, and parts of Chapter 6. Similarly, material from Chapter 7 and parts of Chapters 3 and 4 could be used for a short course on "Multiobjective analysis in management of natural disasters."

My plan is to maintain an active Web site for this book, which will provide additional exercises for each chapter as well as suggested solutions. I will maintain an active software component of the Web site as a platform for the improvement of the enclosed computer programs through exchange of experience, and collecting a larger number of different applications that can be shared among the users of this book.

I and the individuals involved in publishing this book have done our best to make it error free, but it is almost inevitable that there will be some mistakes. I take responsibility for any errors of fact, judgment, or science that may be contained in this book. I would be most grateful if readers would contact me to point out any mistakes or make suggestions for improving this book.

Publishing this book was made possible through the contribution of many people. I would likc to start by acknowledging the publication support provided by the Institute for Catastrophic Loss Reduction and its Director Mr. Paul Kovacs. Most of the knowledge contained in this book came from my numerous interactions with teachers, students, and colleagues throughout the world. They taught me all I know. I would like to thank particularly my former student Dr Sajjad Ahmad whose work is discussed in Chapter 5. A special thank goes to Dr Veerakcudy Rajasekaram, who is the developer of all the computer programs. His attention to detail, love of computer programming, and analytical mind are highly appreciated. Mr. Andrew Patrick Belletti did a great job in preparing all the illustrations for this book in the style required by the publisher. I am very thankful for his effort and good work.

The support of my family, Dijana, Damjan, and Tanja, was of the utmost importance in the development of this book. They provide a very large part of my motivation, my goals, my energy, and my spirit. Without the endless encouragement, criticism, advice, and support of my wife Tanja this book would never have been completed.

SLOBODAN P. SIMONOVIĆ

Spring 2008, London

List of Acronyms and Abbreviations

ABM	Agent-based modeling
A&E	Accident and Emergency
CCBFC	Canadian Commission on Building and Fire Codes
COMPRO	Multiobjective analysis tool of compromise programming
CSA	Canadian Standards Association
DFAA	Disaster Financial Assistance Arrangements
DP	Dynamic programming
DRP	Disaster recovery planning
DRS	Disaster recovery strategy
DSS	Decision support system
DYNAMO	DYNAmic MOdels
ELECTRE	Elimination and Choice Translating Algorithm
EMO	Emergency management organization
EMT	Emergency management team
FDAI	Federal disaster assistance initiative
FEMA	Federal Emergency Management Agency
FIDS	Facility for Intelligent Decision Support
FP	Floodplane
FPT	Federal/Provincial/Territorial
GAMS	General Algebraic Modeling System
GIS	Geographic Information System
GNP	Gross National Product
GOC	Government Operations Center
HAZUS-MH	Hazards US Multi-Hazard
ICLR	Institute for Catastrophic Loss Reduction
IJC	International Joint Commission
INFOHYDRO	Hydrological Information Referral Service
IOM	International Organization for Migration
IPCC	Intergovernmental Panel on Climate Change
IRRBTF	International Red River Basin Task Force
ISDR	International Strategy for Disaster Reduction
JEPP	Joint Emergency Preparedness Program
LINPRO	Linear Programming Optimization Tool
LP	Linear Programming
MAEviz	Earthquake Risk Assessment System
MAS	Multiagent System

MDS	Mennonite Disaster Service
MEMO	Manitoba Emergency Management Organization
MINOS	Modular In-core Nonlinear Optimization System
MOEA	Multiobjective Evolutionary Algorithms
MOFLO	Multiobjective Facility Location
MOP	Multiobjective Optimization Problem
NATO	North Atlantic Treaty Organization
NDMS	National Disaster Mitigation Strategy
NDP	New Democratic Party
NEP	National Exercise Program
NRC	National Research Council
NSGA	Nondominated Sorting Genetic Algorithm
OPTEVAC	Optimal Placement of Casualty Evacuation Assets
PAES	Pareto Archived Evolution Strategy
PLE	Personal Learning Edition
POWERSIM	Powersim Software AC Simulation Software
PSC	Public Safety Canada
RCMP	Royal Canadian Mounted Police
RMS	Risk Management Solutions
RRBDIN	Red River Basin Decision Information Network
SAR	Search and Rescue
SARS	Severe Acute Respiratory Syndrome
SEMOPS	Sequential Multiobjective Problem Solving Method
STELLA	Structural Thinking Experimental Learning Laboratory with Animation
SWT	Surrogate Worth Trade-off Method
UN	United Nations
UNDP	United Nations Development Program
USAR	Urban Search and Rescue Program
VENSIM	Ventana Systems Simulations Software
WCe	World CAT Enterprise
WMO	World Meteorological Organization

PART I
Management of Disasters

1 Introduction

Everyday life is overwhelmed by critical phenomena that occur on specific spatial and temporal scales. Typical examples are floods, bridge collapses, stock market crashes, or the outbreak of diseases. All these phenomena might have, whenever they occur, significant negative consequences for our lives. They often result from complex dynamics involving interaction of innumerable system parts within three major systems: the physical environment; the social and demographic characteristics of the communities that experience them; and the buildings, roads, bridges, and other components of the constructed environment. In nonscientific terms, such events are commonly referred to as *disasters* (Bunde et al., 2002).

The terms *hazard*, *vulnerability*, *disaster*, and *risk* are interpreted and understood by different people in different ways. Before progressing with detailed discussions of many topics related to disaster management, let me provide the meaning of these terms in the context of this book (UN/ISDR, 2004).

Hazard is a potentially damaging physical event, phenomenon, and/or human activity, which may cause loss of life or injury, property damage, social and economic disruption, or environmental degradation. Hazards can include latent conditions that may represent future threats and can have different origins: natural (geological, hydrometeorological, and biological) and/or induced by human processes (environmental degradation and technological hazards). Hydrometeorological hazards include natural processes or phenomena of atmospheric, hydrological, or oceanographic nature, which may cause loss of life or injury, property damage, social and economic disruption, or environmental degradation. Examples of hydrometeorological hazards are floods, debris, and mud flows; tropical cyclones, storm surges, thunder/hailstorms, rain and windstorms, blizzards, and other severe storms; drought, desertification, wildland fires, temperature extremes, and sand or dust storms; and permafrost and snow or ice avalanches.

Vulnerability is susceptibility to suffer loss or a set of conditions and processes resulting from physical, social, economic, and environmental factors, which increase the susceptibility of a community, an individual, an economy, or a structure to the impact of hazards.

Disaster occurs when a hazard triggers vulnerability and disruption of the functioning of a community or a society that is so serious that it causes widespread human, material, economic, or environmental losses, which exceed the ability of the affected community or society to cope with using its own resources. A disaster is a function of

Systems Approach to Management of Disasters: Methods and Applications, By Slobodan P. Simonović

the risk. It results from the combination of hazards, conditions of vulnerability, and insufficient capacity or measures to reduce the potential negative consequences of risk. The distinction between natural and other types of disasters is blurred. Many of the deaths resulting from the Hurricane Katrina, New Orleans, in 2005 were caused by dike collapses. A number of assessment studies following the event found that many parts of the complex flood protection infrastructure were not designed and maintained up to existing standards and regulations. Despite the fact that nature created the hurricane, the disaster was intensified by human action or a lack of it. The term *disaster* in this book will be used in its broadest sense and the distinction between natural and other types of disasters will not play an important role.

Risk combines the notions of hazard and vulnerability. It is the probability of harmful consequences, or expected losses (deaths, injuries, property, livelihoods, economic activity disrupted, or environment damages) resulting from interactions between natural- or human-induced hazards and vulnerable conditions. Conventionally risk is expressed by the notation:

$$\text{Risk} = \text{Hazards} \times \text{Vulnerability} \tag{1.1}$$

One important consequence of the definition (1.1) is that a high probability hazard with small consequences has the same risk as a low probability hazard with large consequences.

The longer time period records (traced back to 1900 while more reliable after 1950) show a relentless upward movement in the number of disasters (Figure 1.1) and their human (Figure 1.2) and economic impact (Figure 1.3).

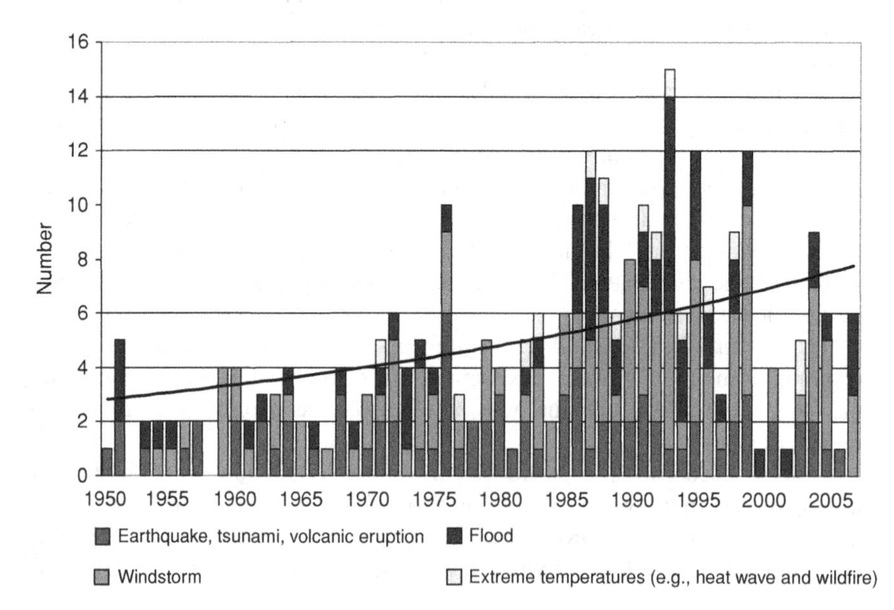

Figure 1.1 Great natural disasters 1950–2007, number of events (after Munich Re, NatCatSERVICE, 2008).

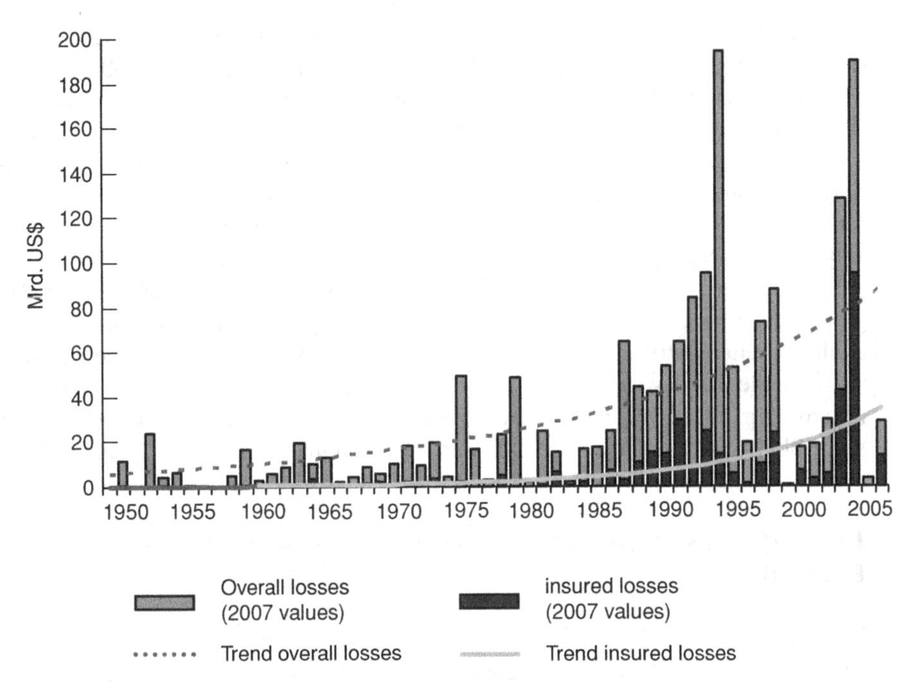

Figure 1.2 Great natural disasters 1950–2007, overall and insured losses (after Munich Re, NatCatSERVICE, 2008).

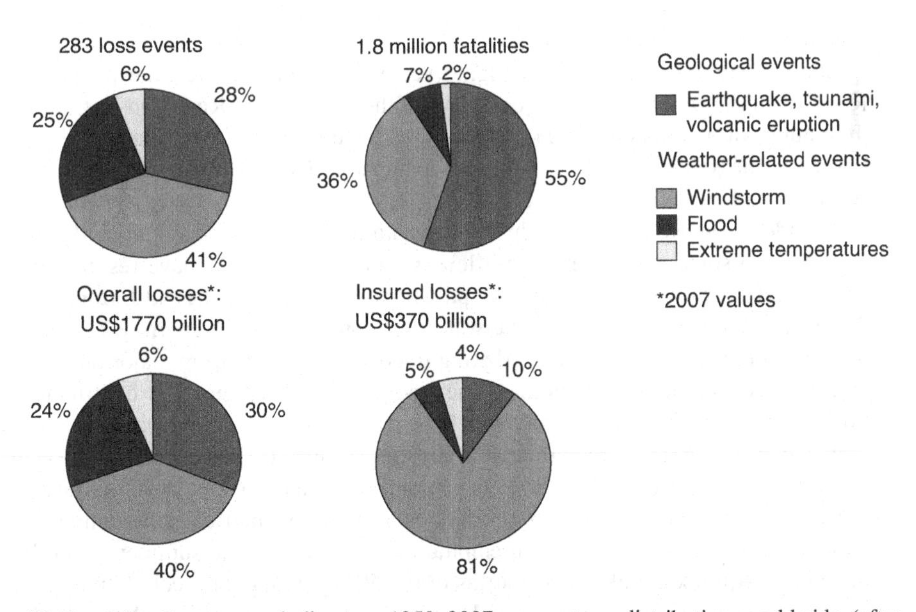

Figure 1.3 Great natural disasters 1950–2007, percentage distribution worldwide (after Munich Re, NatCatSERVICE, 2008).

A comparison of the annual figures verifies the serious increase in great natural disasters. The frequency of these events more than doubled between 1960 and 2005. The 276 great natural disasters in the period under observation are attributed, in almost equal proportions, to earthquake/volcanic eruption, windstorm, and flood. The most fatalities were caused by earthquakes and volcanic eruptions (55%). Economic losses have increased by a factor of 6.7, insured losses by a factor of 13.5, and the trend remains an upward one. As far as insured losses are concerned, windstorm losses are way ahead, accounting for nearly 80% of the US$340 billion.

It is troubling that disaster risk and impacts have been increasing during a period of global economic growth. On the good side, a greater proportion of economic surplus could be better distributed to alleviate the growing risk of disaster. On the bad side, it is possible that development paths are themselves creating the problem: increasing hazards (e.g., through global climate change and environmental degradation), human vulnerability (through income poverty and political marginalization), or both.

1.1 ISSUES IN MANAGEMENT OF DISASTERS—PERSONAL EXPERIENCE

We learn from experience. Here is a personal story of the 1997 flood on the Red River. At the time of "Red River flood of the Century" I lived in Winnipeg, Manitoba, Canada.

1.1.1 Red River Flooding

Situated in the geographic center of North America, the Red River originates in Minnesota and flows north (one of eight rivers in the world that flow north). The Red River basin covers 116,500 km^2 (exclusive of the Assiniboine River and its tributary, the Souris) of which nearly 103,600 km^2 are in the United States (Figure 1.4). The basin is remarkably flat. The elevation at Wahpeton, North Dakota, is 287 m above sea level. At Lake Winnipeg, the elevation is 218 m. The basin is about 100 km across at its widest. The Red River floodplane has natural levees at points both on the main stem and on some tributaries. These levees (some 1.5 m high) have resulted from accumulated sediment deposit during past floods. Because of the flat terrain, when the river overflows these levees, the water can spread out over enormous distances without stopping or pooling, exacerbating flood conditions. During major floods, the entire valley becomes the floodplane. The type of soil in this region also contributes to flooding because, while the topsoil is rich, beneath it lies anywhere from 1 to 20 m of largely clay soil, with characteristic low absorptive capacity. Water tends to sit on the surface for extended periods of time. In general, the climate of southeastern Manitoba is classified as subhumid to humid continental with resultant extreme temperature variations. Annually, most of the precipitation received is in the summer rather than the winter. Approximately three-fourths of the 50 cm of annual precipitation occurs from April to September. Consequently, most years spring melt is well managed by the capacities of the Red River and its tributaries. However, periodically, weather

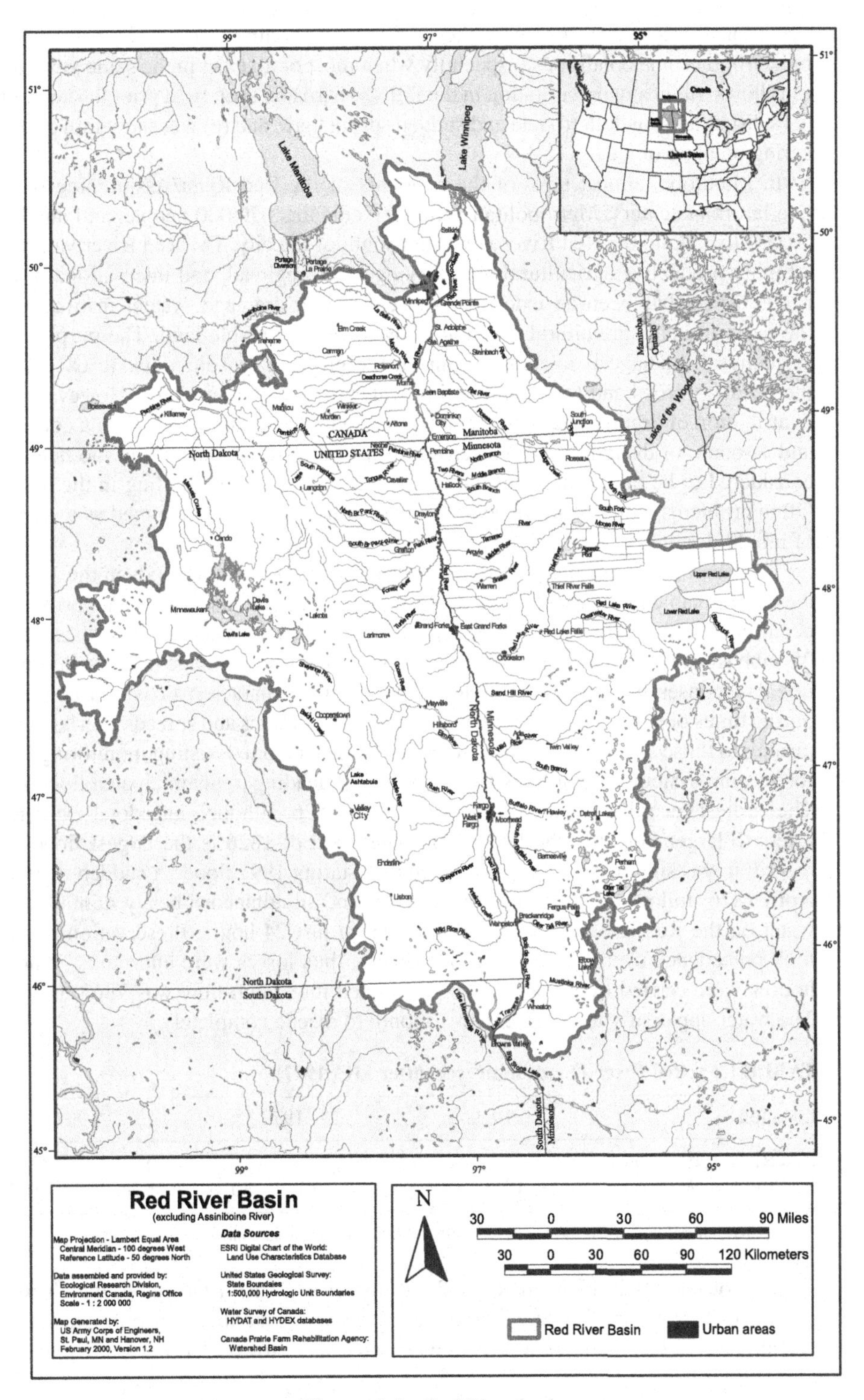

Figure 1.4 Red River basin.

conditions exist that instead promote widespread flooding through the valley. The most troublesome conditions (especially when most or all exist in the same year) are as follows: (a) heavy precipitation in the fall, (b) hard and deep frost prior to snowfall, (c) substantial snowfall, (d) late and sudden spring thaw, and (e) wet snow/rain during spring breakup of ice.

In Manitoba, almost 90% of the residents of the Red River/Assiniboine basin live in urban centers. Metropolitan Winnipeg contains 670,000 people, and another 50,000 live along the Red River north and south of the city. The Red River valley is a highly productive agricultural area serving local, regional, and international food needs. There has been an extensive and expanding drainage system instituted in the basin to help agricultural production by increasing arable land. The purpose of agricultural drainage is to remove, during the growing season, water in excess of the needs of crops and to prevent sitting water from reducing yields. However, the contribution of drainage activities, if any, to flooding and damages is both a concern and a source of disagreement. Faster removal of the spring water from the fields is considered to be one of the contributors to the regular spring flooding in the basin. Often problems with maintenance of drainage infrastructure are claimed as a source of infield flooding.

The basin floods regularly. Early records show several major floods in the 1800s, the most notable being those of 1826, 1852, and 1861. In this century, major floods occurred in 1950, 1966, 1979, 1996, and 1997 (Table 1.1). The Red River basin has 25 subbasins, which have different topography, soils, and drainage that result in different responses during flood conditions. One common characteristic is overland flow during times of heavy runoff. Water overflows small streams and spreads overland, returning to those streams or other watercourses downstream. Existing monitoring and forecasting systems do not track these flows well, leading to unanticipated flooding. The earliest recorded flood in the basin was in 1826, although anecdotal evidence refers to larger floods in the late 1700s. The flood of 1826 is the largest flood on record; it was significantly larger than the devastating 1997 flood. A sudden thaw in April 1826, followed by ice jams on the river and simultaneous heavy rainfall, had water on the Red River rise 1.5 m downtown in just 24 hours. Preservation of life took precedence over preservation of property, thus losses were enormous. Whole houses were carried by the River. The estimated maximum flow was 7362 m^3/sec. The water apparently took more than 1 month to recede completely.

TABLE 1.1 Red River Floods in m^3/sec (after IJC, 1997)

Location	1950		1979		1997	
Red River at Emerson	May 13	2670	May 1	2620	May 2	3740
Red River at Winnipeg	May 19	3058	May 10	3030[a]	May 4	4587[a]

[a] Computed natural flow as would have occurred without existing flood control works.

A pivotal event in the Red River flood history was the 1950 flood, which was classified a great Canadian natural disaster based on the number of people evacuated and affected by the flood. A very cold winter and heavy snowpack in the United States, combined with heavy rain during runoff, were the primary causes. All towns within the flooded area in the upper valley had to evacuate. More than 10,000 homes were flooded in Winnipeg and 100,000 people evacuated. A plan to evacuate all 350,000 people in Winnipeg was prepared, although luckily it did not have to be used.

Most of the flood management planning in Manitoba was initiated after the 1950 flood. This flood was the turning point in the history of flooding and flood control in Manitoba's portion of the Red River basin. Construction of elevated boulevards (dikes) within the City of Winnipeg and associated pumping stations was initiated in 1950. The current flood control works for the Red River valley consist of the Red River Floodway, the Portage diversion and Shellmouth Dam on the Assiniboine River, the primary diking system within the City of Winnipeg, and community diking in the Red River valley (Simonović, 2004). Following the 1950 flood on the Red River, the federal government and the Province of Manitoba set up a fact-finding commission to appraise the damages and make recommendations (Royal Commission, 1958). The commission recommended in 1958 the construction of the Red River Floodway (completed in 1966), the Portage Diversion (completed in 1970), and the Shellmouth Reservoir (completed in 1972). As a consequence of the concern over flood protection for the Red River Valley, a federal-provincial agreement led to the construction in the early 1970s of a series of ring dikes around communities in the Valley. Moreover, financial aid programs encouraged rural inhabitants to raise their homes, as well as to create individual dikes around their properties. All the decisions regarding the capacity of the current flood control works were based primarily on economic efficiency—getting the largest return for the investment.

1.1.2 "Red River Flood of the Century," Manitoba, Canada

Sunday, April 6, 1997, was a day off for most people, including me, but it was not a standard day of rest. Our house on Kirkbridge Drive in the south part of the town was surrounded by drifts of snow, at some places up to the window frames. Our driveway, service road, and the street were covered by snow, at places deeper than 1 m (see Plate 1 in the color plate section). Our plans to do some late shopping and finalize preparations for our daughter's birthday on April 11 ended up in serious snow-moving activities. The city was virtually shut down.

Radio was announcing that the whole Red River valley from North Dakota to Lake Winnipeg was already under the snow varying in depth from more than 2 m along the upper reaches to more than 1.5 m around Winnipeg (more than most people could remember seeing). Temperature just began to peak over the freezing level when this massive snowstorm piled more snow on an already high snow cover. Flooding was an already accepted certainty in the valley. Early forecasts of my colleague and friend Alf Warkentin from the Water Resources Branch were a 10% chance for flood as bad as that of 1979. That flood inundated southern Manitoba and turned it into a lake 90 km long from north to south and 20 km across at its widest point. After the blizzard, Alf's forecast was revised and the Red River valley was facing a flood bigger than the flood of 1950.

Life slowly returned to normal after the weekend. However, the work of Emergency Management Manitoba and Water Resources Branch just started. Preparations for the flood were in full swing. Life, for me and for many citizens of Winnipeg, was kind of unreal. Yes, the big flood was coming, but Winnipeg had resources other smaller municipalities did not. We had engineers and infrastructure and operations departments that had expertise and experience with flooding.

Our home was close to the southern border of the City and my way to University was taking me across the Pembina Hwy—the main north–south artery cutting across Winnipeg. I was going to work to administer the final exams in my courses, meet with students, attend the administrative meetings, and at the same time something serious was going on. Everyone was talking about the flood. The flood was a reality south from Winnipeg. Red River Valley was under siege. My contacts in the Water Resources Branch were providing regular information about the hectic effort to get the best estimate of what is going to hit us and to get prepared as good as we can and as soon as we can. My children were taken from the school to help the sandbagging effort. I offered help to some friends living close to the river. Busloads of school kids, complete strangers, church groups, neighbors, office managers let off work, and anybody able-bodied showed up for sandbagging duty (see Plate 2 in the color plate section).

On my way to work, waiting for the green light at the crossing with Pembina Hwy, I would witness heavy mechanization moving south; later tracks full of soldiers and volunteers; even later school buses full of people being moved from the valley to safer locations. People from the Water Resources Branch like Larry Whitney, emergency flood spokesman (who numerous times delivered lectures in my courses), Rick Bowering, head of the Water Resources, and Doug McNeil from the City became everyday guests in every Winnipeg home through a regular process of updating information about the incoming flood. About 8 million sandbags were laid into ramparts around Winnipeg. It is not known how many sandbags were used outside the City because each municipality took care of those matters. But at one point, the province leased a 747 jet for $225,000 to airlift 3 million sandbags from California to the Red River Valley.

April 19 was a special day. Nearly 2 weeks had passed since the blizzard, and under the bright sun the massive snow blanket had begun to melt. Our neighbors in North Dakota were fighting the flood. Cities were falling to the Red, one after another. All this information was coming to us, but nothing hit us as hard as the front page of the Winnipeg Free Press on Sunday, April 20. It showed the downtown Grand Forks—the Security Building submerged in Red and on fire. It was a strange image showing two forces of nature acting together with destructive power, and nature was winning. Grand Forks was under the water and 35,000 people were rendered homeless. This image, repeated on the TV many times and shown in other local papers, got stuck in my mind. It was real and coming at us. My wife insisted on moving furniture from the basement. I was checking the backup (backflow) valve that for those who did not have it became a valuable commodity. All backup valves were sold out in town and people were ordering them from all over Canada and the United States.

The battle with Red was raging in the Valley (see Plate 3 in the color plate section). Tremendous effort to protect the property and reduce the damage was going on in parallel with the expansion of the water over the land. The "Red Sea" reached up to 37 km wide and covered 1850 km^2 in Manitoba. On April 27, my colleague and friend Prof. Wendy Dahlgrin took me for a flight on her small plane above the southern Manitoba. Our flight route and altitude were under the control of the military. My stomach did not agree with the bumpy flight of a small plane. However, one picture remains in my mind (see Plate 4 in the color plate section). From the altitude we were flying on, all I was able to see was water. We flew from Winnipeg south to thc Canada–US border and back. The river channel could be recognized only by the tops of the trees still above the water level. The picture looked unreal. Farmhouses still above the water and townships protected with ring dikes looked like small islands in the ocean.

The towns of Emerson, Morris, Ste. Agathe, St. Adolphe, Grande Point, and farms around the Valley were receiving help from volunteers, responsible agencies, and Canadian Arm Forces. The ring dikes around communities were raised. Shortly after midnight on Tuesday, April 29, 1997, the Red River struck the small town of Ste. Agathe, 25 km south of Winnipeg (see Plate 5 in the color plate section). It was the first indication that parts of Manitoba thought safe could be vulnerable. The water did not flood from the east side as one might expect, that is, where the Red River flows past the town and where the town dike was built. Instead, the water blindsided the village from the west, flowing overland and crossing Highway 75. All other communities survived. Beside Ste. Agathe, the Red River flood got in one more bite. It took that bite at Grande Pointe, a suburb of Winnipeg bordering southeast city limits. One hundred Grande Pointe homes were flooded. It was time for heroics because, in spite of Winnipeg's and the province's best efforts, the planning and preparations were not complete.

The province introduced the mandatory evacuation of thousands of rural people living outside ring dikes. The order created bands of "outlaws" who ignored the authorities and drove their boats through flooded fields to save their homes and those of others. *Royal Canadian Mounted Police* (RCMP) wanted evacuation and they wanted it in a hurry. They thought there was a grave threat to life, and therefore, pressured other authorities into supporting an evacuation. Shortly after Grand Forks went under on April 18, the province moved out 3400 Red River Valley residents. This was not controversial. Most were people in the ring dikes or who had health or mobility issues. But on April 23, Emergency Management Organization (EMO) dropped the bombshell. It announced a total evacuation of the valley, about 17,000 people. Within days, more than 800 rural homes were reported flooded.

Some residents did not follow the orders. They stayed and raised the height of dikes, plugged leaks in dikes, and made sure pumps were running and properly positioned. They also phoned owners when they discovered problems.

Water was at the doorstep of Winnipeg. The city filled 6.5 million sandbags. But even 6.5 million bags were not enough. City built 14 earth dikes inside the city limits. The floodway was used to maintain the 24.5 ft level at James Avenue. That was considered to be the level that the city's dikes could be expected to hold back.

Maintenance of 24.5 ft level at James Avenue meant almost a week where the Red was at its record high level in Winnipeg and almost two weeks where it was above the level it had reached in any previous year (even the pre-floodway 1950). Emergency dikes were under enormous strain and plugging leaks became a 24-hour-a-day job. Hectic pace to protect the city was confronted with surreal "life as usual" for most of the people leaving and working in the city. My wife was scheduled to have a surgery and the St. Boniface hospital, located very close to the river, was hardly keeping the schedule. Before the date of surgery the hospital was closed for some time. Fortunately, the impact of the Red on the work of St. Boniface hospital did not affect my wife. Surgery was done on time and we learned immediately after about another closure of the hospital. The only similar emotion to what I was experiencing during these days was described in the book *Poplava* (Flood in Serbian language) for those who can read the language of the place where I was born (Nenadic, 1982). I felt anxiety, nervousness, fear, and helplessness, together with a tremendous need to do something, to add some meaning to this waiting time.

The water was still coming up. The last frontier was the extension of Brunkild Z-dike designed to keep the Red River water out of the La Salle River (considered at that time Winnipeg's Achilles' heel). The La Salle is the Red River's last tributary before Assiniboine and it flows into the Red at La Barriere Park in St. Norbert. That is north of the floodway gate and behind Winnipeg's primary diking system. As many as 100,000 Winnipeggers, including my family, would be forced from their homes if enough water got over the high ground and came down the La Salle. Resources were scarce and available time was short. The province put all its energies and earth-moving equipment into a 72-hour dash to build the 24-km Brunkild Z-dike extension (see Plate 6 in the color plate section). When the water reached the critical Brunkild gap on April 29, the Z-dike blocked the way.

The river had crested in Winnipeg on May 1, and all the city's defenses held. But a water elevation of 24.5 ft above winter ice levels at the James Avenue pumping station was considered all Winnipeg could safely handle, so floodway gates were raised to hold water inside Winnipeg to that level.

Not everyone understands exactly how the floodway works (see Plate 7 in the color plate section). Its two gates are actually in the Red River, where the river and diversion channel meet. The two gates are raised to elevate water enough to push it into the diversion channel. The reason the water level has to be raised is because there is a large mound at the opening of the diversion channel to stop ice going into the floodway. Large ice would damage bridges and other structures along the floodway.

But raising those gates caused artificial water levels south of the floodway. On May 2, some 125 of 150 homes in Grande Pointe took on water. The province initially denied the floodway had caused artificial flooding. But a review later determined that the floodway operation caused artificial flooding of 2 ft above what water levels upstream should have been. Many residents of Grande Pointe felt "sacrificed."

With the river crest passing the city on May 1 the flood was not over. Communities north from Winnipeg were just starting their battle with Red and Winnipeg with those south of the city were embarking on a difficult path of recovery. Many homes were bought out because their location made flood-proofing too difficult. For example, on

St. Mary's Road just south of Winnipeg, 25 homes were purchased by the government because the cost of flood-proofing was too high.

Assessment of damage started in May. However, the process was slow and plugged with problems (see Plate 8 in the color plate section). Initially, the province was only going to pay 80% compensation to flood victims, even though 90% of the money came from the federal government. Claimants had to pay 20% deductible, and the maximum government compensation was to be $100,000. The premier of Manitoba was adamant about these terms. His explanation at the time is still being quoted today: "If you live on the floodplane, you have to take some responsibility." Many residents immediately south from the floodway gates were convinced that it was not the floodplane, but the floodway, that caused their homes to be deluged. In the 1999 election, Grande Pointe got its revenge. New Democratic Party (NDP) candidate upset the Conservative incumbent by a mere 111 votes. The roughly 130 voters from Grande Pointe that went to the NDP made the difference. Compensation to flood victims was eventually raised. The province finally eliminated both the $100,000 cap, and the 20% deductible, for Disaster Financial Assistance funds. Compensation covered essentials for living only.

At the end a total of 3747 private homes had claims for flood damage approved according to the province's Emergency Management Organization. Another 633 flood damage claims from full-time farms were approved. Also, claims for 383 full-time businesses were approved. The Disaster Financial Assistance payments for those claims reached $257 million. That does not include business losses.

In addition to the government support, the effort of many volunteers and donations from all over the country made a difference. In the Red River Valley south of Winnipeg, the Mennonite Disaster Service (MDS) built 14 new homes, did major reconstruction on 71 homes, minor reconstruction on 28 homes, relocated 5 homes, and cleaned 802 flooded homes and yards. MDS volunteers put in 21,061 volunteer days, worth an estimated $2.5 million in labor. MDS used donations of nearly $1.9 million for food, transportation, and lodgings for volunteers. They also used donations to buy building materials, for which they were later reimbursed by Emergency Measures Organization, so people did not have to wait for their claims to be settled before they had roofs over their heads.

The most humbling event may have been the donations that poured in to help flood victims. The Canadian Red Cross collected $25 million in donations from more than 144,000 private citizens across the country. But 70% of the $25 million came from other Manitobans. The Red Cross employed 250 people on flood relief, and mobilized another 2200 volunteers in Manitoba. It helped rebuild or restore 230 homes, and plug gaps between government aid and family incomes.

Salvation Army also provided free cleanup supplies, toys for children, tickets for local sporting events, and covered grocery costs. It even took seniors on a 2-day bus trip to Gimli.

Many families in the valley were under stress. The financial bottom line for people just collapsed. There were divorces, there were suicide attempts, and trauma teams were working overtime to help population under stress (Morris-Oswald and Simonović, 1997).

By letters of June 12, 1997, the Governments of Canada and the United States requested the International Joint Commission (IJC) to examine and report on the causes and effects of damaging floods in the Red River basin and to recommend ways to reduce and prevent harm from future flooding. The IJC is a binational Canada–United States organization established by the Boundary Waters Treaty of 1909 that assists the governments in managing waters shared by the two countries for the benefit of both. To assist it with the Red River flood of 1997 binational investigation, the Commission has appointed an International Red River Basin Task Force. The Task Force, composed of members from a variety of backgrounds in public policy and water resources management, was to provide advice to the Commission on matters identified in the letters from governments. The Governments asked the Commission to examine a full range of management options, including structural measures (such as building design and construction, basin storage, and ring dikes) and nonstructural measures (such as floodplane management, flood forecasting, emergency preparedness, and response) and to identify opportunities for enhancement in preparedness and response that could be addressed to improve flood management in the future. I was appointed to serve on the Task Force together with four more members from Canada and five members from the United States. For more information, please consult the IJC International Red River Basin Task Force's Web site at http://www.ijc.org/rel/boards/rrbtf.html (last accessed July 21, 2008).

Work on the Task Force was an experience of a lifetime. I participated in a large number of public hearings across the Canadian and US parts of the basin, literally meeting thousands of people affected by the flood. I had an opportunity to hear horror stories of those who lost everything; listen to the rage of people who felt left without assistance; and meet with those who worked hard to save their families and property from damage. This work brought me in touch with basin managers in Canada and United States too. We had extensive meetings with representatives of all governments (local, provincial/state, and federal). I was part of many technical, social, and environmental studies commissioned by the IJC. For the first time in my professional life I got an opportunity to understand the full extent of the impact my work has on people, environment, and society in general. The Task Force prepared a December 1997 interim report (IRRBTF, 1997) that cautioned against complacency and made 40 recommendations for better flood preparedness in the short term. At the end of our work, we submitted the final report (IRRBTF, 2000).

The International Red River Basin Task Force defined required projects, coordinated the funding and scheduling, exercised quality control, provided oversight of subgroups, synthesized the findings, and prepared the recommendations. The Task Force established three subgroups—database, tools, and strategies—to conduct or direct much of the data collection, model development, program evaluation, and to prepare preliminary recommendations. Each subgroup included experts from both the United States and Canada. The concept for accomplishing the required tasks included three main activities: database development, modeling, and the development of damage reduction strategies. A coordinated database was found to be fundamental, as it supports the development of models and flood damage reduction strategies. Each of these working topics ended up as a key element in the decision support system. The

Task Force's final report (IRRBTF, 2000) drew together the findings of the subgroups and made recommendations on policy, operations, and research issues.

The IJC used the final report as the basis for public hearings in the basin prior to the submission of its report to the governments. Public participation was an important part of the process. Following the distribution of the Interim Report, the IJC and the Task Force conducted a series of public meetings throughout the basin in February and October 1998. The results from these meetings were incorporated into the work plan. Efforts were made to keep people in the basin informed throughout the study using the Internet, news releases, and other means of contact. Public and technical inputs were invited throughout the study period.

The fact that this work involved two countries implied two different ways of doing business, two political systems, two or more ways of collecting, analyzing and storing data, and many other political dichotomies. These dichotomies created a unique challenge for this work, but the reality that floodwaters do not recognize an international border made a basin-wide approach to flood management an imperative. Although this work did not develop a comprehensive basin-wide water management plan, the work of the Data, Tools, and Strategies Groups contributed to more effective and efficient floodplane management, facilitated integrated flood emergency management in the basin, and fostered improved international cooperation and communication.

In investigating what can be done about flooding in the Red River basin, the Task Force examined the issue of storage—through reservoirs, wetlands, small impoundments, or micro-storage—and drainage management. The conclusions (IRRBTF, 2000) are:

> **Conclusion 2**: It would be difficult if not impossible to develop enough economically and environmentally acceptable large reservoir storage to reduce substantially the flood peaks for major floods.
>
> **Conclusion 4**: Wetland storage may be a valued component of the prairie ecosystem but it plays an insignificant hydrologic role in reducing peaks of large floods on the main stem of the Red River.

Since the Task Force concluded that storage options provide only modest reductions in peak flows for major floods, a mix of structural and nonstructural options were examined. Winnipeg, the largest urban area within the basin, was found to remain at risk. The city survived the 1997 flood relatively unharmed, but it cannot afford to be complacent. If it had not been favored with fair weather during late April 1997, it could have suffered the fate of its southern neighbors. The Task Force made a number of recommendations to address the city's vulnerabilities and better prepare it for large floods in the future. To achieve the level of protection sufficient to defend against the 1826 or larger floods, major structural measures on a scale equal to the original Floodway project were found to be needed to protect the city. Two options were suggested: expansion of the Floodway and construction of a water detention structure near Ste. Agathe to control floodwaters for floods larger than 1997. After detailed feasibility studies, the Floodway expansion project was selected as a preferred alternative.

Structural protection measures are only part of the response to living with major floods. The Task Force looked at a wide range of floodplane management issues to see how governments and residents might establish regulatory and other initiatives to mitigate the effects of major floods and to make communities more resilient to the consequences of those floods. It made a number of recommendations on defining the floodplane, and adopting and developing building codes appropriate to the conditions in the Red River basin, education, and enforcement.

In an effort to gain a better understanding of the flooding issues, and in recognition of weaknesses in technological infrastructure within the basin, the Task Force devoted much of its energy and resources to data issues and computer modeling. On reviewing current data availability, the Task Force concluded that further improvement and maintenance of the Red River floodplane management database was required. Federal, state, and provincial governments and local authorities needed to maintain a high level of involvement in further database development and in improving data accessibility. The Red River Basin Decision Information Network (RRBDIN, 2005) now provides information about water management within the basin and links to other relevant resources. While RRBDIN concentrates on information and activities on the US side, the Government of Manitoba has been involved in collecting and disseminating flood information from the Canadian side (Province of Manitoba, 2005). Information from RRBDIN includes databases, references, technical tools, communication tools, and GIS data, as well as the most up-to-date information available on weather and flood forecasting. The Task Force found difficulty in securing public access from Canadian agencies to data and other flood-management-related information. The Task Force recommended that Canadian data be made available at no cost and with no restrictions for flood management, emergency response, and regional or basin-wide modeling activities. The Web site of the Government of Manitoba now provides up-to-date reports on daily flood conditions, in the form of maps and reports, along with miscellaneous information on flood management. A prototype version of the real-time flood decision support for the Red River basin is operational (Province of Manitoba, 2004).

In year 2000, I moved from Winnipeg and accepted a job with the University of Western Ontario and the Institute for Catastrophic Loss Reduction. However, my link to the flood of 1997 did not end there. In May of 2008, I organized the fourth International Symposium on Flood Defense in Toronto (http://www.flood2008.org, last accessed July 21, 2008). One plenary session of the Symposium was devoted to the Flood of the Century: "Red River Flood of the Century—10 Years Later."

Has it really been 10 years? Yes, judging by major improvements in flood protection since 1997. Winnipeg's floodway now provides protection from a one-in-300-year flood, and will be up to one-in-700-year protection by the time it is completed. It is costing $665 million (the largest infrastructure investment in Canada in 2005).

A new system of earthen dikes and preformed concrete walls protecting Grand Forks in North Dakota and Minnesota from a one-in-250 year flood is functionally complete. The Grand Forks system cost about US$400 million.

But outside those centers, the approach to flood-proofing in Manitoba versus North Dakota is quite different. In North Dakota, you will not see any houses elevated against flooding like in Manitoba, and you will see only a handful of personal ring dikes. Instead, North Dakota and Minnesota chose to use federal money from FEMA (Federal Emergency Management Agency) to buy out homeowners on the floodplane, instead of protecting them. Behind the decision to buy out homeowners is a government cost–benefit analysis. Government determined that it would be simply cheaper to buy people out than protect them. That was after looking at such things as the cost to protect a home versus its value, how many times it has flooded, and how many times it may flood in the future.

Minnesota has been especially aggressive about buyouts. But on the west side of the river in North Dakota, which has a much smaller tax base, many rural people have been ignored because there was not enough money. Their homes that were damaged by flooding still sit at the same elevations as in 1997. In North Dakota today, there are still 1100 rural residences on the Red's floodplane. Authorities do not know the level of flood protection for those homes. What is known is that very few have received government assistance to protect themselves. In the ongoing buyout program, FEMA pays 75% of a home's preflood value, the state pays 10%, and the county and homeowner split the remaining 15%. Most of the buyouts in North Dakota were in the cities, and most of those were in Grand Forks. There were 850 homes and 50 businesses bought out in Grand Forks. There were more than 1200 buyouts in that state between federal programs FEMA and the Federal and Urban Development program.

Contrast that with Manitoba where the government has bought out fewer than 75 homes in the Red River Valley since the big flood. Manitoba did a cost–benefit analysis too, but concluded it was better to help people stay on the land. The Red River Valley is an extremely prosperous agricultural area, and people do need to live in that floodplane to do their business. Flood protection allows businesses to develop with a level of security that they are not going to be damaged by a major flood.

North Dakota does not help fund the elevating of houses above flood levels, like in Manitoba. However, North Dakota and Minnesota have run small programs to help rural homeowners build individual ring dikes. Under the program, North Dakota agrees to finance a ring dike 50-50 with the landowner, committing a maximum US$25,000. The ring dikes for farms are costing well more than $50,000, so the farmer must pay much more than $25,000. Minnesota's program is more generous, with the state picking up 75% of costs. However, only a finite amount of funding is available for the American program, and many people have not been approved. Washington does not pay into the program. Since the program began, in 2001, 120 rural landowners have applied for assistance to build a ring dike in North Dakota. Just 16 have received funding so far. About double that number have been approved in Minnesota.

That is a meager number compared to Manitoba. Since 1997, a total 1830 rural homeowners in Manitoba's Red River Valley have received federal and provincial money to protect them from flooding. Today, virtually every home in the Red River Valley is protected to 1997 flood levels, plus 2 ft. Manitoba homeowners received on average $40,000 apiece to elevate their homes, build a dike, or otherwise fortify their residences. The feds and province cost-shared the program 50-50. The total program spending came to $73 million. Under the program, Manitoba homeowners could get up to $60,000 in government funds to build a ring dike or elevate a home, the most common types of flood protection. They had to contribute $10,000. But many landowners reduced their $10,000 share down to a small amount because the province knocked off dollars for labor, like a farmer using his or her tractor to help build the mound (they were paid hourly rates), and compensation for the soil they took from their land to build the mound.

It is always interesting to see how governments spend their money in Canada versus the United States, and how much they spend. Yet direct comparisons are not fair. One should not forget that North Dakota suffered much more damage than Manitoba, and had a much bigger hole to climb out of. The 1997 flood cost the state US$3.7 billion, including estimates of losses to businesses, according to FEMA. Still, public money has not flowed in the United States like in Manitoba.

Manitoba has done a much better job, flood-proofing its towns and villages, too. Every community along the Manitoba portion of the Red River is protected. There are 13 communities with new ring dikes: St. Mary's Road, Grande Pointe, Rosenort, Niverville, Gretna, Aubigny, St. Pierre-Jolys, Lowe Farm, Riverside, Rosenfeld, Ste. Agathe, and Roseau River. The cost of those dikes was shared 50-50 by federal and provincial governments. They have also improved the dikes for Dominion City, Emerson, Letellier, St. Jean Baptiste, Morris, St. Adolphe, and Brunkild. In total, 2133 homes and businesses have received new or upgraded protection in the form of community ring dikes at a cost of $42 million. Federal and provincial governments paid 90% of that, and rural municipalities 10%.

The same cannot be said in North Dakota. The city of Fargo is still waiting on flood-proofing funds from Washington after 1997. The delay in Fargo getting flood protection is important because memory and the urgency for flood protection fade with time. Federal funding in the United States is extremely tight now because of the costs of both the Iraq war and the flooding of New Orleans. In Breckenridge-Wahpeton, a series of diversions and dikes have been constructed and offer better protection but are still only half-finished. Construction has been idle for 2 years because federal funds have dried up. The town of Drayton, 45 km south of the Manitoba border, is very susceptible to flooding but cannot get any flood-proofing dollars.

In North Dakota, various government officials were pleased with how the state withstood the 2006 flood. Only about 10 homes were flooded. In Manitoba in 2006, only one home had serious flood damage, and that was a home in which the owner had refused flood-proofing assistance after 1997.

Today, Grande Pointe has a ring dike. Ste. Agathe has a ring dike too. The Brunkild Z-dike is now permanent. Casings for 500 wells on Manitoba's floodplane also have been raised to 1997 levels, plus 2 ft, so aquifers are not contaminated. Winnipeg also fared well in 2006, which compared to the 1996 flood. With about 200 homes now protected by a ring dike for Kingston Row and Kingston Crescent, the city needed just 20,000 sandbags last year.

The water- and climate-monitoring network in the Red River Valley has been upgraded at a cost of more than $1.5 million. This included activating or establishing

34 monitoring stations and installing 165 new climate stations. Manitoba forecasting office has dozens of satellite water monitors in the Red River, 60 new rain gauges in streams, and computer flood modeling programs.

A total of 1830 Red River Valley homes and businesses outside Winnipeg received individual flood-proofing since the flood at a cost of $73 million from federal and provincial governments. An additional 2133 homes and businesses have been protected by community ring dikes since the flood at a cost of $38 million from federal and provincial governments. Municipalities cost-shared 10%, raising the total to $42 million.

This is the end of my private story of the Red River Flood of 1997. I decided to provide this detailed experience in order to (a) illustrate the level of complexity that one natural disaster can bring, (b) demonstrate the need for a new approach to natural disasters management, and (c) offer the context for the set of tools presented in this book as one potential approach to address complexities in the management of natural disasters. The personal message I took from this experience was written on one temporary sandbag dike at Rosenort—"No Man is an Island" (see Plate 9 in the color plate section).

1.2 TOOLS FOR MANAGEMENT OF DISASTERS—TWO NEW PARADIGMS

Management of natural disasters has a long tradition in many countries around the world including Canada. There is no reason to abandon the approaches that have been used to date and the knowledge that has been accumulated through experience. However, there are some troubling questions about why are the losses shown in Figures 1.2 and 1.3 on such a sharp rise and why more progress does not appear to have been made. There is no shortage of ideas about what can be done to improve the disaster management tools and their implementation to achieve more impressive results on the ground than those realized so far. In the context of this book any empirical, analytical, or numeric procedure used in the process of disaster preparedness, emergency management, disaster recovery, and disaster mitigation and prevention is referred to as a "tool."

The application of various tools for disaster management during the last 50 years shows a pattern of change. Some of the lessons summarized by the National Research Council (1996), the Global Disaster Information Network (1997), Mileti (1999), Woo (1999), Godschalk et al. (1999), Stallings (2002), Bunde et al. (2002), Kohler et al. (2004), and Skipper and Jean Kwon (2007) are noted below.

Domain-specific lessons

1. Disaster losses are increasing. Precise estimate of losses is impossible since there is no systematic reporting method and no single repository for loss data in many countries.
2. Disaster losses are affecting global economies. Currently, disaster losses from natural disasters in developing countries represent only a small fraction

of GNP. However, the situation in the developing world is quite the opposite. With the trend of increasing losses, individual disasters will represent more significant fractions of the GNP of affected countries and may affect global economies.

3. Disasters have a broad social impact. Disasters impose deaths, injuries, and monetary losses. However, they can also redirect the character of social institutions, result in new and costly regulations imposed on future generations, alter ecosystems and disturb the stability of political regimes.
4. The nature of disasters is changing. Disasters are becoming more complex. They are the result of the interaction of, and changes in, the physical environmental systems that produce extreme events, the people and communities that experience those events and the constructed environment that is affected.
5. Climate change is increasing the frequency and magnitude of natural disasters. It is widely accepted that climate change will cause an increase in convective storms, floods, drought, and extreme temperature events.
6. Population increase creates serious disaster management problems. As areas become more densely populated, their exposure to hazards increases. Differences in socioeconomic status, gender, and race, or ethnicity result in a complex system of wealth stratification, power, and status, which in turn results in an uneven distribution of exposure and vulnerability to hazards, disaster losses, and access to aid, recovery, and reconstruction.
7. Diminishing strength of the built environment. The ability of public utilities, transportation systems, communications, critical facilities, engineered structures, and housing to withstand the impacts of extreme natural forces is not as strong as is believed.
8. Environmental impacts. Enabling human settlements in certain areas changes the exposure of environment and its vulnerability to disasters.
9. Postponing catastrophic losses. Some mitigation activities are not really preventing damage but merely postponing it. For example, communities behind a levee get a sense of security that fosters the construction of structures at risk to floods larger than those that were designed for. If the postponement amounts to many years, the accumulated losses could be enormous.
10. An interdisciplinary approach is required for solving disaster management problems.
11. The public must be involved in the management of disasters.
12. Institutional change, education, training, and cooperation are necessary in order to address the disaster management in the future.

Technical lessons

1. Integrated planning and management based on the use of systems analysis is a very efficient approach to finding solutions for complex disaster management problems.

2. Mathematical modeling tools have an application in disaster management.
3. Decision support tools including optimization models can be considered for mitigation, planning as well as emergency operational and recovery applications.
4. Improved tools for planning and decision making are necessary, together with well-coordinated databases.
5. Complex disaster decision-making processes require technical support.
6. Training and institutional development play an important role in the practical application of optimal disaster management strategies.
7. Uncertainty, ambiguity, constant change, and surprise characterize disaster management. Most management strategies unfortunately have been designed for a predictable world and static view of natural disasters.

The existing disaster management framework needs to evolve to begin to cope with the complexity of the factors that contribute to disasters in today's and tomorrow's world. As a start, this book offers shifts in thinking about disasters. Two new paradigms are identified that will shape future tools for management of natural disasters. The first focuses on the complexity of the disaster management domain and the complexity of the modeling tools in an environment characterized by continuous, rapid technological development. The second deals with disaster-related data availability and the natural variability of domain variables in time and space that affect the uncertainty of disaster management process.

1.2.1 The Complexity Paradigm

The first component of the complexity paradigm is that disaster management problems in the future will be more complex. Domain complexity is increasing (Figure 1.5). Further population growth, climate change, and regulatory requirements are some of the factors that increase the complexity of disaster management problems. Disaster management strategies are often conceived as too shortsighted (design life of dams, levees, bridges, etc.). Short-term thinking must be rejected and replaced with disaster management schemes that are planned over longer temporal scales in order to take into consideration the needs of future generations. Planning over longer time horizons extends the spatial scale. If resources for disaster management are not sufficient within the affected region, transfer from neighboring regions should be considered. The extension of temporal and spatial scales leads to an increase in the complexity of the decision-making process. Large-scale disaster management process affects numerous stakeholders. The environmental and social impacts of complex disaster management solutions must be given serious consideration.

The second component of the complexity paradigm is the rapid increase in the processing power of computers (Figure 1.5). Since the 1950s, the use of computers in disaster management has grown steadily. Computers have moved from data processing, through the user's office and into information and knowledge processing.

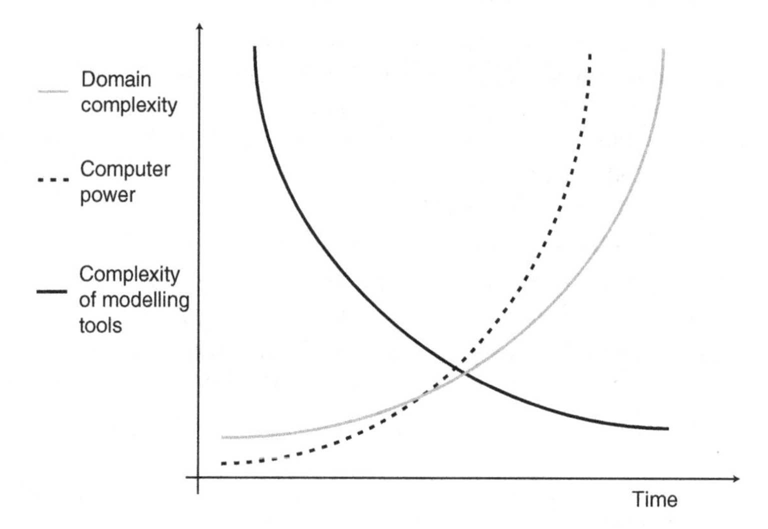

Figure 1.5 Schematic illustration of the complexity paradigm.

Whether the resource takes the form of a laptop PC or a desktop multiprocessing workstation is not important any more. It is important that the computer is used as a partner in more effective disaster management (National Research Council, 1996; Global Disaster Information Network, 1997; Stallings, 2002). The main factor responsible for involving computers in the disaster decision-making processes is the treatment of information as the sixth economic resource (besides people, machines, money, materials, and management).

The third component of the complexity paradigm is the reduction in the complexity of contemporary systems tools (again, see Figure 1.5). The most important advance made in the field of management in the last century was the introduction of systems analysis. Systems analysis is defined here as an approach for representing complex management problems using a set of mathematical planning and design techniques. Theoretical solutions to the problems can then be found using a computer. In the context of this book, systems analysis techniques, often called "operations research," "management science," and "cybernetics," include simulation and optimization techniques that can be used in four-phase disaster management cycle (discussed in detail in Chapter 2). Systems analysis is particularly promising when scarce resources must be used effectively. Resource allocation problems are very common in the field of disaster management and affect both developed and developing countries, which today face increasing pressure to make efficient use of their resources.

Simulation models can play an important role in disaster risk assessment, emergency management, and mitigation planning. Early simulation models were constructed by a relatively small number of highly trained individuals. These models were quite complex, however, and their main characteristics were not readily understood by nonspecialists. Also, they were inflexible and difficult to modify to accommodate site-specific conditions or planning objectives that were not included in the original model. The most restrictive factor in the use of simulation tools is that there is often a large number of feasible solutions to investigate. Even when combined with efficient techniques for selecting the values of each variable, quite substantial computational effort may lead to a solution that is still far from the best possible. Advances made during the last decade in computer software have brought considerable simplification to the development of simulation models (High Performance Systems, 1992; Lyneis et al., 1994; Powersim Corporation, 1996; Ventana Systems, 1996). Simulation models can be easily and quickly developed using these software tools, which produce models that are easy to modify, easy to understand, and that present results clearly to a wide audience of users. They are able to address disaster management problems with highly nonlinear relationships and constraints.

Numerous optimization techniques are available for use in disaster management too. Most resources allocation problems can be effectively addressed using linear programming (LP) solvers introduced in the 1950s (Dantzig, 1963). LP is applied to problems that are formulated in terms of separable linear objective functions and linear constraints. However, neither objective functions nor constraints are in a linear form in most practical disaster management applications. Many modifications can be used in real applications in order to convert nonlinear problems for the use of LP solvers. Examples include different schemes for the linearization of nonlinear relationships and constraints, and use of successive approximations. Nonlinear programming is an optimization approach used to solve problems when the objective function and the constraints are not all in the linear form. In general, the solution to a nonlinear problem is a vector of decision variables that optimizes a nonlinear objective function subject to a set of nonlinear constraints. No algorithm exists that will solve every specific problem fitting this description. However, substantial progress has been made for some important special cases by making various assumptions about these functions. Successful applications are available for special classes of nonlinear problems such as unconstrained problems, linearly constrained problems, quadratic problems, convex problems, separable problems, nonconvex problems, and geometric problems. The main limitation in applying nonlinear programming to disaster management problems is in the fact that nonlinear algorithms generally are unable to distinguish between a local optimum and a global optimum (except by finding another better local optimum). In recent years, there has been a strong emphasis on developing high-quality, reliable software tools for general use such as MINOS (Murtagh and Saunders, 1995) and GAMS (Brooke et al., 1996). These packages are widely used in different fields for solving complex problems. However, the main problem of global optimality remains an obstacle in the practical application of nonlinear programming. Dynamic programming (DP) offers advantages over other optimization tools since the shape of the objective function and constraints do not affect it. DP requires discretization of the problem into a finite set of stages. At every stage, a number of possible conditions of the system (states) are identified, and an optimal solution is identified at each individual stage, given that the optimal solution for the next stage is available. An increase in the number of discretizations and/or state variables would increase the number of evaluations of the objective function and core memory requirement per stage. This problem of rapid growth of computer time and memory requirement associated with multiple-state-variable DP problems is known as "the curse of dimensionality." Some modifications used to overcome this limitation of DP include discrete differential DP (an iterative DP procedure) and differential DP (a method for discrete-time optimal control problems). In the very recent past, most researchers have been looking for new approaches that combine efficiency and ability to find the global optimum. One group of techniques, known as evolutionary algorithms, seems to have a high potential. Evolutionary techniques are based on similarities with the biological evolutionary process. In this concept, a population of individuals, each representing a search point in the space of feasible solutions, is exposed to a collective learning process, which proceeds from generation to generation. The population is arbitrarily initialized and subjected to the process of selection, recombination, and mutation through stages known as generations, such that newly created generations evolve toward more favorable regions of the search space. In short, the progress in the search is achieved by evaluating the fitness of all individuals in the population, selecting the individuals with the highest fitness value and combining them to create new individuals with increased likelihood of improved fitness. The entire process resembles the Darwinian rule known as "the survival of the fittest." This group of algorithms includes, among others, evolution strategy (Back et al., 1991), evolutionary programming (Fogel et al., 1966), genetic algorithms (Holland, 1975), simulated annealing (Kirkpatrick et al., 1983), and scatter search (Glover, 1999). Significant advantages of evolutionary algorithms include:

- no need for an initial solution;
- easy application to nonlinear problems and to complex systems;
- production of acceptable results over longer time horizons; and
- the generation of several solutions that are very close to the optimum.

During the evolution of systems analysis, it has become apparent that more complex analytical optimization algorithms are being replaced by simpler and more robust search tools. Advances in computer software have also led to considerable simplification in the development of simulation models.

1.2.2 The Uncertainty Paradigm

The first component of the *uncertainty paradigm* is the increase in all elements of uncertainty in time and space (Figure 1.6). Uncertainty in disaster management can be divided into two basic forms: uncertainty caused by inherent variability of physical components of the system and uncertainty caused by a fundamental lack of knowledge. Awareness of the distinction between these two forms is integral to understanding uncertainty. The first form is described as *variability* and the second one as *uncertainty*.

Uncertainty caused by variability is a result of inherent fluctuations in the quantity of interest (i.e., the atmosphere, biosphere, hydrosphere, and lithosphere). The three major sources of variability are temporal, spatial, and individual heterogeneity. Temporal variability occurs when values fluctuate over time. Values affected by spatial

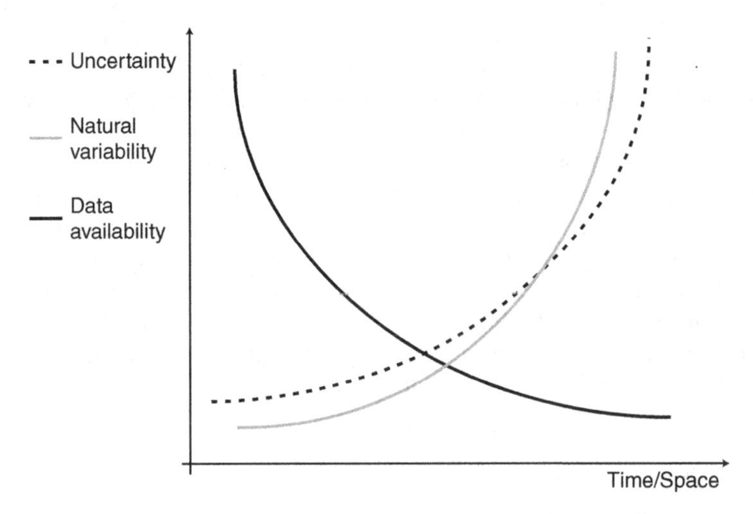

Figure 1.6 Schematic illustration of the uncertainty paradigm.

variability are dependent upon the location of an area. The third category effectively covers all other sources of variability. In disaster management, variability is mainly associated with the spatial and temporal variation of physical variables (precipitation, river flow, wind speed, etc.). The more elusive type of uncertainty is caused by a fundamental lack of knowledge. It occurs when the particular values that are of interest cannot be assessed with complete confidence because of a lack of understanding or limitation of knowledge.

The second component of the uncertainty paradigm is the decrease in disaster data availability (Figure 1.6). Meteorological information on cloud cover, fog, rain, high winds, hail, snow cover, and runoff are necessary for severe weather disaster management. The numbers of gauging stations in operation worldwide, as reported by World Meteorological Organization (WMO), is very impressive. The *INFOHYDRO Manual* (WMO, 1995) estimates that there are nearly 200,000 precipitation gauges operating worldwide and more than 12,000 evaporation stations. For example, monitoring is taking place at more than 64,000 stations for discharge, at nearly 38,000 for water level, at 18,500 for sediment, at more than 100,000 for water quality, and at more than 330,000 for groundwater characteristics. Despite the apparently high global numbers, the stations are not uniformly distributed, and there is a shortage over large areas. The financial constraints of government agencies that are responsible for the collection of disaster-related data have resulted in reductions in the data collection program in many countries. In many countries, disaster data collection activities are very fragmented. A similar fragmentation is observed at the international level. Of particular concern are the gaps in the existing data relative to the informational requirements. Many authors agree that current data collection networks are inadequate for providing the information required to understand and explain changes in natural systems. Given the reductions in the funding of data collection activities, it is clear that a change in the approach to data collection activities is essential.

The third component of the uncertainty paradigm is the increase in natural variability of disaster-related factors of the earth's physical systems (again, see Figure 1.6). Water flow of importance for flooding exhibits both temporal (between years and seasons) and spatial variation. The water flow from the basin is the integrated result of all physical processes in the basin. The topography, the spatial distribution of geological phenomena, and land use are the main causes of spatial variability of flow. Observed natural variability is being affected by climate change. One of the most important aspects of studying the consequences of global warming is estimating possible changes in the extreme conditions (maximum and minimum river discharges, temperatures, etc.). On the one hand, an increase in maximum floods can be expected, and on the other hand, so can a more frequent occurrence of severe droughts. Both could have major economic and ecological consequences.

1.3 CONCLUSIONS

The idea of surprise extremes needs to be broadened (Mileti, 1999). Disaster researchers and practitioners around the world would benefit from embracing a systems approach in their work. Viewed in systems way, disasters can be seen as the anticipated result of interactions among the earth's physical systems, human systems, and the constructed environment. All these systems and their subsystems are dynamic and with constant interactions among them. This complexity is what makes disaster problems difficult to solve.

Various systems tools are available for potential implementation in disaster management. They rely on mathematical modeling of real physical systems and transfer of solutions found to work on the models into real environments. In the past, stakeholders not actively involved in the development of a model tended to mistrust the results of the model. Computer power has increased and costs have fallen to the point that all stakeholders in the resource can play a very important role in disaster management.

Technology is already a facilitating force in political decision making, and will be more so in the future. Spatial decision support systems using object-oriented programming algorithms are integrating transparent tools that will be easy to use and understand. National and international databases, both static and dynamic, now provide much of the necessary information in digital form. The trend will continue for providing public access to all disaster-related data at reasonable cost and in a user-friendly format, and this will play an important role in supporting tools for disaster decision making.

The speed with which data and ideas can be communicated has historically been a control mechanism of scientific progress. The Internet began in 1968 by connecting four hosts. As of March 2008, almost 1.5 billion hosts were connected to multiple computer networks according to the Internet Usage and World Population Statistics (http://www.internetworldstats.com/stats.htm, accessed July 2008). Virtual libraries, virtual databases, virtual forums and bulletin boards, web-enabled software packages, and the use of "write once—run anywhere" languages (such as Java by Sun Microsystems) will create new opportunities for disaster managers.

The future of disaster management will be difficult in both the developing and developed world. My hope is that the tools discussed in this book, supported by good data communicated through powerful networks, will empower people to make wise decisions on how to make best use of limited resources and minimize disaster losses.

REFERENCES

Back, T., F. Hoffmeister, and H.P. Schewel (1991), "A survey of evolution strategies," in *Proceedings of the Fourth International Conference on Genetic Algorithms*, Morgan Kaufmann, San Mateo, CA.

Brooke, A., D. Kendrik, and A. Meeraus (1996), *GAMS: A User's Guide*, Scientific Press, Redwood City, CA.

Bunde, A., J. Kropp, and H.J. Schellnhuber (eds) (2002), *The Science of Disasters: Climate Disruptions, Heart Attacks, and Market Crashes*, Springer, Berlin, Germany.

Dantzig, G.B. (1963), *Linear Programming and Extension*, Princeton University Press, Princeton, NJ.

Fogel, L.J., A.J. Owens, and M.J. Walsh (1966), *Artificial Intelligence Through Simulated Evolution*, John Wiley, Chichester, UK.

Global Disaster Information Network (1997), "*Harnessing information and technology for disaster management*," Report, Disaster Information Task Force, Washington, p. 115.

Glover, F. (1999), "Scatter search and path relinking," in D. Corne, M. Dorigo, and F. Glover (eds) *New Methods in Optimization*, McGraw-Hill, New York.

Godschalk, D.R., T. Batley, P. Berke, D.J. Brower, and E.J. Kaiser (1999), *Natural Hazard Mitigation: Recasting Disaster Policy and Planning*, Island Press, Washington.

High Performance Systems (1992), *Stella II: An introduction to Systems Thinking*, High Performance Systems, Inc., Hanover, NH.

Holland, J.H. (1975), *Adaptation in Natural and Artificial Systems*, University of Michigan Press, Ann Arbor, MI.

International Red River Basin Task Force (IRRBTF) (1997), *Red River Flooding: Short-Term Measures*, Interim Report to the International Joint Commission, Ottawa/Washington, available online, http://www.ijc.org/php/publications/html/taskforce.html, last accessed July 21, 2008, p. 71.

International Red River Basin Task Force (IRRBTF) (2000), *The Next Flood: Getting Prepared*, Final report to the International Joint Commission, Ottawa/Washington, available online, http://www.ijc.org/rel/pdf/nextfloode.pdf, last accessed July 21, 2008, p. 167.

Kirkpatrick, S., C.D. Gelatt, Jr., and M.P. Vecchi (1983), "Optimization by simulated annealing," *Science*, 220(4598): 671–680.

Kohler, A., S. Julich, and L. Bloemertz (2004), "Risk analysis xxx A basis for disaster risk management," *Guidelines*, Federal Ministry for Economic Cooperation and Development, Germany, p. 71.

Lyneis, J., R. Kimberly, and S. Todd (1994), "Professional dynamo: Simulation software to facilitate management learning and decision making," in Morecroft, J. and J. Sterman (eds) *Modelling for Learning Organizations*, Pegasus Communications, Waltham, MA.

Mileti, D.S. (1999), *Disasters by Design*, Joseph Henry Press, Washington.

Morris-Oswald, M. and S.P. Simonović (1997), "Assessment of the social impact of flooding for use in flood management in the Red River Basin," Report prepared for the International Joint Commission, Slobodan P. Simonović Consulting Engineer, Inc., Winnipeg, available online, http://www.ijc.org/php/publications/html/assess.html, last accessed July 21, 2008.

Munich, Re. (2008), NatCatSERVICE, *online database*, http://www.munichre.com/en/reinsurance/business/non-life/georisks/natcatservice/default.aspx, last accessed July 17, 2008.

Murtagh, B.A. and M.A. Saunders (1995), *MINOS 5.4 User's Guide, Technical report SOL 83-20R*, Systems Optimization Laboratory, Department of Operations Research, Stanford University, Stanford, CA.

National Research Council (1996), "Computing and communications in the extreme: Research for crisis management and other applications," Report, the Steering Committee on High Performance Computing and Communications, National Academy Press, Washington, p. 159.

Nenadic, D. (1982), *Poplava*, Narodna knjiga, Beograd (in Serbian language), Serbia.

Powersim Corporation (1996), *Powersim 2.5 Reference Manual*, Powersim Corporation, Inc., Herndon, VI.

Province of Manitoba (2004), "Red River Valley Flood Protection," *online support system*, http://geoapp.gov.mb.ca/website/rrvfp/, last accessed July 21, 2008.

Province of Manitoba (2005), "Flood Information," *online service*, last accessed July 21, 2008, http://www.gov.mb.ca/waterstewardship/floodinfo/index.html.

Royal Commission on Flood Cost-Benefit (1958), *Report*, Winnipeg, Manitoba, Canada.

RRBDIN (2005), "Red River Basin Decision Information Network," *online network*, http://www.rrbdin.org, last accessed July 21, 2008.

Simonović, S.P. (2004), "CANADA—Flood Management in the Red River, Manitoba, Integrated Flood Management Case Study," World Meteorological Organization—The Associated Programme on Flood Management, available online, last accessed July 20, 2008, http://www.apfm.info/pdf/case_studies/cs_canada.pdf.

Skipper, H.D and W. Jean Kwon (2007), *Risk Management and Insurance: Perspectives in a Global Economy*, Blackwell Publishing, Oxford, UK.

Stallings, R.A. (ed.) (2002), *Methods of Disaster Research*, Xlibris Corporation Publishing, Philadelphia, PA.

UN/ISDR (2004), *Living with Risk: A Global Review of Disaster Reduction Initiatives*, United Nations Inter-Agency Secretariat International Strategy for Disaster Reduction (ISDR) report, Geneva, Switzerland, p. 387.

Ventana Systems (1996), *Vensim User's Guide*, Ventana Systems, Inc., Belmont, MA.

WMO (1995), *INFOHYDRO Manual*, Hydrological Information Referral Service, Operational Hydrology Report No. 28, WMO-No. 683, Geneva, Switzerland.

Woo, G. (1999), *The Mathematics of Natural Catastrophes*, Imperial College Press, London, UK.

EXERCISES

1.1 Describe the largest disaster experienced in your region.

(a) What were its physical characteristics?

(b) Who was involved in the management of the disaster?

(c) What is, in your opinion, the most important disaster management problem in your region?

(d) Give some examples of the disaster mitigation works in the region.

(e) What lessons can be learned from the past management of disasters in your region?

(f) What are the most important principles you would apply in future disaster management in your region?

1.2 Review the literature and find a definition of integrated disaster management.

(a) Discuss the Red River example presented in this chapter in the context of this definition.

(b) What would you do, in addition to what has been done, in this case to make flood management decisions in the Red River basin sustainable?

1.3 Discuss characteristics of the disaster from Exercise 1.1 in the context of two paradigms presented in Section 1.2.

(a) What are the complexities of the problem in Exercise 1.1?

(b) Identify some uncertainties in the problem.

(c) Can you find some data to illustrate the natural variability of regional conditions?

(d) How difficult is to find the data? Why?

1.4 For the problem in Exercise 1.1, identify the factors that will provide for sustainable disaster management decisions. What are the spatial and temporal scales to be considered?

2 Integrated Disaster Management

Integrated disaster management requires collaboration and coordination from international and national to local levels. Usually national services are involved in monitoring potential hazards and issuing warnings. This, however, is one component of a much larger *system* that must be in place to prevent and reduce disasters. The details of such systems vary from one country to another. When they are organized efficiently the disaster management plan is clear and unambiguous; the warnings arrive in time and are reliable; people are informed and ready for action; and lives and livelihoods can be protected.

2.1 DEFINITION

Implementation of the disaster management plan involves government departments and agencies at every level, from the national government to local communities, or from local police to fire, health, and social services. It is very important for the success of the plan that each of these bodies has clearly defined authority and responsibility. Ideally, the plan and these responsibilities should be defined by legislation in order to remove all possible ambiguities and to impose on each body an enforceable duty to play its part.

The disaster mitigation plan, in its most general form, follows the disaster management cycle that has four closely integrated components: mitigation, preparedness, response, and recovery (Figure 2.1). Most of the traditional presentations of the integrated disaster management cycle involve graphing these four phases as independent components that follow each other. In this chapter, the Venn diagram presentation is introduced to capture the idea of integrated disaster management where each component of the cycle has some overlapping activities with other components.

This provides an opportunity to introduce the definition of integrated disaster management that will be followed in this book. "Managing," in everyday language, means to handle or direct with a degree of skill. It also can be seen as exercising executive, administrative, and supervisory duties. Managing can involve altering a situation by manipulation, or succeeding in accomplishing something. Management is the act or art of managing.

Systems Approach to Management of Disasters: Methods and Applications, By Slobodan P. Simonović

Figure 2.1 The Venn diagram of integrated disaster management.

Integrated disaster management is an iterative process of decision making regarding prevention of, response to, and recovery from, a disaster. This process provides a chance for those affected by a disaster to balance their diverse needs for protecting lives, property and the environment, and to consider how their cumulative disaster-related actions may affect long-term sustainability of the affected region. The guiding principles of the process are systems view, partnerships, uncertainty, geographic focus, and reliance on strong science and reliable data.

2.2 INTEGRATED DISASTER MANAGEMENT ACTIVITIES

The definition of integrated disaster management includes traditional activities of mitigation, preparedness, response, and recovery. It is more comprehensive and integrates all these activities in an approach to support the decision-making process based on the natural, social, engineering, and other sciences.

2.2.1 Mitigation

Mitigation is long-term planning and it involves identifying the vulnerability of every part of the territory to particular types of hazards, and identification of steps that should be taken to minimize the risks. Disaster mitigation measures are those that eliminate or reduce the impacts and risks of hazards through proactive measures taken before an emergency or disaster occurs. These steps can include modifying building codes to ensure that buildings can withstand earthquake and high winds, forbidding building on land that is prone to flooding and identification of evacuation procedures.

One of the best-known examples of investment in disaster mitigation in Canada is the Red River Floodway. It was built to protect the City of Winnipeg and to reduce the impact of flooding in the Red River Basin. It cost $60 million to build in the 1960s. Since then, the floodway has been used more than 20 times. Its use during the 1997, Red River Flood alone saved an estimated $8 billion.

Mitigation involves government at many levels, emergency services and the military, but also relief agencies. In the mitigation case, all partners in the integrated disaster management will determine what tools and personnel they will require, what training to offer, what outreach products must be prepared and distributed to the communities in the hazard-prone areas.

Many tools and techniques were used for coping with natural hazards and disasters. Disaster mitigation measures may be structural (e.g., flood dikes) or nonstructural (e.g., land use zoning). Mitigation activities should incorporate the measurement and assessment of the evolving risk environment and may include the creation of comprehensive, proactive instruments that enable the prioritization of risk reduction investments. A few examples of mitigation measures include (i) hazard mapping, (ii) adoption and enforcement of land use and zoning practices, (iii) implementing and enforcing building codes, (iv) floodplane mapping, (v) reinforced tornado safe rooms, (vi) burying of electrical cables to prevent ice buildup, (vii) raising of homes in flood-prone areas, (viii) disaster mitigation public awareness programs, and (ix) insurance programs.

The five categories of measures briefly presented here have proven over last three decades to be the most useful for minimizing and/or redistributing losses and reducing social and economic disruption (Mileti, 1999).

Land Use Planning and Management Local governments have many land use management options at their disposal for increasing community resilience: building standards, development regulations, critical and public facilities policies, land and property acquisition, taxation, planning processes, and public information. Land use planning and management leads to communities where people and property are kept out of the areas exposed to hazards, where development is designed to be resilient to natural forces, and where the mitigative capacity of natural environmental systems is not diminished. Many political, social, and economic forces promote development decisions that set the stage for future catastrophes. On the other side, land use planning and management can reduce disasters to a scale that can be managed by the governments, communities, individuals, and businesses exposed to them.

In many countries, land use planning and management are not governed on the federal or state level. These levels of government seem unwilling to take a role in active land use management and they communicate to local governments conflicting signals about exposure to hazards—advocating risk reduction and transfer rather than risk assumption and elimination. On the local level, there is a lack of political will to manage land use. Few local governments are ready to manage disasters by managing development. Like many individuals, local governments view disasters as a minor problem compared to more local pressing concerns such as unemployment, housing, education, and crime. Also, the costs of land use management are immediate while the benefits are uncertain, not easily visible, and may not occur during the tenure of elected officials.

Promoting appropriate land use for local conditions (a) keeps people and property out of hazardous areas, (b) provides more affordable housing and living conditions, (c) protects the environment, and (c) reduces the costs of growth and development.

Engineering Implementation of the state-of-the-art engineering approaches has been successful in reducing mortality rates from hazards and disasters. One, already mentioned, excellent example is the Red River Floodway in Canada—built to protect the City of Winnipeg and to reduce the impact of flooding in the Red River Basin. Used many times since the 1960s, when it was built, it prevented during the 1997 Red River Flood alone an estimated $8 billion in damage.

One of the important aspects of relying on engineering solutions in managing disasters is the potential increase in future damage that may come from ever-more-extensive development in hazardous places. Nevertheless, carefully engineered buildings and infrastructure will be essential to the future disaster resiliency of all localities. The task will be to accurately assess all of the hazards involved and balance them against the benefits to be gained, according to a given community's preferences. Localities will need to reach their own decisions about the level of hazard they believe is appropriate to their situation and balance that with the desired quality of life and protection of the environment.

Part of the decision-making process in integrated management of disasters will be to determine the extent to which an engineering solution will reduce future hazard, for how long, against what magnitude of extreme event, at what economic cost, and with what environmental impact. This integrated disaster management framework is complex, unique to each locality, and changing continuously with time. Tools presented in this book provide support for the practical implementation of systems approach in integrated disaster management.

Engineering codes, standards, and practice have evolved (see the next section). For buildings and critical infrastructure, these codes vary widely on the basis of local perceptions of benefits, costs, and risks. Affordability is a key factor in whether engineering codes and standards are changed. Thus, the establishment and enforcement of codes and standards are a problem of social choice and are not based exclusively on technical feasibility.

Buildings, in systems view, are complex combinations of the basic foundation and structure, plumbing, electrical, heating, ventilation, air conditioning, and ancillary systems. At the same time, buildings are used to provide shelter, comfort for families, place for work, space for entertainment, and a host of other social functions. The structural aspects of a building consider flooding, earthquake, high-wind, and related disaster loads. Social aspects of a building during a disaster consider loss of shelter, comfort, and other functionalities it provides.

Infrastructure includes all of the structural components that provide connections among human developments, such as electric power lines, transportation networks, water supply systems, drainage and water treatment systems, gas lines, communications networks, and hazardous materials storage facilities. For infrastructure, the engineering process integrates all of the hazard impacts as part of the design and management.

Building Codes and Standards The quality of buildings and infrastructure is directly related to loss of life, injuries, and the financial costs of disasters. Therefore, disaster-resilient construction is an essential component of local resiliency. The regulation of building construction in Canada, as in many other countries, is accomplished through building codes—a collection of laws, regulations, ordinances, or other requirements adopted by a government legislative authority related to physical structure of buildings. Building codes provide the minimum acceptable requirements necessary (a) to preserve the public safety, health, and welfare and (b) to protect the property and the built environment.

The primary application of building codes is to regulate new or proposed construction. They have little application to existing buildings (unless they are undergoing reconstruction, rehabilitation, or alterations). The set of codes generally consists of a number of documents divided more for convenience than for specific technical or legal reasons. In Canada (NRC, 2008), these codes are as follows:

- National Building Code of Canada 2005
- National Fire Code of Canada 2005
- National Plumbing Code of Canada 2005
- National Farm Building Code of Canada 1995
- National Housing Code of Canada and Illustrated Guide 1998
- Model National Energy Code of Canada for Houses 1997
- Model National Energy Code of Canada for Buildings 1997

The Canadian Commission on Building and Fire Codes (CCBFC) oversees the work of a number of technical standing committees, whose members apply their experience to develop and improve the codes that protect the health and safety of Canadians. Representing all major facets of the construction industry, Commission members include building and fire officials, architects, engineers, contractors, and building owners as well as members of the public. The input of provincial governments, municipalities and the construction industry is crucial to the strength of Canada's codes. The current code-writing process has one of the most extensive- and involved-public review procedures in the world.

Standards are prescribed sets of rules, conditions, or requirements with definition of terms; classification of components; delineation of procedures; specification of dimensions, materials, performance, design, or operations; description of fit or measurement of size; measurement of quality and quantity of describing materials, products, and systems; and services and practices. There are three basic classifications of standards used in building codes: (i) engineering practice standards, (ii) materials standards, and (iii) test standards.

Canadian Standards Association (CSA) (http://www.csa.ca) is a not-for-profit, nonstatutory, voluntary membership association engaged in standards development and certification activities. CSA has developed more than 2000 standards covering the life sciences, environment, electrical and electronic products, communications systems, building construction, energy, transportation/distribution, materials technology, and quality business management. CSA standards are often incorporated into government regulation.

Even with a nationwide code, the administration and enforcement of all building codes rests with the local governments, with varying degrees of oversight by higher levels of governments. The local government is responsible for creating the organizational structure for the code enforcement. The local enforcement entity, directed by the code enforcement official, can come in any size or shape that are determined by the amount and nature of construction activity, the relative importance of code enforcement in the priorities of the jurisdiction, and the resources (specially financial) that are available to support the activity.

A building permit process is established for the review, inspection, and approval of proposed activities to secure compliance with the building code. A certificate of occupancy is issued when all inspections have been performed, any deficiencies corrected, and the construction work completed. Any code is only as good as the enforcement that goes along with it.

Currently, however, there are significant reasons why building codes may fall short of their disaster loss reduction potential (Mileti, 1999). For example, building codes are for life safety and do not provide for property protection or functionality after a disaster.

Prediction, Forecast, and Warning A complete and effective warning system comprises four elements, spanning knowledge of the risks faced through to preparedness to act on early warning. Failure in any one part can mean failure of the whole system. (1) Risk knowledge phase involves systematic data collection and risk assessments. (2) Monitoring phase involves development of hazard monitoring and early warning services. In this phase questions such as—Are the right parameters being monitored? Is there a sound scientific basis for making forecasts? Can accurate and timely warnings be generated?—should be answered. (3) Dissemination and communication phase provides for communication of risk information and early warnings to all of those at risk in understandable form and on time. (4) Response capability development phase is aimed at building national and community response capabilities.

Good warning systems have strong linkages between the four elements. The major players concerned with the different elements meet regularly to ensure they understand all of the other components and what other parties need from them. Risk scenarios are constructed and reviewed. Specific responsibilities throughout the chain are agreed and implemented. Past events are studied and improvements are made to the warning system. Manuals and procedures are agreed and published. Communities are consulted and information is disseminated. Operational procedures such as evacuations are practiced and tested. Behind all of these activities lies a solid base of political support, laws and regulations, institutional responsibility, and trained people.

Often there are warning signs well ahead of a hazard event. Weather forecasters observe the start of hurricanes and storms and can calculate their likely future strength and tracks; vulcanologists can interpret telltale rumblings below the earth; drought experts estimate the chances of refreshing rains. Good early warning systems also need to consider community vulnerabilities as well as the hazards. What are the early warning signs for vulnerability? Key signs are growing poverty, environmental degradation, populations crowded in risky locations, civil strife, and lack of knowledge and preparedness.

All warning systems, however simple they may be, are based on some idea—model—of how the phenomenon behaves. Models are then used to say what is likely to happen next. In the simplest cases, the model may amount to no more than common sense—for example, the recognition that poor people who have settled in a river valley will lose their dwellings and all their belongings in even a small flood. At the other extreme, models of the physics of the global weather system are immensely complex and require large computers to do all their calculations and produce detailed forecasts for the whole globe. Predictions are never perfect or precise. There is always some uncertainty. Often the warning may be expressed statistically—for example, as a 60% chance of intensive rain hitting a region in the next 24 hours. In addition, there is always a great deal of uncertainty about the social components of a warning and about the specific human impacts that might occur. This can make it difficult for decision makers to act. Decision makers and those at risk must weigh up the chances and consider the implications for their particular situation. Which option to follow—to close down a subway and disrupt a city or let the city keep running and risk chaos? To follow advice to evacuate or stay put and try to protect your home? Each person or community must make its own judgment as appropriate to their own circumstances. In each case, there are costs of acting as well as costs of not acting.

Canada has no national warning strategy that covers all hazards in all places. Public warning practice is decentralized across different governments and the private sector. Public alert systems can be improved with the new hardware and technology, but disseminating warning preparedness knowledge is a much more complicated problem. There is a threat that further technological advances will only increase the gap between the practice and the state of the art. Improvements to local warning systems can be done by (a) improving procedures, training, knowledge sharing and management practices, and (b) investing more seriously in better communications and warning system equipment. Most advances in Canada have come from better monitoring, instrumentation, data collection, and modeling.

Insurance In places where available, many property owners rely on private insurance to protect themselves against losses from natural disasters. Insurance coverage varies significantly among countries and is now available for some but not all natural disasters. Insurance coverage is nearly everywhere available for wildfires, winter storms, volcanoes, tornadoes, lightning, and hail. These perils are covered under most standard property insurance contracts. Generally speaking, these events are sufficiently random and widespread to permit the private insurance mechanism to operate effectively. Landslides are normally not considered an insurable peril by private insurers. They are covered by insurance programs only if the damage is caused by an insured earthquake or flood damage. Hurricane wind damage is included as part of the basic wind coverage in most property insurance policies. Flood damage from hurricanes is not often included but can be purchased separately. Insurance coverage for damage from earthquakes is also not automatically included in homeowners' insurance policies, but it can be purchased at many places as a rider for an additional premium. Earthquake coverage is often included on commercial policies for structures in hazard-prone areas, and it can also be purchased separately. Protection against loss by fires that might follow an earthquake is included in the basic fire peril coverage in most property insurance contracts. Business interruption from an earthquake is covered by a separate policy. Insurance companies have viewed flood risk as uninsurable because of problems of both adverse selection and highly correlated risks.

Insurance providers generally are able to pay for common losses from the annual premiums collected, supplemented by investment income and reinsurance recoveries. In the United States, the recent suffering created by Hurricane Katrina has Congress rethinking strategies for natural disaster insurance with an initiative for a public–private partnership to spread the risk on a national scale. Currently, homeowners need a separate policy for the resulting fires that more often than not spring up as a result of the earthquake. In the case of hurricane insurance purchase, a separate flood insurance policy is required to cover water damage.

In Canada, most of disaster insurance comes through home insurance policy (Bain and Mitchell, 2006). Homeowner insurance policies cover the building and its contents for "direct loss" or damage caused by insured perils, which may be stated individually or merely described as "all risks." Some insurers offer very basic coverage for properties that do not meet their normal underwriting standards.

Insurable perils include aircraft or vehicle impact; electrical current; explosion; falling object (not including objects propelled by snowslide or earth movement); fire; lightning; riot; smoke (released suddenly from malfunctioning cooking or heating devices, but not from fireplaces); theft; transportation (of personal property while it is temporarily away from your home; includes building fixtures and fittings being repaired or in seasonal storage); vandalism (while building is normally occupied); water damage (more details below); wind and hail (applies to the outside of a building except for antennas, satellite dishes, etc.; the interior of a building and its contents are covered only if the storm has first created an opening); and window glass (breakage in a building that is normally occupied; not covered by tenant insurance).

Water damage coverage usually includes damage arising from sudden and accidental escape of water from an indoor plumbing, heating, sprinkler, or air-conditioning system; or from an indoor or outdoor "domestic appliance" on the premises; or from a water main. "Domestic appliance" is usually defined in the policy and includes water heaters, water beds, and swimming pools. Water may be in liquid form, or it may be steam or ice. However, coverage for freezing damage is restricted to property *inside* the home and there are special requirements if the home is unoccupied for more than four consecutive days during the normal heating season. Coverage usually excludes damage arising from (a) floodwater such as that from an overflowing creek, (b) repeated or continuous water seepage (from a cracked basement wall, e.g., or from an unrepaired pipe), and (c) sewer backup.

Coverage for the following perils, not normally included in any type of home insurance policy, may often be purchased separately in Canada: earthquake—particularly worth considering in quake-prone regions of Quebec and British Columbia. Cost varies according to location and type of construction, furnace oil spills, sewer backup—useful in some low-lying areas, particularly those with combined storm and sanitary sewers.

Regulation influences the supply of disaster insurance by controlling insurers' entry to and exit from insurance markets. In principle the job of regulators is to protect the public. However, regulation increasingly has been influenced by public perceptions and preferences on how the cost of risk should be shared among different groups. In many countries, home insurance is currently a minefield of fine print and exclusions when it comes to insuring against natural disasters.

Insurance is not considered a mitigation measure because it redistributes rather than reduces disaster losses. Well-designed insurance programs can, however, encourage the adoption of loss reduction measures by creating incentives through rate discounts, lower deductibles, and higher coverage limits.

To deal with natural disasters of the magnitude now predicted, the new insurance strategies should consider improved risk assessment, documentation of damage resistance, stronger enforcement, equitable approach, and probably public–private partnerships in order to protect the insurance industry against insolvency. Achievement of these goals requires direct involvement of government at all levels to link insurance and mitigation.

Summary The mitigation tools reviewed briefly in this section of the text are not new. They have always been available for, and they are essential for, the future of integrated disaster management. The challenge for future disaster managers and researchers will be to combine these tools in the most effective and economical ways to increase community or regional resiliency. The systems approach to disaster management requires:

- wise use of long-term land-use plans;
- construction practices, engineering approaches, and building codes that go beyond life safety toward protecting the functionality of structures located in disaster-prone areas;
- improved preparation, dissemination, and use of hazard forecasts in disaster management decision making; and
- insurance mechanisms as a vehicle to foster mitigation efforts.

2.2.2 Preparedness

The purpose of preparedness is to anticipate problems in disasters so that ways can be devised to address the problems effectively and so that the resources needed for an effective response are in place beforehand. Preparedness includes such activities as formulating, testing, and exercising disaster plans; providing training for disaster responders and the general public; and communicating with the public and others about disaster vulnerability and what to do to reduce it.

Preparedness also involves the wider community. Public must be educated about the nature of the hazards it faces and how to recognize and respond safely to them. It is important that the citizens become familiar with the way in which integrated disaster management and emergency services communicate information on, and warnings of, severe weather. Individuals and families should have personal action plans, lists of emergency supplies to have on hand, and a clear understanding of how to coordinate their own safety.

Citizens must also know how best to cooperate with each other and with the emergency services. Carrying out exercises and drills can be used to expose any sources of confusion or deficiencies in the way that information is disseminated. Each community should refine its local plan according to its particular circumstances and should consider contingency plans in case any aspect of the emergency plan fails for some unforeseen reason.

Preparedness, as well as recovery, in Canada is coordinated by the federal government that develops national policy, response systems and standards, issues timely alerts, and similar products. Federal government also works closely with emergency management organizations (EMOs) across Canada and supports regional partners and first responders with funds, tools, and training. Provincial and territorial EMOs are active in planning and research, training, response operations and the administration, and delivery of disaster financial assistance programs. EMOs are a good source of information about how to prepare for emergencies in a region. Other partners in preparedness are fire, police, and medical services. All of them, up to a certain level, plan independently for disaster management.

The Government of Canada established the Joint Emergency Preparedness Program (JEPP) to enhance the national capability to manage all types of emergencies and ensure a reasonably uniform emergency response and recovery capacity across Canada.

The government, in consultation and cooperation with provincial and territorial governments, contributes to emergency preparedness and critical infrastructure protection projects and initiatives. Whether it is through training, the purchase of emergency response equipment, for emergency planning or for capacity building, this shared investment is aimed at reducing injuries and loss of human life, property damage, and assuring the continuation of our critical services in an emergency.

The National Exercise Program (NEP) consists of training courses and operation center exercises. Government, first responders, and military officials work together in these exercises, simulating emergency scenarios. The exercises are often conducted with multiple departments, including provincial, territorial, and municipal governments, to ensure efficient and effective responses. Regional exercises entail collaborations between municipal, provincial/territorial, and federal agencies. Public Safety Canada (PSC) is at the highest level of the federal response: (i) coordinating all federal departments that may be implicated in a given emergency, (ii) overseeing the cycle of federal activities during an exercise, (iii) collecting data, (iv) measuring response times, and (v) identifying strengths and weaknesses. As well, PSC works with the provinces and territories to further develop the NEP, creating policies and standards, identifying priorities, and developing other potential exercise scenarios.

In general, disaster emergency preparedness is influenced and constrained by institutional power differences; the nature of intergovernmental system itself; the complexity of agencies, responsibilities, and legislation; and the difficulty of effective interagency coordination.

2.2.3 Response

Disaster response activities include emergency sheltering, search and rescue, care of injured, fire fighting, damage assessment, and other emergency measures. First responders must also cope with response demands such as the need for coordination, communications, ongoing situation assessment, and resource mobilization during emergency period.

The response phase of the integrated disaster management involves the implementation of the measures developed during the mitigation and preparedness phases. Usually, the national service will provide up-to-the-minute weather advisories and warnings. Emergency, health and social services, volunteers, and citizens have their own parts of the overall plan to enact, and should be able to calmly set about ensuring the safety of the community.

Emergencies in Canada are managed first at the local level, for example, hospitals, fire departments, police, and municipalities. If they need assistance, they request it from the provinces or territories. If the emergency escalates beyond their capabilities, the provinces or territories seek assistance from the federal government. The coordination and provisioning of resources can move quickly from the local to the national level.

EMOs in Canada play an important role in disaster response as well as fire, police, and medical services. In responding to disaster they face a number of challenges. They must mobilize; assess the nature of emergency; prioritize goals, tactics, and resources; and coordinate with other organizations and public. All of these activities must be accomplished under conditions of uncertainty, urgency, limited control, and limited access to information. The main challenge of emergency response is to overcome the natural tendency of all organizations involved to maintain their independence and autonomy during the emergency management.

2.2.4 Recovery

The term *recovery* has been used interchangeably with reconstruction, restoration, rehabilitation, and postdisaster redevelopment. Regardless of the term used, the meaning has generally implied putting a disaster-affected community back together. Early views of recovery almost exclusively saw it as reconstruction of the damaged physical environment (Mileti, 1999). Its phases went all the way from temporary measures to restore community functions through replacement of capital stocks to predisaster levels and returning the appearance of the community to normal to the final phase, which involved promoting future economic growth and development. Usually communities try to reestablish themselves in forms similar to predisaster patterns and the resulting continuity and familiarity in postdisaster reconstruction enhance psychological recovery. Others view disasters as opportunities to address long-term material problems in local housing and infrastructure. They recast reconstruction into a developmental process of reducing vulnerability and enhancing economic capability (Anderson and Woodrow, 1989).

The most recent view is that recovery is not just a physical outcome but a social process that encompasses decision making about restoration and reconstruction activities (Berke et al., 1993). This view highlights how decisions are made, who is involved in making them, what consequences those decisions have on the community, and who benefits and who does not. The process approach also stresses the nature, components, and activities of related and interacting groups in a systemic process and the fact that different people experience recovery differently. Rather than viewing recovery as a linear process, this approach views the recovery process as probabilistic and recursive.

Analyses of many disasters have demonstrated that there are patterns in the recovery process. Recovery is a set of actions that can be learned about and implemented accordingly. It is possible to anticipate some of the main concerns, problems, and issues during recovery and to plan for them as a part of mitigation phase. Local participation and initiative must be achieved. Some of the key deterrents to speedy recovery are exclusion of local involvement; poorly coordinated and conflicting demands from various government-assisted programs; staff who are poorly prepared to deal with aid recipients; top-down, inflexible, standardized approaches; and aid that does not meet the needs of the affected population.

With each disaster, new knowledge is gained about how to plan more effectively for recovery and reconstruction. All participants in the integrated disaster management should at this stage assess how well the plan worked, its strengths and weaknesses, and start the process of revising the management plan for greater success the next time. This is where the recovery overlaps with the mitigation phase, and the cycle of integrated disaster management continues.

2.3 DISASTER MANAGEMENT IN CANADA—BRIEF OVERVIEW

Disaster management in Canada is carried out by different levels of government in many departments and ministries. There are some deficiencies in their coordination. However, PSC is established to provide, among other tasks, for the coordination of disaster management process. The broad portfolio of PSC is well suited to effective disaster management.

2.3.1 Emergency Management Act

When an emergency strikes, lives are at stake and effective response means knowing who is in charge. The *Emergency Management Act* (PSC, 2008a) sets out clear roles and responsibilities for all federal ministers across the full spectrum of emergency management. This includes prevention/mitigation, preparedness, response and recovery, and critical infrastructure protection.

The main purpose of the Act is to strengthen emergency management in Canada. The Act reinforces efforts to ensure that Canada is well prepared to mitigate, prepare for, respond to, and recover from natural and human-induced risks to the safety and security of Canadians. The Act:

- gives responsibility to the Minister of Public Safety to provide national leadership and set a clear direction for emergency management and critical infrastructure protection for the Government of Canada;
- clearly establishes the roles and responsibilities of federal Ministers and enhances the Government of Canada's readiness to respond to all types of emergencies;
- enhances collaborative emergency management and improves information sharing with other levels of government as well as the private sector; and
- gives authority to the Minister of Public Safety, in consultation with the Minister of Foreign Affairs, to coordinate Canada's response to an emergency in the United States.

Protecting critical infrastructure is one of the emerging challenges of integrated disaster management. Critical infrastructure consists of physical and information technology facilities, networks, services, and assets that are vital to the health, safety, security, or economic well-being of Canadians or the effective functioning of governments in Canada.

2.3.2 National Disaster Mitigation Strategy

The Government of Canada, together with provincial and territorial governments, launched Canada's National Disaster Mitigation Strategy (NDMS) on January 9, 2008 (PSC, 2008b). This Strategy is based on the recognition by federal, provincial, and territorial governments that mitigation is an important part of a robust emergency management framework, and that all stakeholders are committed to working together to support disaster mitigation in Canada.

Federal/Provincial/Territorial (FPT) governments have worked together to develop the NDMS for Canada. Responding directly to national consultation findings, the NDMS supports all-hazards emergency management, with an initial focus on reducing risk posed by natural hazards, an area that stakeholders agree requires urgent attention.

While the Strategy does not replace existing enterprise risk management programs at all government levels, the incorporation of NDMS principles into FPT initiatives will benefit the management of internal government risks. The goal of the NDMS is *to protect lives and maintain resilient, sustainable communities by fostering disaster risk reduction as a way of life*. The principles that reflect the essence of what the NDMS aims to achieve are as follows:

- Preserve Life—protect lives through prevention.
- Safeguard Communities—enhance economic and social viability by reducing disaster impacts.
- Fairness—consider equity and consistency in implementation.
- Sustainable—balance long-term economic, social and environmental considerations.
- Flexible—be responsive to regional, local, national and international perspectives.
- Shared—ensure shared ownership and accountability through partnership and collaboration.

The proposed Strategy will establish ongoing national disaster mitigation program activity areas. Implementation of program activities will be structured around four key elements:

- leadership and coordination;
- public awareness, education, and outreach;
- knowledge and research; and
- FPT cost-shared mitigation investments.

2.3.3 Joint Emergency Preparedness Program

The Government of Canada established the JEPP to enhance the national capability to manage all types of emergencies and ensure a reasonably uniform emergency response and recovery capacity across Canada. JEPP is administered by PSC. Whether it is through training, the purchase of emergency response equipment, for emergency planning or for capacity building, this shared investment is aimed at reducing injuries and loss of human life, property damage, and assuring the continuation of our critical services in an emergency. Currently, more than $8 million is made available annually for emergency preparedness, urban search and rescue and critical infrastructure protection projects from coast to coast.

JEPP projects are proposed annually by the provincial and territorial governments and selected for funding based on national and regional priorities. Projects are cost shared and the Government of Canada's contribution depends on the nature of the project, other projects under consideration, and the amount of funds available. To be eligible for Government of Canada funding, JEPP projects must (i) have a clear objective that supports priorities aimed at enhancing the national, provincial, and territorial emergency response capability; (ii) have an agreed, identifiable beginning and end; (iii) include a statement of the nature and extent of federal involvement and take into account how federal participation will receive visibility and recognition;

and (iv) include a provincial or territorial commitment to the project through funding or in-kind contribution.

2.3.4 Emergency Response

Emergencies are managed first at the local level, for example, hospitals, fire departments, police, and municipalities. If they need assistance, they request it from the provinces or territories. If the emergency escalates beyond their capabilities, the provinces or territories seek assistance from the federal government. The coordination and provisioning of resources can move quickly from the local to the national level.

PSC's emergency response includes the Government Operations Center (GOC) and Urban Search and Rescue Program (USAR). The GOC is housed at PSC and monitors emerging threats and provides around-the-clock coordination and support in the event of a national emergency. The GOC is Canada's strategic-level operations center. It is the hub of a network of operations centers run by a variety of federal departments and agencies including the Royal Canadian Mounted Police, Health Canada, Foreign Affairs, Canadian Security Intelligence Service, and National Defense. The GOC also maintains contact with the provinces and territories as well as international partners such as the United States and NATO. It operates 24 hours a day, 7 days a week, gathering information from other operations centers and a wide variety of sources, both open and classified, from around the world. The GOC deals with anything—real or perceived, imminent or actual, natural disaster or terrorist activity—that threatens the safety and security of Canadians or the integrity of Canada's critical infrastructure.

The USAR is the capacity to rescue victims from major structural collapse or other entrapments. Heavy USAR teams locate trapped persons in collapsed structures and other entrapments using search dogs and electronic search equipment. Heavy USAR involves work to breach, shore, lift and remove structural components, the use of heavy construction equipment to remove debris, and the medical treatment and transfer of victims. USAR is a general term for a group of specialized rescue skills that are integrated into a team that includes search, medical, and structural assessment resources.

PSC provides national leadership for USAR development to ensure that program development is coordinated and appropriately shared among the federal government, provinces and territories, major urban centers, and other national and international stakeholders. The USAR program is one aspect in the enhancement of Canada's National emergency response capacity.

The main elements of a national USAR program are based on operational readiness and capacity to deploy at short notice in response to domestic disasters. PSC has identified the five following priorities for the development of a national USAR program: (i) plans, policies and protocols to outline responsibilities of the federal government, and of USAR teams deployed in afflicted areas outside home jurisdictions; (ii) standard equipment designed for USAR operations; (iii) training in technical skills and joint operations with other teams; (iv) national guidelines or standards, where required; and (v) exercises to improve capability and develop interoperability. Four Canadian cities (Vancouver, Calgary, Toronto, and Halifax) and the Province of Manitoba currently have, or are developing, interoperable USAR capacity.

PSC works with provincial and territorial EMOs to ensure first responders and emergency management personnel are well-prepared through education, support, and exercises.

For example, Emergency Management Ontario (EMO) devotes a great deal of time and effort to the citizens of Ontario to help prevent or ease the effects of emergencies by making sure they are better prepared. It is our strong belief that the better prepared we are—as a family, as a community, as a province—the more confident we become when responding to an actual disaster. EMO has developed a variety of public education materials tailored to different needs. But, EMO goes beyond simply publishing materials. They have experts assigned to specific areas of the province who are in regular contact with municipal officials responsible for emergency management in their communities. Their focus is to get the message of prevention and preparedness out to as many Ontarians as possible. EMO is involved in developing emergency scenarios, as well as practicing such response scenarios with many federal, provincial, municipal and private sector partners.

2.3.5 The Role of Federal Government in Disaster Recovery

In the event of a large-scale disaster where costs exceed regional resources, PSC provides financial assistance to provincial and territorial governments. Assistance is paid to the province or territory—not directly to individuals or communities. The provincial or territorial governments design, develop, and deliver disaster financial assistance, determining the amounts and types of assistance that will be provided to those who have experienced losses. The following programs for disaster recovery are delivered through PSC.

Disaster Financial Assistance Arrangements The Disaster Financial Assistance Arrangements (DFAA) program, designed in consultation with the provinces and territories and administered by PSC, details how the federal government should respond to a natural disaster. Provinces and territories are responsible for designing, developing, and delivering financial assistance to the victims of emergencies and disasters as they see fit, with no restrictions placed on them by the federal government. The federal government does not directly provide disaster relief funding to individuals or businesses. It is up to the provincial or territorial government affected by the disaster to request assistance from the federal government, in accordance with DFAA guidelines.

Under the DFAA program, provincial/territorial governments can ask the federal government for disaster relief when eligible expenditures surpass $1 per capita (based on provincial/territorial population). The program sets out guidelines respecting what expenses resulting from a disaster qualify for relief, following a graduated funding formula based on the size of the disaster. Generally speaking, DFAA guidelines stipulate that the federal government will not provide funding to the province to cover costs already insured or where insurance was available at a reasonable price but was not purchased. There is nothing in the law, however, preventing the federal government from covering any cost it wishes.

From a theoretical perspective, the DFAA rule against providing aid to those who choose not to purchase insurance helps protect the federal government against the danger of moral hazard. In short, should the government agree to pay for all damages regardless of whether insurance was available, people would have a strong incentive not to buy insurance. Such a result would increase the cost of a natural disaster borne by the taxpayer. Moreover, it would also decrease the pool in insured risks, thus increasing the risk of insolvency for insurers.

Federal Disaster Assistance Initiative The goal of the FDAI is to provide provincial and territorial governments with a comprehensive suite of disaster recovery programs that respond to the needs of Canadians. This will be accomplished in three ways: (1) revision of the existing DFAA; (2) development of new or improved disaster assistance instruments to complement the DFAA; and (3) creation of an inventory of existing federal and provincial programs, policies, and legal tools for disaster assistance and recovery.

2.4 DECISION MAKING AND INTEGRATED DISASTER MANAGEMENT

It is well documented in the disaster-related literature that many factors influence implementation of integrated disaster management activities that will minimize loss of life, injury, and/or material damage. It seems that governments, businesses, organizations, and individuals do not implement large-scale disaster management decisions that would enable them to avoid long-term losses from hazards (Mileti, 1999). Making integrated disaster management a reality requires overcoming many social, political, and financial obstacles, as well as changing many human behaviors. One of the most important factors that influences integrated disaster management is the decision-making process itself.

Decisions to (a) make an initial commitment of resources to take precautionary measures and (b) continue allocation of resources to follow through on them can be made at personal, organizational, or government levels. The decision-making process differs with the different level of decision making. In the most general way, *decision making is* defined as *the process of making an informed choice among the alternative actions that are possible*. The process may involve establishing objectives, gathering relevant information, identifying alternatives, setting criteria for the decision, and selecting the best option. Decision theory can be used to assist in the process of decision making. Specific techniques used in decision making include heuristics and decision trees. Computer systems designed to assist decision making are known as decision support systems.

2.4.1 Individual Decision Making

People are typically unaware of the hazards they are exposed to. They also underestimate hazards they area aware of and overestimate their ability to cope when disaster strikes. It is also very common that people blame others for their losses, do not utilize properly preimpact measures and rely heavily on emergency relief when the need arises. Individual decision making is characterized with imperfect assessment of the situation, lack of knowledge of the full range of alternative options available to them, imperfect use of information to assess likely states of nature, and the consequences of their actions.

There is a considerable body of work focused on finding correlations between disaster decision making and a broad range of human socioeconomic (age, ethnicity, gender, income, education), geographic (recency and frequency of disaster experience, proximity to the disaster impact area), and psychological characteristics (Morris-Oswald and Simonović, 1997). The results are not uniform. In some cases, age is a determinant while in another situation community connections are more important. In general, people think about disaster in terms of usefulness, cost, time requirements, and protection implementation obstacles. There is a consensus that motivating people to take disaster-mitigative action is a process of overcoming social conformity and encouraging innovation.

2.4.2 Decision Making in Organizations

Not much is known about the process that formal organizations follow in responding (or failing to respond) to disasters. In general, disaster management decisions are preceded by awareness of a hazard, awareness of alternative options, adoption of one or more options, implementation of the options adopted, and subsequent evaluation (Burton et al., 1978).

2.4.3 Decision Making in Government

Governments adopt and implement large-scale disaster management options such as dams and levees, mandate that lower levels of government and individuals engage in risk reduction activities such as adopting building codes, and regulate development and land use. A variety of models exist to help understand natural hazard adoption and implementation by governments.

Models of organizational decision making are often applicable to public policy decisions. Political science identifies behavioral models such as political systems theory, group theory, elite theory, institutionalism, and rational choice theory as available forms of government decision making (Mileti, 1999).

It is expected that federal, state, and local policymakers' processing of disaster information is systematic rather than heuristic. Therefore, information programs could emphasize scientific information and modeling of estimated losses and impacts of alternative mitigation measures. The degree to which such tools would increase local policymakers' adoption of disaster mitigation remains to be determined. No information is available on (a) perception of state and local policymakers of credibility of loss estimation models; (b) whether their past experience inclines them to have confidence in computer models; and (c) whether the models would provide all of the information they need to make disaster protection decisions.

Provincial/state and federal governments have a strong motivation to encourage local governments to reduce potential losses—for example, to reduce province/state and federal disaster relief. Many studies point to the reluctance of local governments to adopt on their own, or to adequately enforce, strong measures for managing land use and development in hazardous areas (Mileti, 1999). State requirements for local disaster management vary according to the nature of the hazards, differing perceptions of the seriousness of hazards, and differing beliefs about role of government.

It is very common that disaster management is not very high on the agenda of many province/state-level agencies. National policymakers face pressure to reduce federal disaster relief. However, they are also aware of the political power of the property rights movement. Limited knowledge is available about the extent to which government agencies attempt to minimize their own losses and to assure continued provision of routine government services after a disaster.

2.5 SYSTEMS VIEW OF INTEGRATED DISASTER MANAGEMENT

Integrated disaster management involves complex interactions within and between the natural environment, human population (actions, reactions, and perceptions), and built environment (type and location). A different thinking is required to address the complexity of disaster management. Mileti (1999, p. 26) strongly suggests adaptation of a global systems perspective.

Systems theory is based on the definition of a system—in its most general sense as a collection of various structural and nonstructural elements that are connected and organized in such a way as to achieve some specific objective through the control and distribution of material resources, energy, and information. The basic idea is that all complex entities (biological, social, ecological, or other) are composed of different elements linked by strong interactions, but a system is greater than the sum of its parts. This is a different view from the traditional analytical scientific model based on the law of additivity of elementary properties that view the whole as equal to the sum of its parts. Because complex systems do not follow the law of additivity, they must be studied differently.

A systemic approach to problems focuses on interactions among the elements of a system and on the effects of these interactions. Systems theory recognizes multiple and interrelated causal factors, emphasizes the dynamic character of processes involved, and is particularly interested in a system change with time—be it a flood, hurricane, or a disaster-affected community. The traditional view is typically linear and assumes only one, linear, cause-and-effect relationship at a particular time. A systems approach allows a wider variety of factors and interactions to be taken into account.

Using a systems view, Mileti (1999, p. 107) states that disaster losses are the result of interaction among three systems and their many subsystems:

- the earth's physical systems (the atmosphere, biosphere, cryosphere, hydrosphere, and lithosphere);
- human systems (e.g., population, culture, technology, social class, economics, and politics); and
- the constructed systems (e.g., buildings, roads, bridges, public infrastructure, and housing).

All of the above systems and subsystems are dynamic and involve constant interactions between and among subsystems and systems. All human and constructed systems and some physical ones affected by humans are becoming more complex with time. This complexity is what makes national and international disaster problems difficult to solve. The increase in the size and complexity of the various systems is what causes increasing susceptibility to disaster losses. Changes in size and characteristics of the population and changes in the constructed environment interact with changing physical systems to generate future exposure and define future disaster losses. The world is becoming increasingly complex and interconnected, helping to make disaster losses greater (Homer-Dixon, 2006).

REFERENCES

Anderson, M. and P. Woodrow (1989), *Rising from the Ashes: Development Strategies in Times of Disaster*, Westview Press, Boulder, CO.

Bain, T. and J. Mitchell (2006), *Home Insurance Explained: Including Tenant and Condominium Insurance*, 2nd edn, Insurance Bureau of Canada, Toronto, Canada, p. 24.

Berke, P.R., J. Kartez, and D. Wenger (1993), "Recovery after disaster: Achieving sustainable development, mitigation and equity," *Disasters*, 17(2):93–109.

Burton, I., R.W. Kates, and G.F. White (1978), *The Environment as Hazard*, Oxford University Press, NJ.

Homer-Dixon, T. (2006), *The Upside of Down: Catastrophe, Creativity, and the Renewal of Civilization*, Alfred A. Knopf, Canada.

Mileti, D.S. (1999), *Disasters by Design*, Joseph Henry Press, Washington, DC.

Morris-Oswald, M. and S.P. Simonović (1997), "Assessment of the Social Impact of Flooding for Use in Flood Management in the Red River Basin," *Report* prepared for the International Joint Commission, Slobodan P. Simonović Consulting Engineer Inc., Winnipeg, available online, http://www.ijc.org/php/publications/html/assess.html, last accessed August 11, 2008.

National Research Council (NRC) (2008), *Canadian Code Center*, online resources, http://irc.nrc-cnrc.gc.ca/codes/index_e.html, last accessed August 5, 2008.

Public Safety Canada (2008a), *Emergency Management Act*, available online, http://laws.justice.gc.ca/en/ShowTdm/cs/E-4.56/, last accessed August 5, 2008.

Public Safety Canada (2008b), *Canada's National Mitigation Strategy*, available online, http://www.ps-sp.gc.ca/prg/em/ndms/strategy-eng.aspx, last accessed August 5, 2008.

EXERCISES

2.1 For the largest disaster experienced in your region (from Exercise 1.1 in Chapter 1 in Section 1.5)

- **(a)** Describe mitigation actions taken.
- **(b)** Describe preparedness actions taken.
- **(c)** Describe response actions taken.
- **(d)** Describe recovery actions taken.
- **(e)** Identify, for each set of actions, their advantages and disadvantages.

2.2 For the Red River flood of 1997 example presented in Section 1.1.1

- **(a)** Identify all mitigation, preparedness, response, and recovery activities.
- **(b)** What would you do, in addition to what has been done, in this case to make flood management decisions in the Red River basin sustainable?

2.3 For the largest disaster experienced in your region (from Exercise 1.1 in Section 1.5)

- **(a)** Identify all systems and subsystems.
- **(b)** What are the elements of identified systems and subsystems?
- **(c)** What are the main relationships among the elements of identified systems and subsystems?
- **(d)** Sketch all the relationships from part (c) using graph paper.

2.4 Implement the general definition of a system from Section 2.5 to

- **(a)** A flood.
- **(b)** A hurricane.
- **(c)** An earthquake.

PART II
Systems Analysis for Integrated Management of Disasters

3 Systems Thinking and Integrated Disaster Management

Systems are as pervasive as the universe around us. At one extreme, they are as large as the universe itself, while at the other, they are as small as the atom. Systems first existed in natural forms, but since the appearance of human beings on Earth, a variety of human-made systems have come into existence. Only recently have we come to understand the underlying structure and characteristics of natural and human-made systems in a scientific sense. Following the recommendation of Mileti (1999, p. 26) in this book I am trying to emphasize that the concept of systems thinking should play a dominant role in a disaster management as it is already playing in a wide range of fields, from engineering to different topics of natural and social sciences.

The roots of this development are complex. Two lines of systems thinking emerge from these beginnings. They differ in a number of significant ways that have serious implications for social focus on the role of systems view in communication and control in society, while the other emphasizes the role of systems thinking in dynamic behavior. Each of the two lines of systems thinking that emerge in the social sciences is a set of authors and ideas that are interconnected sociologically and/or methodologically. Richardson (1991) labels them as the "servomechanistic thread" and the "cybernetics thread." Both threads are influenced by ideas associated with the feedback concept (explained later) in engineering and by loop ideas from the social sciences. The servomechanistic thread has greater similarities, and stronger sociological connections, to mathematical modeling traditions in biology and economics. The cybernetic thread can be more associated with ideas in formal logic and psychology. The differences that emerge and become accentuated in these two threads create different understandings about the role of systems thinking in social systems and the appropriate role of feedback thinking in the social science.

The first main characteristic of all systems approaches is that they are committed to holism—to looking at the world in terms of "wholes" that exhibit emergent properties, rather than believing in a reductionist fashion that learning should proceed by breaking wholes down into their fundamental elements. The strength of this commitment varies from seeing holism as a replacement for reductionism to regarding it as simply complementary. Holism is, however, a distinctive feature of systems thinking and systems advocacy of holism in their approaches provides a useful contrast to the prevailing emphasis on reductionism in many disciplines.

Systems Approach to Management of Disasters: Methods and Applications, By Slobodan P. Simonović

Second, it might be argued along with Jackson (2000) that human beings inevitably organize their knowledge in "cognitive systems." These cognitive systems are structured frameworks linking various elements of our knowledge into cohesive wholes. The cognitive systems lie at the very heart of the scientific method itself. From this it could follow that science is much closer to the systems view than has so far been believed. The success of systematizing science in predicting and controlling the real world must mean that the world itself is orderly. I do not insist on following these, somewhat "hard" systems conclusions. It is enough that we use them as justification for using the concept of a system as the fundamental element of methodological approach to solving real-world problems of integrated disaster management.

The final and most developed argument in favor of systems approaches must, however, rest upon the diversity, range, effectiveness, and efficiency of the approaches themselves in relation to real-world problem management. On the basis of extensive social science systems literature, it seems that systems ideas and concepts have a resonance with real-world practice, which is sadly lacking in much social theory. For this reason, systems methodologies can assist in the task of translating social theory into a practical form and aggregating its findings in practical approaches to intervention.

Integrated disaster management involves complex interactions within and between the natural environment, human population (actions, reactions, and perceptions), and built environment (type and location). Interdisciplinary character and complexity of problems in this domain make a systems approach based on systems analysis necessary. For example, in order to find the best (or even a good) way to evacuate people from the hazard exposed area, a disaster management specialist (or team of specialists) must consider alternative solutions, and choose one that promises the optimal outcome at maximum efficiency in a very complex network of interactions. This requires problem solving capacity and power that are far beyond the unaided capacity of a single individual. It should not be taken that systems analysis is merely one of many changes in our contemporary technological society. Rather, it involves a change in our basic thought processes. In one way or another, we are forced to deal with complexities, with wholes or systems, in all fields of knowledge. This implies a basic reorientation in scientific thinking across almost all disciplines, from subatomic physics to history.

3.1 SYSTEM DEFINITIONS

From this point on, I will be approaching systems ideas in a much more applied fashion to develop methodologies for solving real-world disaster management problems. This approach is sometimes referred to as "hard" systems thinking (Checkland, 1999). My applied approach, as presented in this book, includes simulation, optimization, and multiobjective analysis and therefore cannot be considered only as "hard" systems thinking. All the methods and tools that are going to be presented here share the basic assumption that *the problem task they tackle is to select an efficient means of achieving a known and defined end.*

3.1.1 What is a System?

Some kind of system is inherent in all but the most trivial mitigation, preparedness, and response and recovery problems. To understand a problem, the analyst must be able to recognize and understand the system that surrounds and includes it. This has not always been done effectively in the past. Some reasons for poor system definition are poor communications, lack of knowledge of interrelationships, politics, limited objectives, and transformation difficulties. What then is a system? There are many variations in definitions of a system; for example, one dictionary alone provides no less than 15 ways to define the word. However, all of them share common traits. In the most general sense, a system is a collection of various structural and nonstructural elements that are connected and organized in such a way as to achieve some specific objective through the control and distribution of material resources, energy, and information. A more formal definition of a system can be stated as:

$$S : \boldsymbol{X} \rightarrow \boldsymbol{Y} \tag{3.1}$$

where $\boldsymbol{X}$ is an input vector and $\boldsymbol{Y}$ is an output vector. To put this differently, a system is a set of operations that transforms input vector $\boldsymbol{X}$ into output vector $\boldsymbol{Y}$.

A schematic representation of the system definition is shown in Figure 3.1. This sees the system in terms of input, output, a transformation process, feedback, and a restriction. *Input* energizes the operation of a given *transformation process*. The final state of the process is known as the *output*. *Feedback* is the name given to a number of operations that compare the actual output with an objective and identify the discrepancies between them.

A more comprehensive definition may look at a system as an assemblage or combination of elements or parts forming a complex or unitary whole, such as an

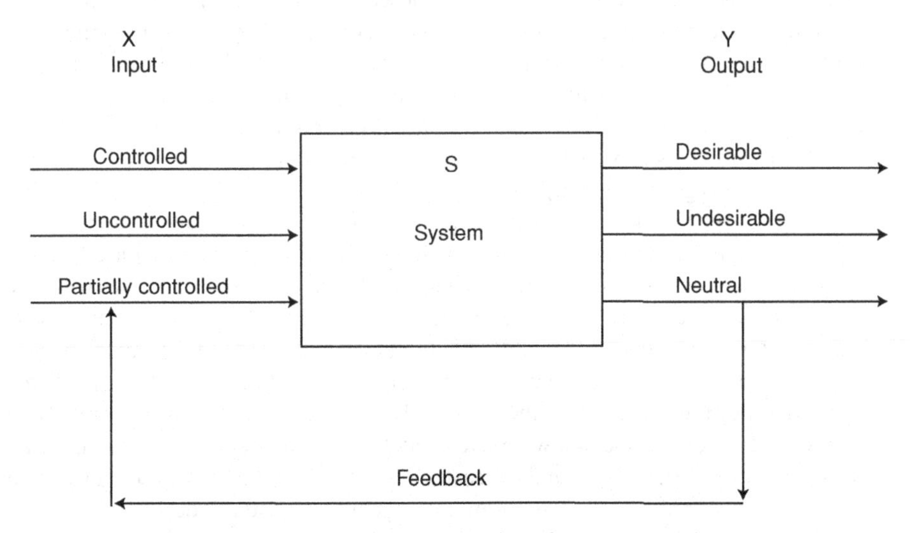

Figure 3.1 Schematic presentation of a system definition.

emergency transportation system; any assemblage or set of correlated members, such as a system of currency; an ordered and comprehensive assemblage of facts, principles, or doctrines in a particular field of knowledge or thought, such as a system of philosophy; a coordinated body of methods or a complex scheme or plan of procedure, such as a system of organization and management; or any regular or special method or plan of procedure, such as a system of marking, numbering, or measuring.

3.1.2 Systems Thinking

The problems that we currently face in disaster management have been stubbornly resistant to solution, particularly with a one-sided solution (i.e., one that looks at the problem in a narrow linear way and "solves" it). As we are discovering, there is no way to completely solve the problem of providing, timely and efficiently, disaster relief to a disaster-affected community, and the lack of one is steadily affecting a rising number of people around the globe. The problems of disaster recovery cost and increase in natural disasters as a consequence of climate change (the buildup of carbon dioxide in the atmosphere, ozone depletion, etc.) also fall into the category of "resistant to simple solutions."

It has probably never been possible to solve problems such as these, but the situation was less critical when it was possible to shift them out of the area of immediate concern. In an era when the connections among the various subsystems were less tight, it was possible to score a temporary victory by essentially pushing a problem into the future or into "someone else's backyard" (Richmond, 1993). Unfortunately, there is less and less space to do that in our modern world, and temporary victories of this nature do not add up to a viable long-term strategy. In an interconnected world we do not have places that we can treat as holders for our "garbage," and it is necessary to face the consequences of our present decisions.

Human beings are quick problem solvers. From an evolutionary standpoint, quick problem solvers were the ones who survived. We can often quickly determine a cause for a situation that we identify as a problem. For example, if a river overflows its banks, we might conclude that this is because it was raining a lot during the previous couple of days. This approach works well for simple problems, but it works less well as the problems get more complex.

If we accept the argument that the primary source of the growing intractability of problems in disaster management is a tightening of the links between the various physical and social subsystems that make up our disaster systems reality, we will agree that systems thinking provides us with tools for a better understanding of these difficult disaster management problems. The fundamental tools of the systems approach have been in use for over 30 years (Forrester, 1990) and are now well established. However, they require a shift in the way we think about disaster management system. In particular, they require that we move away from looking at isolated situations and their causes, and start to look at the disaster system as a system made up of interacting parts. Systems thinkers use diagramming languages to visually depict the feedback structures of these systems. They then use simulation to play out the associated dynamics. These tools give us the ability to see into a "neighbor's backyard," even if that backyard is thousands of kilometers away. They also confer the ability to experience the consequences of our decisions, even if they are somewhere in the future.

The important question is, how can the framework, the process, and the technologies of systems thinking be transferred to future disaster managers in a reasonable amount of time? According to Richmond (1993), if we view systems thinking within the broader context of critical thinking skills and recognize the multidimensional nature of the thinking skills involved in systems thinking, we can greatly reduce the time it takes to pick up this framework. As this framework increasingly becomes the context within which we think, we shall gain much greater leverage in addressing the pressing disaster management issues that await us in the future. A switch must occur from teacher-directed learning to learner-directed learning. Open classrooms, computer-aided instruction, and offering interdisciplinary courses are but a few of the initiatives in the right direction. It has also become apparent to me that good systems thinking means operating on multiple thought tracks simultaneously. This would be difficult even if these tracks comprised familiar ways of thinking. Familiarity with the following aspects could be of assistance.

Dynamic Thinking Dynamic thinking involves acquiring the ability to see behavior patterns rather than focusing on, and seeking to predict, individual events or situations. It means thinking about phenomena as resulting from ongoing circular processes unfolding through time, rather than in terms of events and causes. Dynamic thinking skills are based on the ability to trace out patterns of behavior that change over time. They call for thinking through the underlying closed-loop processes that cycle around to produce particular situations.

Closed-Loop Thinking The second type of thinking process, closed-loop thinking, is closely linked to dynamic thinking. When we think in terms of closed loops, we see the problem as a set of ongoing, interdependent processes rather than as a list of one-way relations between a group of causes and another of effects. In addition when exercising closed-loop thinking, we look to the loops themselves (i.e., the circular cause–effect relations) as being responsible for generating the behavior patterns exhibited by a system. This is in contrast to holding a set of external forces responsible. In this model, external forces tend to be viewed as precipitators rather than as causes.

Generic Thinking Just as most of us are captivated by events, we are generally locked into thinking in terms of specifics. The notion of thinking generically rather than specifically can be applied to disaster management systems. For example, it is useful to appreciate the similarities in the underlying feedback loop relations that generate a flood–drought swing or an oscillation in drinking water quality.

Structural Thinking Structural thinking is one of the most disciplined of the strands of systems thinking. Here, we must think in terms of units of measure or dimensions. The laws of physical conservation are rigorously adhered to in this domain.

Operational Thinking Operational thinking goes hand in hand with structural thinking. Thinking operationally means thinking in terms of how things really work; not how they should work in theory, or how a model can be created by manipulating a bit of algebra and generating some convincing-looking output.

Continuum Thinking Continuum thinking is usually present when we work with simulation models that have been built using a continuous, as opposed to discrete, modeling approach. Discrete models are distinguished by their containing many "if, then, else"-type equations. In such models, for example, we might find that disaster damage compensation is governed by some logic of the form "IF Disaster Relief Budget > A THEN Normal Disaster Compensation ELSE 0." In contrast, the continuous version of this relation would begin with an operational specification of the disaster damage compensation process (e.g., disaster relief budget = population affected × damage compensation). Damage compensation (per person, per event) would then be a continuous function of disaster relief budget. Unlike its discrete analogue, the continuous formulation indicates that damage compensation would be continuously affected as disaster relief budget became depleted. That is, it allows for different measures such as moratoriums on new construction in the disaster-prone area coming into play as it becomes apparent that there are less than adequate relief funds available. The discrete formulation, by contrast, implies "business as usual." Although from a mechanical standpoint the differences between the continuous and discrete formulations may seem unimportant, the implications for thought processes are quite dramatic.

Scientific Thinking The final component of systems thinking that Richmond (1993) identified is scientific thinking. Thinking scientifically means being rigorous about testing hypotheses. This process begins by always ensuring that you do in fact have a hypothesis to test. If there is no hypothesis, the experimentation process can easily degenerate into a game. The hypothesis-testing process itself also needs to be informed by scientific thinking. When we think scientifically we modify only one thing at a time and hold all else constant. We also test our models from a steady state using idealized inputs. We defined the term *system* to mean an interdependent group of items forming a unified pattern. Since our interest here is in disaster management, we shall focus on systems of people and technology that are intended to support mitigation, preparedness, response, and recovery processes. Almost everything that goes on in disaster management is part of one or more such systems. As noted above, when we face a management problem we tend to assume that some external event caused it. With a systems view, we take an alternative viewpoint, namely, that the internal structure of the system is often more important than external events in generating the problem. This is illustrated in Figure 3.2. Many people try to explain aspects of performance by showing how one set of events causes another, or when they study a problem in depth, how a particular set of events is part of a longer-term pattern of behavior. The difficulty with this cause–effect orientation is that it does not lead to very powerful ways to alter the undesirable performance. We can continue

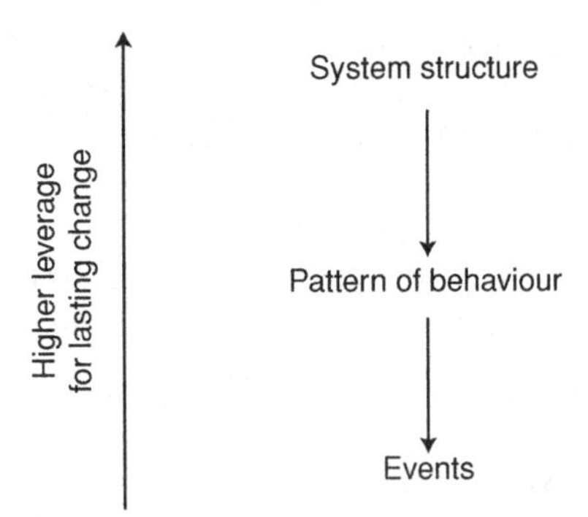

Figure 3.2 Looking for problem solution (high leverage).

this process almost forever, and thus it is difficult to determine what to do to improve performance.

If we shift from this event orientation to focusing on the internal system structure, we improve the possibility of finding the causes of the problem. This is because the system structure is often the underlying source of the difficulty. Unless the deficiencies in the system structure are corrected, it is likely that the problem will resurface, or be replaced by an even more difficult problem.

3.1.3 Systems Analysis

Systems analysis involves the use of rigorous methods to help determine preferred plans and designs for complex, often large-scale, systems. It combines knowledge of the available analytical tools, understanding of when each is most appropriate, and skill in applying them to practical problems. It is both mathematical and intuitive, as is all planning and design.

It is a relatively new field. Its development parallels that of the computer, the computational power of which enables us to analyze complex relationships, involving many variables, at reasonable cost. Most of its techniques depend on the use of the computer for practical application. Systems analysis may be thought of as the set of computer-based methods essential for the planning of major activities. Therefore, it should be central to the education of disaster professionals.

Systems analysis covers much of the same material as operations research, in particular linear and dynamic programming (LP and DP) and decision analysis. However, the two fields differ substantially in direction. Operations research tends to be interested in specific techniques and their mathematical properties. Systems analysis focuses on the use of the methods.

Systems analysis includes the topics of economy, but goes far beyond them in its depth of concept and the scope of its coverage. Now that both personal computers and efficient financial calculators are available, there is little need for professionals to spend much time on detailed calculations. It is more appropriate to understand the concepts and their relationship to the range of techniques available to deal with complex problems.

This approach emphasizes the kinds of real problems to be solved; considers the relevant range of useful techniques, including many besides those of operations research; and concentrates on the help they can provide in improving preparedness and recovery plans, mitigation design options, and emergency management decisions. The use of systems analysis instead of the more traditional set of tools generally leads to substantial improvements in efficiency and reductions in cost. Gains of 30% are not uncommon. These translate into an enormous advantage when we consider disasters that cost tens and hundreds of millions of dollars.

3.1.4 The Systems Approach

The systems approach is a general problem-solving technique that brings more objectivity to the disaster management processes. It is, in essence, concerned with good design: a logical and systematic approach to problem solving in which assumptions, goals, objectives, and criteria are clearly defined and specified. Emphasis is placed on relating system performance to specified goals. A hierarchy of systems is identified, and this makes it possible to handle a complex system by looking at its component parts or subsystems. Quantifiable and nonquantifiable aspects of the problem are identified, and the immediate and long-range implications of suggested alternatives are evaluated.

The systems approach establishes the proper order of inquiry and helps in the selection of the best course of action that will accomplish a prescribed goal by broadening the information base of the decision maker; by providing a better understanding of the system, and the interrelatedness of the complete system and its component subsystems; and by facilitating the prediction of the consequences of several alternative courses of action.

The systems approach is a framework for disaster management and decision making. It does not solve problems, but does allow the decision maker to undertake resolution of a problem in a logical, rational manner. While there is some art involved in the efficient application of the systems approach, other factors play equally important roles. The magnitude and complexity of decision processes requires the most effective use possible of scientific (quantitative) methods of systems analysis. However, we should be careful not to rely too heavily on the methods of systems analysis. Outputs from simplified analyses have a tendency to take on a false validity because of their complexity and technical elegance.

3.1.5 Systems "Engineering"

In this context, word "engineering" does not carry traditional meaning. Systems engineering may be defined as the art and science of selecting from a large number of feasible alternatives, the particular set of actions that will best accomplish the overall objectives of the decision makers, within the constraints of law, morality, economics, resources, politics, social life, nature, physics, and so on.

Another definition sees systems engineering as a set of methodologies for studying and analyzing the various aspects of a system (structural and nonstructural) and its environment by using mathematical and/or physical models. Systems engineering is currently the popular name for the management processes of planning and design used in the creation of a system or project of considerable complexity.

3.1.6 Feedback

Within the last 50 years, the engineering concept of feedback has entered the social sciences. The essence of the concept, defined more precisely later, is a circle of interactions, a closed loop of action and information. The patterns of behavior of any two variables in such a closed loop are linked, each influencing, and in turn responding to, the behavior of the other. Thus the concept of the feedback loop is intimately linked with the concepts of interdependence and mutual or circular causality, ideas very much present in disaster management.

Loop concepts seem to be central to an emerging view of social reality (Richardson, 1991). For some time the unidirectional view of cause and effect has been giving *way* to the circular, looping perspective of mutual causality. For decades, we have applied variations of a research paradigm that strives to derive the causal connection between a "dependent" variable Y and "independent" variables $X_1, X_2, \ldots, X_n$. What characteristics of an instructor correlate with instruction effectiveness? What will the demand for a company's product be next year? How does a manager's leadership style affect worker productivity? Increasingly, social scientists are taking the point of view that the dependent–independent variable view is inadequate for some such questions. Instruction effectiveness is the result of an interaction of the characteristics of instructor and student. The demand for a company's product is not simply the result of consumer behavior but is also determined by the company's own behavior in setting price, maintaining quality, being early to market and ship, and setting a balance among such goals. A manager's behavior is not independent of worker characteristics: worker productivity can turn around and affect a manager's leadership style. In each of these examples, and in a vast range of others, causality appears to be circular. Social scientists, public policymakers, businesspeople, and ordinary folk have to learn to deal with the emerging fact of circular causality in social systems.

The classification of systems in Section 3.1.8 provides a basic differentiation between open and closed systems. An open system is one characterized by outputs that respond to inputs, but where the outputs are isolated from and have no influence on the inputs. An open system is not aware of its own performance. In an open system, past action does not control future action (Forrester, 1990). However, a closed system (also known as a feedback system) is influenced by its own past behavior. A feedback system has a closed-loop structure that brings results from past action of the system back to control future action. A broad purpose may imply a feedback system with many components, but each component can itself be a feedback system in terms of some subordinate purpose. We can then recognize a hierarchy of feedback structures, where the broadest purpose of interest determines the scope of the pertinent system. There is a simplified graphical representation of these two types of system in Figure 3.3.

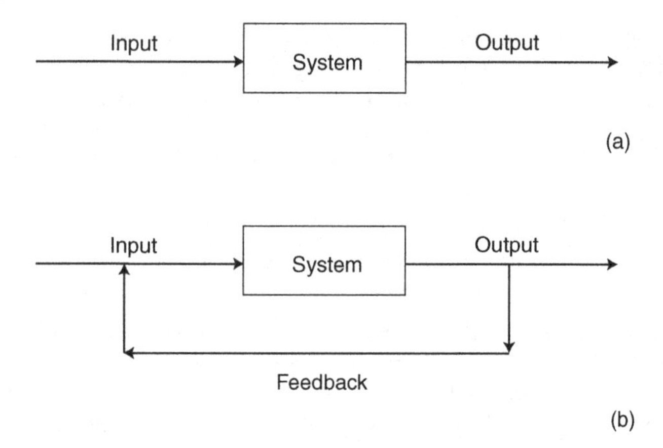

Figure 3.3 Schematic presentation of (a) an open and (b) a closed system.

A basic example of a feedback system is a simple thermostat and the maintenance of constant room temperature. The thermostat senses a difference between desired and actual room temperature, and activates the heating unit. The addition of heat eventually raises the room temperature to the desired level. Then the thermostat automatically shuts off the heater. The system description used for the thermostat applies equally well to many systems: the electric eye of a camera, a thermostatically controlled oven, the cruise control of a car, and the speed governor of a turbine all follow this pattern. Although these are all mechanically controlled systems, there are also equivalents in the biological world. The human body contains numerous self-regulating physiological processes that enable it to maintain a relatively constant internal environment. This self-regulation, called *homeostasis*, maintains, for example, a normal body temperature through the continual alteration of metabolic activities and blood flow rates. Once the concepts of feedback and homeostasis became known to social scientists, they became central organizing concepts for modeling human behavior. Engineering mathematical methods are widely used for the analysis of economic dynamics. Analogy is explored between communication and feedback in neural networks with their counterparts in governmental networks. Forrester (1990) initiated a modeling approach for management science that is explicitly centered on the feedback concept. In addition, the feedback concept is used as a foundation for a form of psychological counseling called reality therapy. Thus psychotherapy, psychology, sociology, anthropology, social psychology, economics, political science, and management have all been touched and changed by the discovery of the potential significance of the feedback loop in human affairs.

The feedback concept is one of the most important aspects of general system theory. The basic model is a circular process (as shown in Figure 3.3b), where part of the output is monitored back into the input, as information on the preliminary outcome of the response, thus making the system self-regulating. This self-regulation can involve either the maintenance of the value of certain variables or steering toward a desired goal.

Feedback systems and homeostatic control are a significant but special class of self-regulating systems and phenomena of adaptation. The following appear to be essential criteria of feedback control systems:

- Regulation is based on a preestablished system structure.
- Causal chains within the feedback system are linear and unidirectional.
- Typical feedback or homeostatic phenomena are "open" with respect to incoming information, but "closed" with respect to matter and energy.

If we compare the flow diagrams of feedback (Figure 3.3b) and open systems (Figure 3.3a), the difference should be apparent. Thus dynamics in open systems and feedback mechanisms are two different model concepts, each of which has a place in its proper sphere. The open-system model is basically nonmechanistic, and supports not only conventional thermodynamics, but also one-way causality, as is basic in conventional physical theory. The cybernetic approach retains the Cartesian machine model of the organism, unidirectional causality, and closed systems; its novelty lies in the introduction of concepts that transcend conventional physics, especially those of information theory.

Unfortunately, up to now the field of disaster management paid very limited attention to feedback thought. By "feedback thought" I mean a powerful way of thinking, linking the concepts of control and self-reinforcement, stability and instability, structure and behavior, mutual causality, interdependence, and many more of the deepest ideas in the natural, social and behavioral sciences.

Usually implicitly, but sometimes explicitly, feedback thought is embedded in the foundations of much of management science and systems theory. It is a building block. The literature shows that feedback thought is both old and new. It shows that in the modern era the concept of feedback moved into prominence in the 1940s and 1950s, and has since appeared to wane. Some consider it an outmoded idea, a metaphor for the management sciences that has had its day and has been replaced by new metaphors. Others, perhaps as a consequence, have not encountered it at all. Still others see it not as a metaphor, but as a natural and crucial property of management systems. Here, I argue that feedback is the most important property of integrated disaster management systems.

The Feedback Loop A basic premise of this investigation is that there is a unifying loop concept underlying a number of superficially diverse ideas in the social sciences. Engineering work that came eventually to influence the social sciences captures feedback systems in terms of differential equations. Econometricians express interdependencies in an economy using difference equations. Other social scientists use words to paint verbal pictures of circular causal processes: vicious circles, self-fulfilling prophecies, homeostatic processes, and invisible hands. Underlying all these representations is the concept of a closed loop of causal influences.

Since the concept of a loop is fundamentally visual, I will use visual terms to give an intuitive definition of the loop concept underlying feedback and mutual causality.

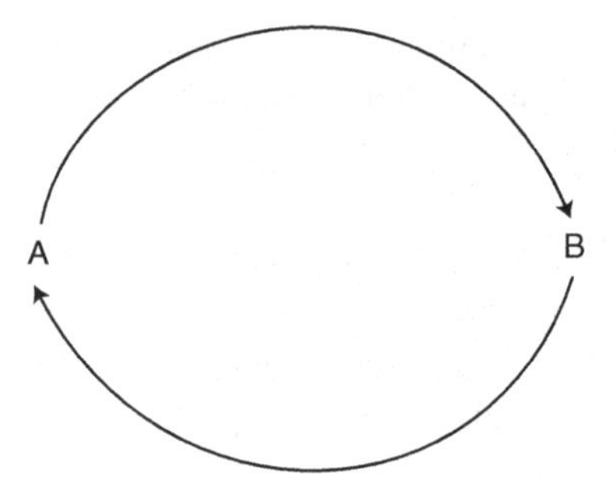

Figure 3.4 A feedback loop.

An arrow drawn from A to B will be taken to mean A "causally influences" B. Thus, if A influences B, and B in turn influences A, we would have a pair of arrows that form a loop of mutual or circular causality (Figure 3.4). To this elemental loop an additional idea is attached, referred to as the *polarity* of the loop. It is the concept of polarity that gives the causal loop its perceived analytic and explanatory power.

The polarity of a circular causal loop reflects the loop's tendency either to reinforce or to counteract a change in any one of its elements. Following the terminology that has become common since the emergence of the feedback concept from engineering into the social sciences, loops are characterized as "positive" or "negative." A causal loop that characteristically tends to reinforce or amplify a change in any one of its elements is called a positive loop. In a positive loop (Figure 3.5a), an increase in an element A feeds around the loop and tends to cause A to increase still further; likewise a decrease in A tends to cause A to decrease still further. Similarly, a causal loop that characteristically tends to diminish or counteract a change in any one of its elements is called a negative loop (Figure 3.5b). In a negative loop, an increase in A feeds around the loop and tends to cause A to slow or reverse its increase; likewise a decrease in A tends to cause A to slow or reverse its decrease.

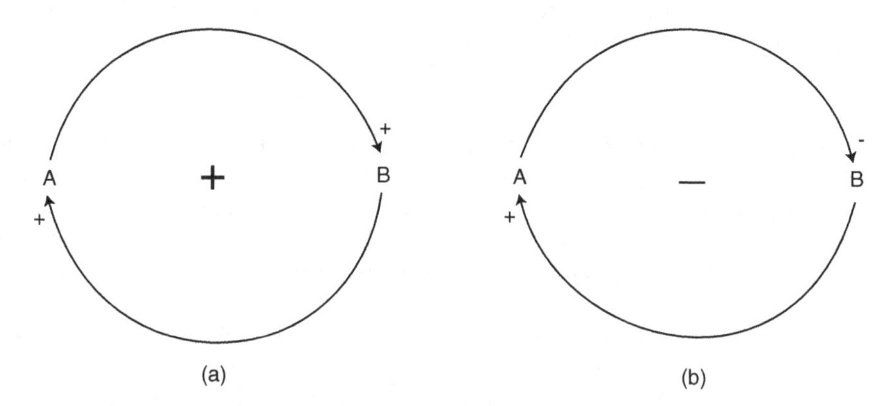

Figure 3.5 Positive (a) and negative (b) feedback loop.

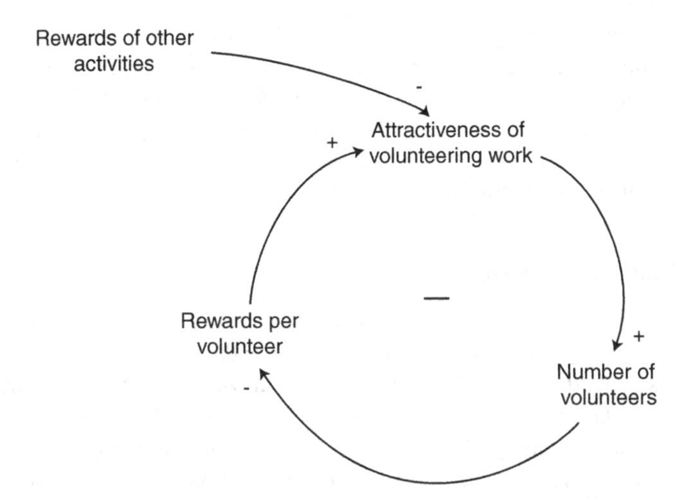

Figure 3.6 Negative feedback loop representing rewards of volunteering activity.

The motivation for the positive and negative labels comes from the way loop polarities can be obtained from the polarities of the individual causal links that combine to form the loop. The "arithmetic" of causal links parallels the arithmetic of multiplying signed numbers. To establish the parallelism, define a causal influence from A to B to be positive if a change in A tends to produce a change in B in the same direction; that is, an increase in A tends to produce an increase in B, or a decrease in A tends to produce a decrease in B. Similarly, define a causal influence from A to B to be negative if a change in A tends to produce a change in B in the opposite direction; that is, an increase in A tends to produce a decrease in B, or a decrease in A tends to produce an increase in B. It is then easy to argue that the polarity of a causal loop is the product of the polarities of its causal links. Therefore, a causal loop is positive if it contains an even number of negative links, and is negative if it contains an odd number of negative links.

An example best illustrates this arithmetic of link and loop polarities. Consider the circular causal loop shown in Figure 3.6, which happens to be abstracted from the classic work of Adam Smith that provides an argument that the reward in all occupations must tend to be the same. The loop part of the figure contains two positive links and one negative link. The causal link from *Attractiveness of volunteering work* to *Number of volunteers* is labeled positive to capture the assumption, in this causal argument, that an increase in attractiveness tends to produce an increase in the number of volunteers, or, a decrease in attractiveness tends to produce a decrease in the number of volunteers. A change is in the same direction. Hence, the link is positive. The causal link from *Number of volunteers* to the *Rewards per volunteer* is labeled negative to indicate an inverse sort of relationship. A change in the number of volunteers is assumed to lead to an opposite change in rewards per volunteer, and so on. Thus, the link is negative by our definition. The remaining links in the diagram can be similarly explained. These influences form together the positive polarity of the closed causal loop. This polarity can be obtained in two ways. Blindly applying the "multiplication rule" for loop polarities, we see that the loop should be called a negative loop because it contains an odd number of negative links, and the product of an odd number of negatives is a negative. More insightfully, we see that the loop describes a stabilizing process in which individual rewards will determine the number of people selecting a volunteering work.

3.1.7 Mathematical Modeling

In general, to obtain a way to control or manage a system, we use a mathematical model that closely represents the system. The mathematical model is solved and its solution is applied to the system. Models, or idealized representations, are an integral part of everyday life. Common examples of models include model airplanes, portraits, and cartographic globes. Similarly, models play an important role in science and business, as illustrated by models of the atom, models of genetic structure, mathematical equations describing physical laws of motion or chemical reactions, graphs, organization charts, and industrial accounting systems. Such models are invaluable for abstracting the essence of the subject of enquiry, showing interrelationships and facilitating analysis (Hillier and Lieberman, 1990).

Mathematical models are also idealized representations, but they are expressed in terms of mathematical symbols and expressions. Such laws of physics as $F = ma$ and $E = mc^2$ are familiar examples. Similarly, a mathematical model of a business problem is a system of equations and related mathematical expressions that describe the essence of the problem. Thus, if there are n related quantifiable decisions to be made, they are represented as *decision variables* (e.g., x_1, $x_2, \ldots, x_n$) whose values are to be determined. The appropriate measure of performance (e.g., profit) is then expressed as a mathematical function of these decision variables (e.g., $P = 3x_1 + 2x_2 + \cdots + 5x_n$). This function is called the *objective function*. Any restrictions on the values that can be assigned to these decision variables are also expressed mathematically, typically by means of inequalities or equations (e.g., $x_1 + 3x_1x_2 + 2x_2 \leq 10$). Such mathematical expressions for the restrictions are often called *constraints*. The constants (e.g., coefficients or right-hand sides) in the constraints and the objective function are called the *parameters* of the model. We might then say that the problem in using a mathematical model is to choose the values of the decision variables so as to maximize the objective function, subject to the specified constraints. Such a model, and minor variations of it, typify the models used in systems analysis.

Mathematical models have many advantages over a verbal description of the problem that is much more common in social sciences. One obvious advantage is that a mathematical model describes a problem much more concisely. This tends to make the overall structure of the problem more comprehensible, and it helps to reveal important cause-and-effect relationships. In this way, mathematical modeling indicates more clearly what additional data are relevant to the analysis. It also facilitates dealing with the problem in its entirety and considering all its interrelationships simultaneously. Finally, a mathematical model forms a bridge to the use of high-powered mathematical techniques and computers to analyze the problem. Indeed, packaged software for both microcomputers and mainframe computers is becoming widely available for many mathematical models.

The procedure of selecting the set of decision variables, which maximizes/minimizes the objective function subject to the systems constraints, is called the *optimization procedure*. The following is a general optimization problem. Select the set of decision variables $x_1^*, x_2^*, \ldots, x_n^*$ such that

$$\begin{aligned}
&\text{Min or Max} f(x_1, x_2, \ldots, x_n) \\
&\text{subject to} \\
&g_1(x_1, x_2, \ldots, x_n) \leq b_1 \\
&g_2(x_1, x_2, \ldots, x_n) \leq b_2 \\
&\vdots \\
&g_m(x_1, x_2, \ldots, x_n) \leq b_m
\end{aligned} \tag{3.2}$$

where $b_1, b_2, \ldots, b_m$ are known values.

A very common claim is that in management sciences optimization fails due to system complexity or computational difficulty. In that case a reasonable attempt at a solution may be obtained by *simulation*. Apart from facilitating the trial and error solution process, simulation is a valuable technique for studying the sensitivity of system performance to changes in model parameters or operating procedure. Simulation will be presented later in the book (see Chapter 5).

According to Equation (3.2), our main goal is the search for an *optimal*, or best, *solution*. Some of the techniques developed for finding such solutions are discussed in this book. However, it needs to be recognized that these solutions are optimal only with respect to the model being used. Since the model necessarily is an idealized rather than an exact representation of the real problem, there cannot be any utopian guarantee that the optimal solution for the model will prove to be the best possible solution that could have been implemented for the real problem. There just are too many uncertainties associated with real problems. However, if the model is well formulated and tested, the resulting solution should tend to be a good approximation to the ideal course of action for the real problem. Therefore, rather than demanding the impossible, the test of the practical success of a systems analysis study should be whether it provides a better guide for action than can be obtained by other means.

The eminent management scientist and Nobel Laureate in Economics, Herbert Simon, points out that *satisficing* is much more prevalent than optimizing in actual practice. In coining the term *satisficing* as a combination of the words *satisfactory* and *optimizing*, Simon is describing the tendency of analysts to seek a solution that is "good enough" for the problem at hand. Rather than trying to develop various desirable objectives, a more pragmatic approach may be used. Goals may be set to establish minimum satisfactory levels of performance in various areas. In this process, we can use past levels of performance or what is expected to be achieved in future. If a solution is found that enables all of these goals to be met, it is likely to be adopted without further ado. Such is the nature of satisficing. The distinction between optimizing and satisficing reflects the difference between theory and the realities frequently faced in trying to implement that theory in practice.

3.1.8 A Classification of Systems

Systems may be classified for convenience and to provide insight into their wide range. There are several classification systems that focus on aspects of the similarities and dissimilarities of different systems (Blanchard and Fabrycky, 1990). This section looks at some common dichotomies between natural and human-made systems, physical and conceptual systems, static and dynamic systems, and closed and open systems. Section 3.2 of the book will extend systems typology to cover social phenomena and in particular integrated disaster management problems.

Natural and Human-Made Systems The origin of systems gives a most important classification opportunity. Natural systems are those that came into being through natural processes. Human-made systems are those in which human beings have intervened by introducing or shaping components, attributes, or relationships. However, once they have been brought into being, all human-made systems are embedded in the natural world. Important interfaces often exist between human-made systems and natural systems. Each affects the other in some way. The effect of human-made systems on the natural world has only recently become a keen subject for study by concerned people, especially in those instances where the effect is undesirable.

Natural systems exhibit a high degree of order and equilibrium. This is evidenced in the seasons, the food chain, the water cycle, and so on. Organisms and plant life adapt themselves to maintain equilibrium with the environment. Every event in nature is accompanied by an appropriate adaptation, one of the most important being that material flows are cyclic. In the natural environment there are no dead ends and no wastes, only continual recirculation. Natural systems adapt to change of any intensity by changing the composition of the system (i.e., eliminating, adding, or rearranging elements as necessary to reestablish equilibrium). This evolution is the price paid for maintaining stability and system integrity. Natural systems adapt (evolve) to survive; systems that fail to adapt will become extinct.

Only recently have significant human-made systems appeared. These systems make up the human-made world. The rapid evolution of human beings is not adequately understood, but their coming upon the scene has significantly affected the natural world, often in undesirable ways. Primitive beings had little impact on the natural world, for they had not yet developed a potent and pervasive technology, but the impact of humanity has steadily increased over time.

Physical and Conceptual Systems Physical systems are those that manifest themselves in physical terms. They are composed of real components, and may be contrasted with conceptual systems, where symbols represent the attributes of components. Ideas, plans, concepts, and hypotheses are examples of conceptual systems. A physical system consumes physical space, whereas conceptual systems are organizations of ideas. One type of conceptual system is the set of plans and specifications for a physical system before it is actually brought into being.

A proposed physical system may be simulated in the abstract by a mathematical or other conceptual model. Conceptual systems often play an essential role in the operation of physical systems in the real world. The system of elements encompassed by all components, attributes, and relationships focused on a given objective employs a process in guiding the state of a system. A process may be mental (thinking, planning, learning), mental-motor (writing, drawing, testing), or mechanical (operating, functioning, producing). Processes exist equally in physical and conceptual systems.

Static and Dynamic Systems Another system dichotomy is the distinction between static and dynamic systems. A static system has a structure but no activity: a dike, for example, is a static system. A dynamic system combines structural components with activity. An example is a school, combining a building, students, teachers, books, and curricula. For centuries we have viewed the universe of phenomena as unchanging. A mental habit of dealing with certainties and constants developed. The substitution of a process-oriented description for the static description of the world is one of the major characteristics separating modern science from earlier thinking.

A dynamic conception of the world has become a necessity, yet a general definition of a system as an ongoing process is incomplete. Many systems would not be included under this broad definition because they lack motion in the usual sense. A highway system is static, yet contains the system elements of components, attributes, and relationships.

It should be recognized that any system can be seen as static only in a limited frame of reference. A dike is constructed over a period of time, and this is a dynamic process. It is then maintained and perhaps altered to serve its intended purpose more fully. These are clearly dynamic aspects, which would need consideration if the field of reference was the dike over a long period of time.

Systems may be characterized as having random properties. In almost all systems in both the natural and human-made categories, the inputs, process, and output can only be described in statistical terms. Uncertainty often occurs in both the number of inputs and the distribution of these inputs over time. For example, it is difficult to predict exactly the peak flood flow that will arrive at a particular location in a river basin, or the exact time it will arrive. However, each of these factors can be described in terms of probability distributions, and system operation is then said to be probabilistic.

Closed and Open Systems A closed system is one that does not interact significantly with its environment. The environment only provides a context for the system. Closed systems exhibit the characteristic of equilibrium resulting from internal rigidity, which maintains the system in spite of influences from the environment. An example is the chemical equilibrium eventually reached in a closed vessel when various chemicals are mixed together. The reaction can be predicted from a set of initial conditions. Closed systems involve deterministic interactions, with a one-to-one correspondence between initial and final states.

An open system allows information, energy, and matter to cross its boundaries. Open systems interact with their wider environment. Examples are plants, ecological systems, and business organizations. They may exhibit the characteristics of steady state, wherein a dynamic interaction of system elements adjusts to changes in the environment. Because of this steady state, open systems can be self-regulatory and are often self-adaptive.

It is not always easy to classify a system as either open or closed. Systems that have come into being through natural processes are typically open. Human-made systems have characteristics of both open and closed systems. They may reproduce natural conditions not manageable in the natural world. They are closed when designed for invariant input and statistically predictable output, as in the case of an aircraft in flight.

3.1.9 A Classification of Mathematical Models

Now that we have outlined some common classifications of systems, we can classify mathematical models according to the nature of the objective function and the constraints introduced in Section 3.1.7.

Linear and Nonlinear Models If the objective function and all the constraints are linear in terms of the decision variables, a mathematical model is considered to be linear. Similarly, if some or all of the constraints and/or the objective function are nonlinear, a mathematical model is described as nonlinear.

Deterministic and Probabilistic Models If each variable and parameter can be assigned a definite fixed number or a series of fixed numbers for any given set of conditions, a model is a deterministic one. If it contains variables, the values of which are subject to some measure of randomness or uncertainty, it is called probabilistic or stochastic.

Static and Dynamic Models Models that do not explicitly take time into account are static, and those that involve time-dependent interactions are dynamic.

Distributed and Lumped Parameter Models Models that take into account detailed variations in behavior from point to point throughout the system space are called distributed parameter models. In contrast, models that ignore the variations, in which the parameters and dependent variables can be considered to be homogeneous throughout the entire system space, are known as lumped parameter models.

3.2 SYSTEMS VIEW OF INTEGRATED DISASTER MANAGEMENT

Integrated disaster management is defined as an iterative process of decision making regarding prevention of, response to, and recovery from a disaster in Section 2.1 of the book. It involves complex interactions within and between the natural environment (represented by natural systems), human population (represented by human activity systems that frame actions, reactions, and perceptions), and built environment (represented by human-made systems). A systems perspective is recommended (Mileti, 1999, p. 26) to address the complexity of the disaster management.

3.2.1 A Systems Typology in Integrated Disaster Management

The main objective of this book is to present the applicability of systems ideas within all of the activities of integrated disaster management: mitigation, preparedness, response, and recovery. Using systems view, Mileti (1999, p. 107) states that disaster losses are the result of interaction among three systems and their main subsystems: (a) the earth's physical systems (e.g., the atmosphere, biosphere, cryosphere, hydrosphere, and lithosphere); (b) human systems (e.g., population, culture, technology, social class, economics, and politics); and (c) the constructed systems (e.g., buildings, roads, bridges, public infrastructure, and housing).

Natural Systems In this systems typology, natural systems are the evolution-made, irreducible wholes that can be *observed* and described as such, being made up of other entities having mutual relationships. They are "irreducible" in the sense that meaningful statements can be made about them as wholes, and this remains true even if we can describe their components and the relationships between the components with some precision (Checkland, 1999). Carbon dioxide is not reducible in this sense to carbon and oxygen—it remains a higher level whole having properties of its own.

Many of the entities that appear in a hierarchy of natural systems are of course the subject matter of scientific disciplines. Most scientific work takes as given the wholes existing at some level in the natural hierarchy and tries to describe them and the laws that govern their behavior. Most science is therefore concerned with the behavior of particular systems, even though interest is not usually centered on system hierarchies or on the property of "wholeness."

As an example of a natural science describing a particular kind of system Checkland (1999) suggests an example from chemical thermodynamics. The prime question that chemistry faces is: Does A react with B? In general, the natural world will contain in close proximity only substances that do not react together under normal conditions. The air could not contain oxygen and nitrogen if they reacted easily with each other. Beaches would not exist if salt water reacted with sand. On the other hand, zinc and hydrochloric acid, if brought together at room temperature, do react spontaneously, producing hydrogen gas. The science of chemistry tackles the problem: In a system consisting of substances A and B, what decides whether or not they react together to produce C? Thermodynamics has answered this question. Initially, it might be thought that such a system would suffer a spontaneous change if the product of reaction, C, represented a lower energy state. This is a good way to an answer—in very many instances A reacts with B to create the reaction product and to release energy in the form of heat. But this is not the whole answer because in some cases a spontaneous reaction is accompanied by an absorption of heat. It is now known that the behavior of a system consisting of $A + B$, as regards its tendency to produce C, is determined by a combination of both energy and system entropy. The system's behavior is described by a now well-tested equation:

$$\Delta G = \Delta H - T\Delta S \tag{3.3}$$

where ΔG measures the tendency to react, ΔH is the heat change in the reaction, T is temperature, and ΔS is the entropy change in the reaction.

Both closed and open systems as defined in Section 3.1.8 exhibit the property of *entropy*. Entropy is defined here as the degree of disorganization in a system. In the thermodynamic usage, entropy is the energy unavailable for work resulting from energy transformation from one form to another. In systems, increased entropy means increased disorganization. A decrease in entropy takes place as order occurs. Life represents a transition from disorder to order. Atoms of carbon, hydrogen, oxygen, and other elements become arranged in a complex and orderly fashion to produce a living organism.

Human Activity Systems There are many examples of sets of human activities related to each other so that they can be viewed as a whole using systems view. The fact that they form an entity is often emphasized by the existence of other systems (often human-made systems) that are associated with them; the activities that make Canadian Rail a human activity system, for example, are associated with the human-made physical system that is the railway network, with its stations, tracks, engine depots, and so on. Even if there are no closely associated systems to emphasize the grouping of the activities, it is difficult to deny the right of an observer to choose to view a set of activities as a system if he/she wishes to do so. What is less obvious is that human activity systems (and, for that matter, human-made systems) are fundamentally different in kind from natural systems.

The difference lies in the fact that such systems could be very different from how they are. Natural systems, without human intervention, could not be different. And the origin of this difference is what distinguishes the human being from other natural systems. Human activity systems depend on the observer. Since we all cannot be observers (some of us are being observed) the restriction on membership in the group of observers means that whatever is observed cannot acquire the full status of public knowledge. Hence there cannot in principle be a strict science of human activity exactly similar to a science of a natural phenomenon like magnetism. Also, the freedom of human actors means that there can never be accounts of human activity systems similar to, and having the same logical status as, accounts of natural systems. The well-established methods of science are entirely appropriate for the study of natural systems, with some generalization based on systems terminology. In the case of human activity systems the way to proceed is less obvious. The research on real-world problem solving suggests that it is essential to include with a description of human activity system an account of the observer and the point of view from which his/her observations are made.

Human-Made Systems Human-made systems exist because a need for them in some human activity system has been identified. Humans as designers are able to create physical systems to serve particular purposes. Similarly, humans can create structured sets of thoughts that can be called "human-made abstract systems."

3.2.2 Systems View of Disaster Management

Integrated disaster management is a part of all social and environmental processes aimed at minimizing loss of life, injury, and/or material damage. Mileti (1999) advocates systems view of disaster management processes in order to address their complexities, dynamic character, and interdisciplinary needs of management options. A primary emphasis of systems analysis in disaster management is on providing an improved basis for effective decision making. A large number of systems tools, from simulation and optimization to multiobjective analysis, are available for formulating, analyzing, and solving disaster management problems. Two paradigms are identified, in Section 1.2 of the book, that are shaping contemporary disaster management. The first focuses on the complexity of the disaster management domain, and the complexity of the modeling tools, in an environment characterized by continuous, rapid technological development. The second deals with disaster-related data availability and the natural variability of domain variables in time and space that affect the uncertainty of disaster management decision making.

The question I would like to answer in this section is: What are we trying to manage? We keep trying to manage environments (water, land, air, etc). We keep trying to manage people within environments. It seems that every time we push at one point, it causes unexpected change elsewhere—first fundamental systems rule. Perhaps it is time to sit back and rethink what we are trying to manage.

A Model In order to apply a continuous improvement approach to disaster management, it is essential to have a way or thinking—a model—of what is being managed. Without this, it is not possible to see where energy or resources are being wasted, or might significantly alter outcomes. Up to now, no such model has been proposed, let alone accepted, as a basis for predicting outcomes from different management interventions and their combinations.

The system in our focus is a social system. It describes the way disasters affect people. The purpose of describing the system is to help clarify the understanding and determine best points of systems intervention.

Management Systems Principles The disaster management system comprises four linked subsystems: *individuals*, *organizations* and *society*, nested within the *environment*. Individuals are the actors that drive organizations and society to behave in the way they do. They are decision makers in their own right, with a direct role in disaster mitigation, preparedness, response, and recovery. Organizations are the mechanism people use to produce outcomes that individuals cannot produce. Organizations are structured to achieve goals. Structure defines information and/or resource flows and determines the behavior of the organization. The concept of society is different from those of individuals and organizations, being more difficult to put boundaries around. In general, society itself is a system, of which individuals and organizations are subsets, and contains the relationships people have with one another, the norms of behavior, and the mechanisms that are used to regulate behavior. The environment includes concrete elements such as water and air, raw materials, and natural systems. It also encompasses the universe of ideas, including the concept of "future." This concept is important in considering disaster management—it is the expectation of future damages and future impacts that drives concern for sustainable management of natural disasters.

Management principle 1: To achieve sustainable management of disasters, interactions between the four subsystems, individual, organization, society, and environment, must be appropriately integrated.

A second principle we can use in developing our framework is that we can order systems inputs and outputs into three categories: resources, information, and values. These connect individuals, organizations, society, and environment, linking the four subsystems. Only information and resources flow link people and organizations. Value systems are influenced by these two flows, but operate in a different way. Value systems are generated within the individual or organization but feed of information and resource flows.

Management principle 2: Two flows, resource flow and, information flow, link the individual, organization, society, and environment subsystems. Value systems are the means through which different values are attached to information and resource flows.

All open systems require input of energies—resources—to produce outputs. The need to constantly access resources is a major mechanism for the operation of subsystems. Each subsystem relies on other subsystems and on the environment for its resources. In an ideal state, the goals of each subsystem, and performance relative to those goals, must represent a gain for other subsystems for all to continue to receive resources. The physical environment exerts passive pressure on the subsystems to ensure fit. In addition, the environment can limit the action by running out of a resource, or by changing circumstances to make the resource more precious—for example, changing climate.

Management principle 3: The ongoing need of subsystems for resources from one another sets the limits of their exploitation of one another and of the environment, and is a determinant of behavior within the system.

Information is used by each of the subsystems to make decisions required to ensure fit with other subsystems and the environment. Without flows of information from outside the system—or subsystem—the system must rely on its own internal information (knowledge) to make decisions. Such a circumstance increases the risk that the subsystem will drift out of fit with its context. Regardless, it constantly receives signals from the outside world, and it is itself sending signals to other systems. Well-functioning systems have structures built into them, which capture relevant information and use that information to maximize chances of utilizing resources to achieve their systems goals.

Management principle 4: Information is used by subsystems to make decisions intended to ensure fit with the needs of other subsystems and the environment.

Data does not in itself have meaning. A process of interpretation occurs between information and meaning, and this process is driven by existing values. Value systems determine what individuals, organizations, and societies find important: (a) the sorts of resources they will pursue and (b) the interpretation of information received and used. Value systems are embedded in the culture of society and organizations, and in the values held by individuals, and they determine how subsystems behave. Use of value systems may be triggered by information, and shaped by flow of resources.

Management principle 5: Values provide meaning to information flows, which are then used to determine resource use by subsystems.

Reality of linking mechanisms indicates that it is the availability of resources that largely conditions choice. It is information about availability that signals to the decision maker (individuals, organizations, or society) whether it is implementing appropriate management strategies. It is through the process of optimizing resource access that learning takes place and significant changes in culture and values are achieved. So, the most powerful management strategies will go directly to resource access and will initiate signals that show which social or environmental performance will allow for access to resources on improved terms.

Management principle 6: The most effective management strategies for sustainable management of disasters are those that condition access to resources.

Each subsystem utilizes different mechanisms for minimizing negative impacts of natural disasters. Within each subsystem there are many different interactions and many different options to optimize resource use.

No "Right" Management Strategy Disaster management is a process of managing behavior. There is no one strategy that will be optimal for any situation. Neither regulation nor economic incentives, nor education or shifts in property rights, is the "right" management strategy. What will work will vary with the social system being managed in response to three variables: the information and resource flows, and the value systems that are in place. The challenge for the disaster manager is to manage these three elements, across individuals, organizations and the society, and within the environment, to achieve the most effective outcome that is possible.

> *Management principle 7: More intensive focus on the systems view of disaster management will accelerate understanding of what management strategies work, and particularly why they might work.*

For example, when one program deals with economic incentives, another deals with improving information flows, and a third is focused on regulatory enforcement, it is very easy to believe that they are focused on different aspects with fragile links. What is necessary is a systems model to make sense of the interactions and dynamics that are being managed. This will allow us to learn from what we are so "clumsily" doing, so that eventually we can do it better.

3.2.3 Systems View of Disaster Management Activities

Integrated disaster management is defined as an iterative process of decision making that combines mitigation, preparedness, emergency response, and recovery.

Mitigation is long-term planning, and it involves identifying the vulnerability of every part of the territory to particular types of hazards and identification of steps that should be taken to minimize the risks. Section 2.2.1 of the book elaborates on difference between structural and nonstructural mitigation measures and identifies the five categories of measures seen as the most useful in practice. For the purpose of this discussion, I will focus my attention on mitigation as a long-term planning process.

The purpose of preparedness, as presented in Section 2.2.2, is to anticipate problems in disasters so that ways can be devised to address the problems effectively and so that the resources needed for an effective response are in place beforehand. Preparedness includes such activities as formulating, testing, and exercising disaster plans; providing training for disaster responders and the general public; and communicating with the public and others about disaster vulnerability and what to do to reduce it.

Disaster response activities (see Section 2.2.3 of the book) include emergency sheltering, search and rescue, care of injured, fire fighting, damage assessment, and other emergency measures. First responders are involved with coordination, communications, ongoing situation assessment, and resource mobilization during emergency period. The response phase of the integrated disaster management involves the implementation of the measures developed during the mitigation and preparedness phases.

Recovery deals with putting a disaster-affected community back together (see Section 2.2.4 of the book). Recovery includes mainly reconstruction of the damaged physical environment. However, the broader view includes temporary measures to restore community functions through replacement of capital stocks to predisaster levels and returning the appearance of the community to normal to the final phase, which involved promoting future economic growth and development.

Generalization of characteristics of all four disaster management activities implies that systems planning and systems design are involved with each and all of them. They are so closely related that it is difficult to separate one from the other.

Systems Planning The planning process closely follows the systems approach, and may involve the use of sophisticated analysis and computer tools. Basically, planning is the formulation of goals and objectives that are consistent with political, social, environmental, economic, technological and aesthetic constraints, and the general definition of procedures designed to meet those goals and objectives. Goals are the desirable end states that are sought. They may be influenced by the actions or desires of government bodies, such as legislatures or courts, of special interest groups, or of administrators. Goals may change as the interests of the concerned groups change.

Objectives relate to ways in which the goals can be reached. Planning should be involved in all types of mitigation activities. A good plan will bring together diverse ideas, forces, or factors, and combine them into a coherent, consistent structure that when implemented will improve target conditions without affecting nontarget conditions. Effective use of the systems approach will help to ensure that mitigation studies address the true problem at hand. Mitigation planning studies that do not do this could not, if implemented, produce useful and desirable changes.

Systems Design The design of a system represents a decision about how resources should be transformed to achieve some objectives. The final design is a choice of a particular combination of resources and a way to use them; it is selected from other combinations that might accomplish the same objectives. For example, the design of a flood protection dike involves decisions about the dike location (height, length, slopes), the type of materials used, and so on; the same result could be achieved in many different ways.

A design must satisfy a number of technical considerations. It must conform to the laws of the natural sciences; only some things are possible. To continue with the example of the dike, there are limits to the available strength of material used for dike design, and this constrains what will be the shape of the dike as a function of different material. The creation of a good design for a system thus requires solid competence in the matter at hand. System managers may take this fact to be self-evident, but it often needs to be stressed to industrial or political leaders, who are motivated by their hopes for what a proposed system might accomplish.

Economics and values must also be taken into account in the choice of design; the best design must take into consideration various groups involved in disaster management as well as variations in their value systems. Moreover, economics and other issues tend to dominate the final choice between many possible designs, each of which appears equally effective technically. The selection of a design is then determined by the costs and relative values associated with the different possibilities. The choice of material and design in dike construction is generally a question of cost and benefits. For more complex systems, political or other values may be more important than costs. In planning a flood control reservoir, for instance, it is usually the case that several sites can be made to perform technically; the final choice hinges on societal decisions about, for example, the relative importance of ease of access and the environmental impacts of the reservoir, in addition to its cost and flood protection.

3.3 SYSTEM FORMULATION EXAMPLES

General systems theory offers a new way for formulating various disaster management problems. A number of simple examples are developed in this section to illustrate the problem formulation phase, which calls for transfer of the conceptual system definitions into the practical problem domain. I have found from my experience that this is the most difficult phase for novices in the field.

3.3.1 Dynamics of Epidemics

Epidemics of infectious diseases can be nicely modeled using systems approach. The cumulative number of cases follows an S-shaped curve—the rate at which new cases occur rises exponentially, peaks, and then falls as the epidemic ends. The epidemics usually start with a single infected person (patient zero). The disease mostly spreads through contact and by inhalation of aerosols released when infected individuals cough and sneeze. The disease spreads slowly at first, but as more and more people get ill and become infectious, the number they infect grows exponentially. Because of the various factors (e.g., close quarters), and thus, a high rate of exposure, a large portion of the population eventually becomes ill, and the epidemic ends because of the depletion of the pool of susceptible people. A few examples that confirm this typical epidemic spread include the (a) epidemic of influenza at an English boarding school in 1978 (News and Notes, 1978), (b) epidemic of plaque in Bombay in 1905–1906 (Kermack and McKendrick, 1927), and the more recent (c) international SARS epidemic in 2003 (Dye and Gay, 2003). It is interesting that the pathogen does not have to be a biological agent—epidemics of computer viruses follow similar dynamics.

Let us consider the region with a total population N. Formulate a systems model of a disease epidemic using the following information: S number of people susceptible to the disease and I number of people who are infectious (therefore the name SI model). Use the following simplifying assumptions: (i) ignore births, deaths, and migration; and (ii) once people are infected, they remain infectious indefinitely (in other words, this model applies only to chronic infections).

Solution The SI model contains two loops (Figure 3.7): a positive loop describing the spread of infection and a negative loop describing depletion of the number of people who can be infected. Infectious diseases spread as those who are infectious come into contact with and pass the disease to those who are susceptible, increasing

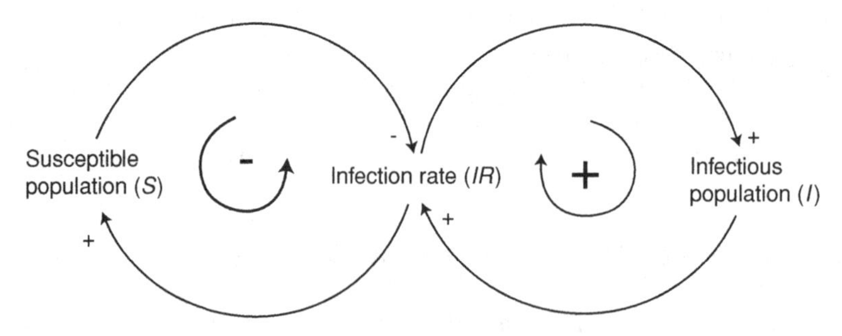

Figure 3.7 Causal structure of a simple model of an epidemic.

the infectious population still further (the positive loop) while at the same time depleting the pool of susceptibles (the negative loop).

Mathematical Formulation The infectious population I is increased by the infection rate IR, while the susceptible population S is decreased by it.

$$I = \int (IR, I_0)\,\mathrm{d}t \tag{3.4}$$

$$S = \int (-IR, N - I_0)\,\mathrm{d}t \tag{3.5}$$

where N is the total population in the community and I_0 is the initial number of infectious people (a small number or even a single individual). To formulate the infection rate, consider the process by which susceptible people become infected.

People in the community interact at a certain rate (the contact rate c measured in people contacted per person per time period, or 1/time period). Thus, the susceptible population generates Sc encounters per time period. Some of these encounters are with infectious people. If infectious people interact at the same rate as susceptible people (they are not quarantined or confined to bed), then the probability that any randomly selected encounter is an encounter with an infectious individual is I/N. Not every encounter with an infectious person results in infection. The infectivity i of the disease is the probability that a person becomes infected after contact with an infectious person. Therefore, the infection rate is the total number of encounters, Sc, multiplied by the probability that any of those encounters is with an infectious individual, I/N, multiplied by the probability that an encounter with an infectious person results in infection:

$$IR = (c \times i \times S) \times \left(\frac{I}{N}\right) \tag{3.6}$$

The dynamics can be determined by noting that without births, deaths, or migration, the total population is fixed:

$$S + I = N \tag{3.7}$$

Though the system contains two state variables, it is actually a first-order system because one of the variables is completely determined by the other. Substituting $N - I$ for S in Equation (3.6) yields:

$$IR = ((c \times i) \times I) \times \left(I - \frac{I}{N}\right) \tag{3.8}$$

Equation (3.8) is known as the logistic model. The infection rate follows a bell-shaped curve, and the total infected population follows the classic S-shaped pattern of the logistic curve. The higher the contact rate or the greater the infectivity, the faster the epidemic progresses. More on the mathematical solution is provided in Chapter 5 of the book.

The SI model captures the most fundamental feature of infectious diseases: the disease spreads through contact between infected and susceptible individuals. It is the interaction of these two groups that creates the positive and negative loops and the nonlinearity responsible for the shift in loop dominance as the susceptible population is depleted. The nonlinearity arises because the two populations are multiplied together in Equation (3.6); it takes both a susceptible and an infectious person to generate a new case.

3.3.2 Shortest Supply Route

An emergency manager is responsible for providing emergency supply to groups of disaster-hit communities. He has m supply storage locations throughout the region. Storage location i can supply at most S_i units of emergency supply (blankets, generators, boats, or similar) for $i = 1, 2, \ldots, m$. Within the shortest possible time, the manager must provide its n communities with at least D_j units of emergency supply, for $j = 1, 2, \ldots, n$ to meet demand requirements. The logistics problem facing the disaster manager is to designate which storage locations are to provide to which community so that the total distribution time is a minimum.

Solution Decision variables in this problem are x_{ij}, the number of emergency supply units transported from storage i to community j. Let us assume that c_{ij} is the associated transportation time between the storage location i and the community j.

Objective function—minimize the total distribution time. A mathematical description of the manager's problem is:

$$\text{Minimize} \sum_{i=1}^{m} \sum_{j=1}^{n} x_{ij} \tag{3.9}$$

Constraints—restrictions on the supply and on the demand side.

$$\sum_{j=1}^{n} x_{ij} \leq S_i \tag{3.10}$$

for $i = 1, 2, \ldots, m$ (supply restrictions).

$$\sum_{i=1}^{m} x_{ij} \geq D_j \tag{3.11}$$

for $j = 1, 2, \ldots, n$ (demand restrictions) and each $x_{ij} \geq 0$.

If we look at the mathematical structure of this problem, it is evident that both the objective function (Eq. 3.9) and all constraints (Eq. 3.10 and 3.11) are linear functions of decision variables. Therefore, this mathematical model can be classified as a linear optimization problem. Further discussion of linear programming and algorithms for the solution of linear programming problems are presented in Chapter 6 of the book.

3.3.3 Resources Allocation

A concrete manufacturing company is hired to provide material required for repairs of the municipal flood control system. Repairs require (a) cement blocks and (b) cement pipes. A manufacturing company has limited resources (machine and manpower) that are required for manufacturing of both products. The production process generates some pollution and the manufacturing company would like to maximize profits and at the same time minimize pollution. Table 3.1 provides the necessary information for the formulation of this problem using the systems approach.

TABLE 3.1 Input Data for the Resources Allocation Problem

	Cement Blocks	Cement Pipes
Number of units to be produced	x_1	x_2
Units of cement required	1	5
Machine time required (hour)	0.5	0.25
Manpower required (man-hour)	0.2	0.2
Direct cement cost ($)	0.25	0.75
Direct labor cost ($)	2.75	1.25
Sales price per unit	4	5

In addition, the following assumptions are made: one construction period is involved (time $t = 0, 1$); machine capacity is 8 hours per period; manpower capacity is 4 man-hours per period; material is limited to 72 units; and the amount of pollutant emitted is 3 units per cement block produced and 2 units per cement pipe produced.

Solution On the basis of the provided information, the objective functions and constraints of the problem can be formulated. The contribution margin (selling price per unit less variable cost per unit) of each product can be calculated:

Cement block:

$$\$4.00 \text{ (sales price)} - \$0.25 \text{ (direct cement cost)} - \$2.75 \text{ (direct labor cost)}$$
$$-\$1.00 \text{ per cement block manufacturer}$$

Cement pipe:

$$\$5.00 \text{ (sales price)} - \$0.75 \text{ (direct cement cost)} - \$1.25 \text{ (direct labor cost)}$$
$$-\$3.00 \text{ per cement pipe manufacturer}$$

The objective function for profit becomes:

$$\max_{x_i} \; Z_1(\mathbf{X}) = x_1 + 3x_2 \tag{3.12}$$

The objective function for pollution is:

$$\min_{x_i} \; Z_2(\mathbf{X}) = 3x_1 + 2x_2 \tag{3.13}$$

The technical constraints due to machine capacity, man-power capacity, and cement availability are respectively:

$$0.5x_1 + 0.25x_2 \leq 8 \tag{3.14}$$

$$0.2x_1 + 0.2x_2 \leq 4 \tag{3.15}$$

$$x_1 + 5x_2 \leq 72 \tag{3.16}$$

In the final form, the multiobjective problem of resources allocation is represented by objective functions (Eqs. 3.12 and 3.13) subject to constraints (Eqs. 3.14–3.16). If we look at the mathematical structure of this problem, it is evident that both the objective functions and all constraints are linear functions of decision variables. Therefore, this mathematical model can be classified as a multiobjective linear optimization problem.

REFERENCES

Blanchard, B.S. and W.J. Fabrycky (1990), *Systems Engineering and Analysis*, Prentice Hall, Inglewood Cliffs, NJ.

Dye, C. and N. Gay (2003), "Modeling the SARS epidemic," *Science*, 300(5627):1884–1885.

Checkland, P.B. (1999), *Systems Thinking, Systems Practice*, Wiley, Chichester, UK.

Forrester, J.W. (1990), *Principles of Systems*, Productivity Press, Portland, OR, first published in 1968.

Hillier, F.S. and G.J. Lieberman (1990), *Introduction to Operations Research*, McGraw Hill, New York.

Jackson, M.C. (2000), *Systems Approaches to Management*, Kluwer Academic/Plenum Publishers, New York; Boston, MA.

Kermack, W. and A. McKendrick (1927), "Contributions to the mathematical theory of epidemics," *Proceedings of the Royal Society*, 115A:700–721.

Mileti, D.S. (1999), *Disasters by Design*, Joseph Henry Press, Washington, DC.

News and Notes (1978), "Influenza in a boarding school," *British Medical Journal*, March 4, p. 587.

Richardson, G.P. (1991), *Feedback Thought in Social Science and Systems Theory*, University of Pennsylvania Press, Philadelphia, PA.

Richmond, B. (1993), "Systems thinking: Critical thinking skills for the 1990s and beyond," *System Dynamics Review*, 9(2):113–133.

EXERCISES

3.1 What is a system?

3.2 Identify and contrast a physical and a conceptual system.

3.3 Identify and contrast a closed and an open system.

3.4 Define in your own words and give examples of:

(a) uncontrolled system input;

(b) neutral system output; and

(c) feedback.

3.5 What is a mathematical model? Why do we develop mathematical models? List what are, in your opinion, the main purposes of mathematical models.

3.6 Describe one disaster management system consisting of various interdependent components. What are the inputs to the system and what are its outputs? How did you decide what to include in the system and what not to include? How did you decide on the level of spatial and temporal detail to be included?

3.7 For the following problems, specify in words the possible objectives, the unknown decision variables whose values need to be determined, and the constraints that must be met by any solution of the problem.

(a) Disaster evacuation.

(b) Locating and deciding the capacity of an emergency water treatment plant.

(c) Determining the size of a reservoir for a flood control of a small community.

(d) Locating new sites for location of fire hydrants in the urban community.

(e) Allocating funds to disaster recovery programs.

(f) Selecting a most efficient flood control alternative.

(g) Determining the number and location of fire stations for a large urban region.

3.8 A city has just approved the use of its municipal budget for building a new emergency evacuation facility. You have been hired as a consultant to assist in the selection of a location for this new facility. List all the objectives to be optimized in selecting the location. Prioritize these objectives from the perspective of:

(a) the mayor of a city and

(b) the residents of a city.

3.9 According to the mathematical structure of each problem in Exercise 3.7, classify the models using the classification presented in Section 3.1.9.

3.10 The emergency blood supply system that serves a city needs a review. The city is starting to explore alternatives to avoid disruption in its emergency blood supply, which is essential for the management of disasters. Options that have been proposed include: (i) expanding the current blood supply system; (ii) developing a new blood supply system; (iii) importing blood when needed from another region; and (iv) modifying the blood collection and storage processes in order to conserve the blood and, at the same time, expanding the existing blood collection program. Develop an optimization model to help evaluate these alternatives.

3.11 Give an example of a disaster mitigation process that exhibits exponential growth.

3.12 Give an example of a disaster recovery process that exhibits exponential decay.

3.13 More evacuees in a region are causing an increase in the food demand. This increased demand prompts the expansion of the food supply system capacity. Develop a simple feedback loop representation of the problem. Identify the type of feedback relationship.

3.14 Develop a small disaster emergency management problem that includes a minimum of two feedback loops. Describe the problem and develop a causal influence diagram for it.

4 Introduction to Methods and Tools for a Systems Approach to Management of Disaster

Integrated disaster management, as defined in this text, is an iterative process of decision making regarding prevention of, response to, and recovery from, a disaster. It relies on the application of a systems approach to formulating disaster management problems, and the use of systems analysis in finding their solutions. To use a systems approach calls for a change in our basic categories of thought about the physical reality under consideration. In contemporary disaster management we are forced to deal with complexities, with wholes or systems. This implies a basic reorientation in thinking.

Systems analysis is the use of rigorous methods to help determine preferred plans, design, and operations strategies for complex, often large-scale, systems. Its methods depend on the use of the computer for practical application. This part of the book provides a basic introduction to some of the techniques used: simulation, optimization, and multiobjective analysis.

Many disaster management problems are subject to uncertainty that has important implications for what can be achieved by integrated disaster management. Systems analysis techniques are available in deterministic form—using the assumption that the data required by a systems model are known exactly; and stochastic form—when uncertainties of various forms impinge on the decision. Stochastic systems techniques are based on two different paradigms—probability and fuzziness. Stochastic systems techniques are beyond the scope of the book and will be addressed in more advanced text in the future. However, in this section an introduction to disaster management under uncertainty is presented in order to provide the reader with basic definitions and point out the difference between probabilistic and fuzzy set approaches.

4.1 SIMULATION

Simulation models describe how a system operates, and are used to predict what changes will result from a specific course of action. Such models are sometimes referred to as cause-and-effect models. They describe the state of the system in

Systems Approach to Management of Disasters: Methods and Applications, By Slobodan P. Simonović

response to various inputs, but give no direct measure of what decisions should be taken to improve the performance of the system. Therefore, simulation is a problem-solving technique. It contains the following phases:

1. Development of a model of the system,
2. Operation of the model (i.e., generation of outputs resulting from the application of inputs), and
3. Observation and interpretation of the resulting outputs.

The essence of simulation is modeling and experimentation. Simulation does not directly produce the answer to a given problem.

Simulation includes a wide variety of procedures. In order to choose among them, and use them effectively, the potential user must know how they operate, how they can be expected to perform, and how this performance relates to the disaster management problem under consideration. The generic simulation procedure involves decomposition of the problem in order to aid in the system description. When the main elements of the system are identified, the proper mathematical description is provided for each of them. The procedure continues with computer coding of the mathematical description of the model. Each model parameter is then calibrated, and the model performance is verified using data that has not been seen during the calibration process. The completed model is then operated using a set of input data. Detailed analysis of the resulting output is the final step in the simulation procedure.

A completed model can be reused many times with alternative input data. If there is a need for modification of the system description or model structure, the whole process starts again and the model has to be recoded, calibrated, and verified again before its use.

The major components of a simulation model are:

Inputs: quantities that "drive" the model. (In disaster forecasting models, for example, a principal input is the set of streamflows, rainfall sequences, extreme temperature, pollution loads, wind speeds, etc.)

Physical relationships: mathematical expressions of the relationships among the physical variables of the system being modeled (continuity, energy conservation, routing equations, etc.).

Nonphysical relationships: those that define economic variables, political conflicts, public awareness, and so on.

Operation rules: the rules that govern operational control.

Outputs: the final product of operations on inputs by the physical and nonphysical relations in accordance with operating rules.

Simulation models play an important role in integrated disaster management. They are widely accepted within the disaster management community and are usually designed to predict the response of a system under a particular set of conditions.

Computer modeling and simulation offer a valuable tool to help disaster managers plan for and respond to disasters. By permitting repeated examination of diverse disasters under multiple parameter settings, computer simulations can project worst-case scenarios, evaluate the capabilities of the entire response system, and explore suitable mitigation strategies. Altay and Green (2006) show that only about 10% of research effort included in their review deals with simulation.

Examples of traditional simulation models available in disaster management literature are covering a broad range of disaster management problems. Smith et al. (2009) present a novel simulation approach to multihazard modeling to support medical and public health disaster planning and response using a sarin release scenario in a metropolitan environment.

An interdisciplinary team at the University of Pittsburgh has developed the Dynamic Discrete Disaster Simulation System to simulate the large-scale disaster responses (Wu et al., 2008). This project incorporates principles and approaches from various disciplines into a framework to assist decision makers in planning and managing responses to disaster events. Although the system is built mainly for emergency response management, the architecture has broad applications in military operations, global supply chain management, and financial portfolio management. This project was motivated by catastrophic events such as the terrorist attacks of September 11, 2001. Hurricane Katrina, and the tsunami in the Indian Ocean. Because the time, location, and scale of the events are impossible to predict, it is increasingly clear that pre-event planning and preparedness are imperative in order to limit the damage to human lives and property and restore the affected communities to a semblance of normal operations. The authors envision the sophisticated simulation model as an efficient approach for three aspects of disaster management: (i) assisting key decision makers in developing and improving emergency response policies; (ii) training emergency personnel on how to respond by simulating various disaster scenarios; and (iii) assisting disaster management team in making response decisions in real time. The overall objective is to provide a circumstance-independent laboratory for testing how the type and scale of the event, situational state, and command decisions affect responders' efficiency and effectiveness in dealing with complex, evolving disasters.

Another example includes a simulation system that can be used during an emergency situation inside a building (Filippoupolitis and Gelenbe, 2009). The system is able to function in real time, adapt to the changes of the environment, and provide reliable suggestions to the evacuees regarding the direction of the best available exit. This approach has been inspired by some work in the transportation area. The system operates in a building environment, where civilians are taking part in an evacuation in the presence of a spreading hazard (fire, earthquake, terrorist attack, or similar). By following the directions provided by the simulation system, they can evacuate the building using the best available paths and avoiding the hazardous areas.

Research related to pandemic disasters introduced another interesting view of system simulation (Steward and Wan, 2007). In their work, simulation is offered as a form of tabletop exercise. What is a tabletop exercise? Tabletop exercises are often employed as an act of preparation for emergency response management. Generally, they consist of a gathering of decision-making actors who are presented with a narrative by facilitators and subsequent attempt to emulate thoughts, decisions, and actions that would be taken in response to the hypothetical facts faced in the narrative. Steward and Wan (2007) propose a formal model of tabletop exercise maturity and fit for disaster preparation. Fundamentally, a tabletop exercise is a simulation using human actors in place of abstract data objects moving across a network of pathways from the start state to end or steady state. In their work, tabletop exercises for emergency and disaster preparedness are, in fact, simulations of systems of behavior of social, governance, agency, and service systems. Given the fact that the main interest of emergency management is in the interaction of events distributed throughout societal infrastructure, such modeling supplies foundational support for design of response, procedural analysis, and performance assessment.

There are few commercial software companies, mostly for the support of insurance industry, that offer a range of simulation products. The Risk Management Solutions (2009) offers the RMS Simulation Platform, which is based on precompiled data. Instead of generating simulation years in real time, for each region and peril (earthquake, flood, hurricane, wildfire, etc.), a presimulated event dataset is used to generate year loss tables. Simulation datasets are developed using a stratified sampling methodology, which reduces the number of simulation years required and guarantees consistent results.

The AIR Worldwide Corporation (2009) offers a similar set of risk models that covers earthquakes, winter storms, floods, severe thunderstorms, tropical cyclones, and wildfires. Each of these models contains some innovative modeling aspects and uses up-to-date computer simulation technology.

The EQECAT (2009) offers one integrated platform WorldCATenterprise (WCe) for catastrophe risk management. WCe is a new generation model optimized for today's computing environments. It offers hazard coverage for 89 countries and territories worldwide. The latest release includes wind, earthquake, tornado, hail, and wildfire models.

The Federal Emergency Management Agency (FEMA) in the United States has developed the Hazards US MultiHazard (HAZUS-MH), a nationally applicable standardized simulation software that estimates potential losses from earthquakes, hurricane winds, and floods. HAZUS-MH uses state-of-the-art Geographic Information Systems (GIS) software to map and display hazard data and the results of damage and economic loss estimates for buildings and infrastructure. It also allows users to estimate the impacts of earthquakes, hurricane winds, and floods on populations. Estimating losses is essential to decision making at all levels of government, providing a basis for developing mitigation plans and policies, emergency preparedness, and response and recovery planning. In the newest version of HAZUS-MH, HAZUS-MH MR3 (HAZUS-MH Version 1.3), the Flood Model will run building analyses and users may import user-supplied flood maps and flood depth grids. The Hurricane Model includes an updated and revised historical database that includes storms from 2004 and 2005. New vulnerability functions will permit calculation of additional losses to manufactured housing due to tree blowdown. Changes to the Earthquake Model include adjustable population distribution parameters in the casualty module and the elimination of partial ignitions in the fire-following module.

Despite the elegance of some disaster computer simulation models, several limitations remain: environments without realistic geographical topography (except commercial products), inadequate scalability, deterministic or linear-response assumptions, avoidance of a population-based systemic approach, inattention to human psychosocial characteristics and individual movement, disregard of learning ability or altered responses, or inability to integrate disaster prediction with management. Many of the examples mentioned are quite complex, however, and their main characteristics are not readily understood by nonspecialists. Also, they are inflexible and difficult to modify to accommodate site-specific conditions or planning objectives that were not included in the original model. The most restrictive factor in the use of simulation tools is that there are often a large number of feasible solutions to investigate. Even when combined with efficient techniques for selecting the values of each variable, quite substantial computational effort may lead to a solution that is still far from the best possible.

4.2 SYSTEM DYNAMICS SIMULATION

System dynamics is an academic discipline introduced in the 1960s by researchers at the Massachusetts Institute of Technology. System dynamics was originally rooted in the management and engineering sciences but has gradually developed into a tool useful in the analysis of social, economic, physical, chemical, biological, and ecological systems (Forrester, 1990; Sterman, 2000). In the field of system dynamics, as in the context of this book, a system is defined as a collection of elements that continually interacts over time to form a unified whole. The underlying pattern of interactions between the elements of a system is called the *structure* of the system. One familiar disaster management example of a system is a disaster relief. The structure of a disaster relief is defined by the interactions between inflow of funds, available funds, outflow of funds, and other variables specific to a particular disaster event and location (extent of disaster, number of affected people, damage level, etc.). The structure of the disaster relief includes the variables important in influencing the system.

The term *dynamics* refers to change over time. If something is dynamic, it is constantly changing in response to the stimuli influencing it. A dynamic system is thus a system in which the variables interact to stimulate changes over time. System dynamics is a methodology used to understand how systems change over time. The way in which the elements or variables comprising a system vary over time is referred to as the *behavior* of the system. In the disaster relief example, the behavior is described by the dynamics of relief fund growth and decline. This behavior is due to the influences of inflow, outflow, and damage level, which are elements of the system.

One feature that is common to all systems is that a system's structure determines its behavior. System dynamics links the behavior of a system to its underlying structure. It can be used to analyze how the structure of a physical, biological, or any other system can lead to the behavior that the system exhibits. By defining the structure of a disaster relief, it is possible to use system dynamics simulation to trace out the behavior over time of the disaster relief.

The system dynamics simulation approach relies on understanding complex interrelationships among different elements within a system. This is achieved by developing a model that can simulate and quantify the behavior of the system. The simulation of the model over time is considered essential to understanding the dynamics of the system. The major steps that are carried out in the development of a system dynamics simulation model include:

- understanding the system and its boundaries,
- identifying the key variables,
- describing the processes that affect variables through mathematical relationships,
- mapping the structure of the model, and
- simulating the model for understanding its behavior.

Advances made during the last two decades in computer software provide considerable simplification in the development of system dynamics simulation models. Software tools like STELLA (High Performance Systems, 1992), DYNAMO (Lyneis et al., 1994), VENSIM (Ventana Systems, 1996), and POWERSIM (Powersim Corp., 1996) use the principles of object-oriented programming for the development of system dynamics simulation programs. They provide a set of graphical objects with their mathematical functions for easy representation of the system structure and the development of computer code. Simulation models can be easily and quickly developed using these software tools. The resulting models are easy to modify, easy to understand, and present results clearly to a wide audience of users. They are able to address disaster management problems with highly nonlinear relationships and constraints.

So, what are the advantages of system dynamics simulation over the classical simulation discussed earlier?

- The power and simplicity of use of system dynamics simulation applications is not comparable with those developed in functional algorithmic languages. In a very short period of time, the users of the system dynamics simulation models can experience the main advantages of this approach. The power of simulation lies in the ease of constructing what-if scenarios and tackling big, messy, real-world problems.
- The general principles upon which the system dynamics simulation tools are developed apply equally to social, natural, and physical systems. Using these tools in disaster management allows for the enhancement of models by adding social, economic, and ecological sectors to the model structure.
- The structure–behavior link of system dynamics models allows analysis of how structural changes in one part of a system might affect the behavior of the system

as a whole. Perturbing a system allows one to test how the system will respond under varying sets of conditions. To return to the example of a disaster relief, someone can test the impact of a delay in providing financial resources on the disaster relief, or analyze the impact of the elimination of a particular group of disaster victims on the behavior of the entire system. The manipulation of graphical objects in the system dynamics model that describes the structure of a system is as easy as a click of the computer mouse button.

- For well-defined systems with sufficient and good data, the system dynamics simulation offers predictive functionality, determining the behavior of a system under particular input conditions. However, the ability to use system dynamics simulation models and extend disaster simulation models to include social, ecological, economic, and physical system components offers learning functionality—the discovery of unexpected system behavior under particular input conditions. This is one of the main advantages of system dynamics over traditional simulation.
- In addition to relating system structure to system behavior and providing users with a tool for testing the sensitivity of a system to structural changes, system dynamics requires an active participation of various stakeholders in the rigorous process of modeling system structure. Since the use of system dynamics software is very simple, the modeling process can be done directly by most experienced stakeholders. Modeling a system structure forces a user to consider details typically glossed over within a mental model. System dynamics simulation can easily become a group exercise, providing for the active involvement of all stakeholders and an interactive platform for the resolution of conflicts among them. In the water resources management literature, this process has been named *shared vision modeling* (Palmer et al., 1999). Disaster management offers plenty of opportunities for adoption of shared vision modeling approach. Initial effort is already documented in the work of Lane et al. (2003).

In the review of Altay and Green (2006), it is shown that system dynamics simulation is just at the beginning of making entry into the field of disaster management. Less than 2% of research effort included in their review deals with system dynamics simulation.

Some examples of systems dynamics simulation include modeling of an accident and emergency department (Lane et al., 2000). Accident and emergency (A&E) units provide a route for patients requiring urgent admission to acute hospitals. This paper discusses the formulation and calibration of a system dynamics model of the interaction of demand pattern, A&E resource deployment, other hospital processes, and bed numbers; and the outputs of policy analysis runs of the model that vary a number of key parameters. One significant policy finding is that while some delays to patients are unavoidable, reductions can be achieved by selective augmentation of resources within, and relating to, the A&E unit. This suggests that basing A&E policy solely on any single criterion will merely succeed in transferring the effects of a resource deficit to a different patient group.

Lane et al. (2003) extended their systems dynamics modeling to active involvement of clients—mentioned earlier as shared vision modeling. This paper describes the collaborative process of building a systems dynamics simulation model in order to understand patient waiting times in an accident and emergency department. The purpose is to explore the issues that arise when involving clients, in this case health-care professionals, in the process of model building. Given this study's first promising results, further collaborative studies are encouraged.

Fawcett and Oliveira (2000) present a new approach to the casualty treatment problem following a large-scale disaster, based on a systems dynamics simulation model of how a regional health-care system responds to an earthquake event. The numbers and locations of casualties rescued alive, the scale of prehospital care, the postearthquake hospital capacity, and the transport system are inputs to the model. The model simulates the movement of casualties from the stricken areas to hospitals. It predicts the number of casualties that die as well as other statistics about the health-care system response, such as waiting time before treatment. The model can be run with varying input assumptions to simulate alternative disaster response strategies. Preliminary runs demonstrate the potential of the model as a tool for planning and training.

Problems of disaster management include multiple actors, multiple perspectives, conflicting interests, important tangibles, and many uncertainties. Systems dynamics simulation is a tool that can take advantage of the inherent structure of disaster problems. Use of comprehensive simulation models, preferably of interdisciplinary nature, is suggested to capture some aspects of natural phenomena and their interaction with people and human-built environment. For example, meteorological or geological principles and technical details can be integrated using engineering and social behavior simulation models.

The rest of this book focuses on system dynamics simulation as one of the methods for integrated disaster management. A detailed description of system dynamics simulation modeling and its application to disaster management follows in Part III.

4.3 OPTIMIZATION

Prospective research on optimal disaster management has often been characterized as "difficult, if not impossible," providing challenges to the establishment of evidence-based guidelines for disaster planning. Systems approach, however, offers numerous optimization tools that can be applied to the investigation of a wide range of disaster management problems.

The procedure of selecting the set of decision variables that maximizes/minimizes the objective function, subject to the systems constraints, is called the *optimization procedure*. A general mathematical form of an optimization problem is:

$$\text{Min or Max } f(\boldsymbol{x})$$

subject to

$$g_j(\boldsymbol{x}) \leq b_j \quad j = 1, 2, \ldots, m$$
$$\boldsymbol{x} \geq 0 \quad i = 1, 2, \ldots, n \tag{4.1}$$

where $\boldsymbol{x}$ is a vector of decision variables, n is the total number of decision variables, g is a constraint, b is the known right-hand-side value, j is the constraint number, and m is the total number of constraints.

Consideration of optimization in management of disasters will be limited in this book to linear programming (LP). LP is applied to problems that are formulated in terms of separable objective function and linear constraints, as in:

$$\text{Min or Max } x_o = \sum_{i=1}^{n} c_i x_i$$

subject to

$$\sum_{i=1}^{n} a_{ij} x_i = b_j \quad j = 1, 2 \ldots, m$$
$$x_i \geq 0 \quad i = 1, 2, \ldots, n \tag{4.2}$$

where x_i is a decision variable, n is the total number of decision variables, x_o is the objective function, b_j is the right-hand-side coefficient, c_i is the objective function coefficient, a_{ij} is the technological coefficient, j is the constraint number, and m is the total number of constraints.

The objective is usually to find the best possible allocation of resources (human, material, financial, etc.) within a given time period in complex disaster management systems. For most practical disaster management applications, the nonlinearity of the objective function and/or constraints means that many modifications have been used to convert nonlinear problems for the use of LP solvers.

Nonlinear programming is an optimization approach used to solve problems when the objective function and the constraints are not all in the linear form. In general, the solution to a nonlinear problem is a vector of decision variables that optimizes a nonlinear objective function subject to a set of nonlinear constraints. Successful applications are available for some special classes of nonlinear programming problems such as unconstrained problems, linearly constrained problems, quadratic problems, convex problems, separable problems, nonconvex problems, and geometric problems. The main limitation in applying nonlinear programming to disaster management problems is in the fact that it is generally unable to distinguish between a local optimum and a global optimum (except by finding another better local optimum).

Dynamic programming (DP) offers advantages over other optimization tools since the shape of the objective function and constraints do not affect it, and as such, it has some valuable application in management of disasters. For example, Chen et al. (2009) offer an interesting application of DP in emergency recovery. A fast and efficient recovery is capable of reducing the disaster loss and prevent the public panic. In this paper, an optimization model of the recoverability process is established to minimize the recovery time and optimize the allocation of resources.

DP requires discretization of the problem into a finite set of stages. At every stage a number of possible conditions of the system (states) are identified, and an optimal solution is identified at each individual stage, given that the optimal solution for the next stage is available. An increase in the number of discretizations and/or state variables would increase the number of evaluations of the objective function and core memory requirement per stage.

However, the complexity of real disaster management problems today exceeds the capacity of traditional optimization algorithms. Altay and Green (2006) show that mathematical programming, including heuristics, is the most frequently utilized method, accounting for 50% of surveyed examples, in disaster management. About 44% of all papers address mitigation and only 11.9% of papers are on natural disasters.

In the recent past, most researchers have been looking for new approaches that combine efficiency and an ability to find the global optimum for complex disaster management systems. Among them the special place belongs to *agent-based modeling* (ABM) as a computational approach for simulating the actions and interactions of autonomous individuals with a view to assessing their effects on the system as a whole. It combines elements of game theory, complex systems, emergence, computational sociology, multiagent systems, and evolutionary programming. The models simulate the simultaneous operations of multiple agents in an attempt to recreate and predict the actions of complex phenomena. The process is one of emergence from the lower (micro) level of systems to a higher (macro) level. As such, a key notion is that simple behavioral rules generate complex behavior; that is, the whole is greater than the sum of the parts. Individual agents are typically characterized as boundedly rational, presumed to be acting in what they perceive as their own interests, such as reproduction, economic benefit, or social status, using heuristics or simple decision-making rules. ABM agents may experience "learning," adaptation, and reproduction. Most agent-based models are comprised of (1) numerous agents specified at various scales (typically referred to as agent-granularity), (2) decision-making heuristics, (3) learning rules or adaptive processes, (4) an interaction topology, and (5) a nonagent environment.

A very important group of optimization techniques associated with ABM, known as *evolutionary algorithms*, seems to have a high potential for disaster management. Evolutionary techniques are based on similarities with the biological evolutionary process. In this concept, a population of individuals, each representing a search point in the space of feasible solutions, is exposed to a collective learning process, which proceeds from generation to generation. The population is arbitrarily initialized and subjected to the process of selection, recombination, and mutation through stages known as *generations*, such that newly created generations evolve toward more favorable regions of the search space. In short, the progress in the search is achieved by evaluating the fitness of all individuals in the population, selecting the individuals with the highest fitness value, and combining them to create new individuals with an increased likelihood of improved fitness. The entire process resembles the Darwinian rule known as the *survival of the fittest*.

Evolutionary algorithms are becoming more prominent in the disaster management field. Significant advantages of evolutionary algorithms include:

- no need for an initial solution,
- ease of application to nonlinear problems and to complex systems,
- production of acceptable results over longer time horizons, and
- generation of several solutions that are very close to the optimum (and that give added flexibility to a disaster manager).

This book focuses on only one optimization methods, LP. It is presented for its academic and practical significance. The other method will be the topic of more advanced text on disaster management. A detailed description of LP, together with its practical implementation, follows in Part III.

4.4 MULTIOBJECTIVE ANALYSIS

Disaster management is by nature multiorganizational, but organizations are only loosely connected leading to managerial confusions and ambiguity of authority. Public sector problems, such as disaster management, are generally ill defined, have high behavioral content, and are overlaid with strong political implications. The multifunctional nature and political hierarchy in emergency response organizations are well suited for hierarchical planning and multiattribute multiobjective approaches, as various groups have different priorities before, during, and after a disaster hits.

A multiobjective programming problem is characterized by an r-dimensional vector of objective functions:

$$Z(\boldsymbol{x}) = Z_1(\boldsymbol{x}), Z_2(\boldsymbol{x}), Z_3(\boldsymbol{x}), \ldots, Z_r(\boldsymbol{x})$$

subject to

$$\boldsymbol{x} \in \boldsymbol{X} \tag{4.3}$$

where $\boldsymbol{X}$ is a feasible region:

$$\boldsymbol{X} = \{\boldsymbol{x} : \boldsymbol{x} \in \boldsymbol{R}^n, g_i(\boldsymbol{x}) \leq 0, x_j \geq 0 \forall i, j\}$$

where $\boldsymbol{R}$ is a set of real numbers, $g_i(\boldsymbol{x})$ is a set of constraints, and $\boldsymbol{x}$ is a set of decision variables.

The word "optimization" has been purposefully kept out of the definition of a multiobjective programming problem since one cannot, in general, optimize a priori a vector of objective functions. The first step of the multiobjective problem consists of identifying the set of nondominated solutions within the feasible region $\boldsymbol{X}$. So instead of seeking a single optimal solution, a set of nondominated solutions is sought. More detailed mathematical definitions are provided in Section 7.2.

The essential difficulty with multiobjective analysis is that the meaning of the optimum is not defined as long as we deal with multiple objectives that are truly different. For example, suppose we are trying to determine the best design of a system of dikes on a river, with the objectives of promoting national income, reducing deaths by flooding and increasing employment. Some designs will be more profitable, but less effective at reducing deaths. How can we state which is better when the objectives are so different, and measured in such different terms? How can we state with any accuracy what the relative value of a life is in terms of national income? If we resolved that question, then how would we determine the relative value of new jobs and other objectives? The answer is, with extreme difficulty. The attempts to set values on these objectives are, in fact, most controversial.

To obtain a single global optimum over all objectives requires that we either establish or impose some means of specifying the value of each of the different objectives. If all objectives can be valued on a common basis, the optimization can be stated in terms of that single value. The multiobjective problem disappears and the optimization proceeds relatively smoothly in terms of a single objective.

In practice it is frequently awkward if not indefensible to give every objective a relative value. The relative worth of profits, lives lost, the environment, and other such objectives is unlikely to be established easily by anyone, or to be accepted by all concerned. We cannot hope, then, to be able to determine an acceptable optimum analytically.

The focus of multiobjective analysis in practice is to sort out the mass of clearly dominated solutions, rather than determine the single best one. The result is the identification of a small subset of feasible solutions that is worthy of further consideration. Formally, this result is known as the *set of nondominated solutions*.

Multiple-objective decisions do not have an optimal solution, unless one solution completely dominates every other solution for every objective. This does not usually happen in disaster management. As a result, methods are developed for assessing tradeoffs between alternatives based on using more than one objective. In the last three decades of multiobjective research, efforts have been made in objective quantification, the generation of alternatives, and selection of the preferred alternative.

Very limited application of multiobjective analysis has been reported in disaster management. Altay and Green (2006) show that only about 10% of research effort included in their review deals with multiobjective analysis. Most of the examples deal with evacuation planning (Chow and Lui, 2002; Georgiadou et al., 2007; Saadatseresht et al., 2009).

Contemporary research into multiobjective analysis has shifted away from continuous theoretical models, and explored issues in evaluating a discrete set of alternatives. There are plenty of options when it comes to choosing a multiobjective method.

The consensus is that simpler transparent methods, or no formal method at all, were preferred by experienced disaster managers.

The shortcoming of most multiobjective methods is that they rely on an a priori articulation of preferences—an expression of the importance of each objective to a decision maker. The difficulty for group decision-making is that conflicts arise and complicate the evaluation process by tying decision makers to their articulation of preference. Prior articulation methods are typified by an effort to aggregate the objectives of decision makers and reduce the problem to a multiple-participant multiple-objective problem. Exceptions to prior articulation are methods that employ a progressive articulation of preferences. These are the true interactive conflict capable multiobjective methods.

This book provides in Chapter 7 the introduction of (a) one method for generating nondominated solutions and (b) one efficient discrete multiobjective method with a progressive articulation of preferences, known as *Compromise Programming* (originally introduced by Zeleny (1973)).

4.5 DISASTER RISK MANAGEMENT

Integrated disaster management includes measures for before (prevention, preparedness, risk transfer), during (humanitarian aid, rehabilitation of the basic infrastructure, damage assessment), and after disaster (disaster response and reconstruction). Disaster risk management is part of integrated disaster management, focusing on the before (risk analysis, prevention, preparedness) of the extreme event, and relating to the during and after of disaster only through risk analysis. Disaster risk management is an instrument for reducing the risk of disaster primarily by reducing vulnerability and strengthening self-protection capabilities (Kohler et al., 2004; UNDP, 2004). The components of the disaster risk management are risk analysis, prevention, and preparedness.

Risk analysis consists of hazard analysis and vulnerability analysis, together with analysis of protective capabilities. Some authors treat the analysis of the protective capabilities of the local population (coping strategies) as part of vulnerability analysis, others as a third component of risk analysis, others see it as an additional activity, and as such a component of risk assessment and not risk analysis. Here, the analysis of self-protection capabilities is treated as part of vulnerability analysis.

Disaster prevention includes those activities that prevent or reduce the negative effects of extreme natural events, primarily in the medium to long term. These include political, legal, administrative, planning, and infrastructural measures such as spatial and land use planning, urban development planning, building codes; sustainable resource management and river basin management; establishment of social organizational structures for preventive measures and to improve the response to extreme natural events (disaster risk management structures); training and upgrading of population and institutions; and infrastructural improvements.

Preparedness for disasters is intended to avoid or reduce loss of life and damage to property if an extreme natural event occurs. The participating institutions and the population at hazard are prepared for the situation that might arise, and precautions are taken. In addition to increasing the alert level, mobilizing the self-help resources of the population for the emergency and operating a monitoring system, preparedness may include the following measures: participatory formulation of emergency and evacuation plans; coordination and deployment planning; training and upgrading; logistical measures, such as emergency accommodation and stockpiling food and drugs; establishing and/or strengthening local and national disaster protection structures and rescue services; disaster protection exercises; early warning systems. Preparedness and prevention measures also include designing and implementing risk transfer concepts.

Altay and Green (2006) show that probability theory and statistics are very frequently utilized methods in the research literature (about 20% of research effort included in their review) but much less frequently in practice. Disaster risk management techniques include both paradigms, probability theory and fuzzy set theory. This book will not be presenting methods for risk management. The following discussion provides a brief introduction of both theoretical concepts and a much more advanced text, to be prepared in the future, will include methods and tools for a systems approach under uncertainty.

4.5.1 Sources of Uncertainty

Uncertainty is defined in plain language as lack of certainty. It has important implications for what can be achieved by integrated disaster management. All disaster management decisions should take uncertainty into account. Sometimes the implications of uncertainty involve *risk*, in the sense of significant potential unwelcome effects. Then disaster managers need to understand the nature of the underlying threats in order to identify, assess, and manage the risk. Sometimes, the implications of uncertainty involve an opposite form of risk, significant potential welcome effects. Then managers need to understand the nature of the underlying opportunities in order to identify and manage the associated decrease in risk. Failure to do so can result in a failure to capture good luck, which can increase the risk.

Uncertainty is in part about variability in relation to the physical characteristics of the problem under consideration. But uncertainty is also about ambiguity (Ling, 1993). Both variability and ambiguity are associated with a lack of clarity because of the behavior of all system components, a lack of data, a lack of detail, a lack of structure to consider disaster management problems, working and framing assumptions being used to consider the problems, known and unknown sources of bias, and ignorance about how much effort it is worth expending to clarify the management situation. Uncertainty caused by variability is a result of inherent fluctuations in the quantity of interest (for example in case of flood disasters hydrologic variables). The three major sources of variability are temporal, spatial, and individual heterogeneity. *Temporal variability* occurs when values fluctuate with time. Other values are affected by *spatial variability*; that is, they are dependent on the location of an area. The third category of *individual heterogeneity* effectively covers all other sources of variability. In disaster management, variability is mainly associated with the spatial and temporal variation.

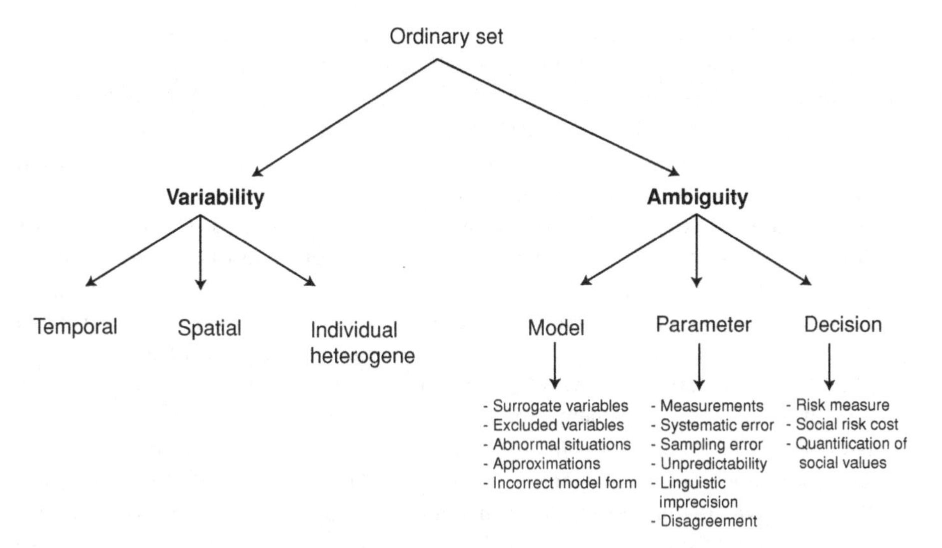

Figure 4.1 Sources of uncertainty.

The more elusive type of uncertainty is *ambiguity*, which is caused by a fundamental lack of knowledge. It occurs when the particular values that are of interest cannot be presented with complete confidence because of a lack of understanding or limitation of knowledge. The main sources of uncertainty because of a lack of knowledge are depicted in Figure 4.1.

Model and structural uncertainties refer to the knowledge of a process. Models are simplified representations of real-world processes, and model uncertainties can arise from oversimplification or from the failure to capture important characteristics of the process under investigation. Addressing this type of uncertainty is the coarse tuning function of the analysis. This type of uncertainty is best understood by studying its major sources. In disaster management, the modeling process includes *surrogate variables* (i.e., the substitution of variables for quantities that are difficult to assess). They are an approximation of the real value. The second source of model uncertainty stems from *excluded variables* (variables deemed insignificant in a model). The removal of certain variables or factors introduces large uncertainties into the model. For example, many disaster risk assessment methods do not consider how hazardous chemicals can be propagated through vegetation. Attempting to address excluded variables raises a paradox: we do not know when we have forgotten something until is too late. The *impact of abnormal situations* on models is the third source of uncertainty. The very nature of a disaster management model requires model calibration and verification using a set of broad circumstances. The problem occurs when a model is used for a situation that lies outside the set of situations used in the process of model calibration and verification. *Approximation uncertainty* is the fourth source of model uncertainty. This source covers the remaining types of uncertainty as a result of model generalizations. Examples of approximation uncertainty in disaster management can be found in the use of discrete probability distributions to represent a continuous process. The final type of model uncertainty, *incorrect form* (the correctness of the model being used to represent the real world) is initially the most obvious. To properly address this source, we must remember that all results are directly dependent on the validity of the model being used as a representation of the true process.

The next general category of uncertainty is *parameter uncertainty*. It involves the fine-tuning of a model, and cannot cause the large variations found in model uncertainty. The most common uncertainty in this category is caused by random error in direct measurements. It is also referred to as metric error, *measurement error*, random error, or statistical variation. This error occurs because no measurement of variables relevant for various disasters can be exact. Imperfections in the measuring instruments and observation techniques lead to imprecision and inaccuracies of measurements. The second, and largest, source of parameter uncertainty is *systematic error* (error as a result of subjective judgment). Measurements involve both random and systematic error. The latter is defined as the difference between the true value and the mean of the value to which the measurements converge. The third type of error is *sampling error* (error in drawing inferences from a limited number of observations). Sampling causes uncertainty in the degree to which the sample represents the whole. Well-developed statistical techniques such as confidence intervals, coefficient of variation, and sample size are used in disaster management to quantify this type of uncertainty. The fourth type of parameter uncertainty is caused by the unpredictability of an event. Limitations in knowledge and the presence of inherent *unpredictability* in the process make it impossible to predict, for example, wind direction and velocity at a future date. The fifth source of uncertainty is *linguistic imprecision*. Everyday language and communication is rather imprecise. It is possible to reduce linguistic uncertainty through clear specifications of events and values. The final source of uncertainty is derived from *disagreement* (conflicting expert opinion).

The third category of uncertainty is *decision uncertainty*, which arises when there is controversy or ambiguity over how to compare and weigh social objectives. It influences disaster decision making after parameter and model uncertainties have been considered. The first decision uncertainty includes uncertainty in the selection of an index to *measure risk*. The measure must be as technically correct as possible while still being measurable and meaningful. The second source of decision uncertainty lies in deciding the *social cost of risk* (transforming risk measures into comparable quantities). The difficulties in this process are clearly illustrated in the concept of developing a monetary equivalent for the value of life. The *quantification of social values* is the third source of uncertainty. Once a risk measure and the cost of risk are generated, controversy still remains over what level of risk is acceptable. This level is dependent upon society's attitude to risk.

4.5.2 Conceptual Risk Definitions

An attempt by risk analysis experts in the late 1970s to come up with a standardized definition of risk concluded that a common definition is perhaps unachievable, and that authors should continue to define risk in their own way. As a result, numerous definitions can be found in recent literature, ranging from the vague and conceptual to the rigid and quantitative. At a conceptual level, we defined risk above as a significant potential unwelcome effect of an event occurring, or the predicted or expected likelihood that a set of circumstances over some time frame will produce some harm that matters. More pragmatic treatments view risk as one side of an equation, where risk is equated with the probability of failure or the probability of load exceeding resistance.

Other symbolic expressions equate risk with the sum of uncertainty and damage, or the quotient of hazards divided by safeguards (Lowrance, 1976). Here we shall start with a risk definition based on the concept of load and resistance, terms borrowed from engineering. In the field of disaster management, these two variables have a more general meaning. For example, in water resources management the physical system may be a dike: load—flood water elevation, resistance—dike elevation, and type of failure—flooding. Another example in the health area may consider human organism as the physical system: load—exposure to harmful substance, resistance—human capacity, and type of failure—health damage. As one more example, the physical system could be an economic system: load—investment needs, resistance—money supply, and type of failure—fiscal failure. In social sciences, a physical system can be a social system: load—change of system, resistance—acceptance level, and type of failure—change of population.

Load l is a variable reflecting the behaviors of the system under certain external conditions of stress or loading. Resistance r is a characteristic variable that describes the capacity of the system to overcome an external load. When the load exceeds the resistance ($l > r$), there should be a failure or an incident. A safety or reliability state is obtained if the resistance exceeds or is equal to the load ($l \leq r$). From the examples above it can be seen that load and resistance may take different meanings, depending on the specific problem domain.

Perhaps the most expressive definition of risk is the one that conveys its multidimensional character by framing risk as the set of answers to three questions: What can happen? How likely is it to happen? If it does happen, what are the consequences? (Kaplan and Garrick, 1981). The answers to these questions emphasize the notion that risk is a prediction or expectation which involves a hazard (the source of danger), uncertainty of occurrence and outcomes (the chance of occurrence), adverse consequences (the possible outcomes), a timeframe for evaluation, and the perspectives of those affected about what is important to them. The answers to these questions also form the basis of conventional quantitative risk analysis methodologies.

Three cautions surrounding risk must be taken into consideration: risk cannot be represented objectively by a single number alone, risks cannot be ranked on strictly objective grounds, and risk should not be labeled as real. Regarding the caution of viewing risk as a single number, the multidimensional character of risk can only be aggregated into a single number by assigning implicit or explicit weighting factors to various numerical measures of risk. Since these weighting factors must rely on value judgments, the resulting single metric for risk cannot be objective. Since risk cannot objectively be expressed by a single number, it is not possible to rank risks on strictly objective grounds. Finally, since risk estimates are evidence-based, risks cannot be strictly labeled as real. Rather, they should be labeled *inferred* at best.

A major part of the risk management confusion according to Slovic (2000) relates to an inadequate distinction between three fundamental types of risks:

- Objective risk (real, physical) R_o and objective probability p_o, which is the property of real physical systems.
- Subjective risk R_s and subjective probability p_s. Here, probability is defined as the degree of belief in a statement. R_s and p_s are not properties of the physical systems under consideration (but may be some function of R_o and p_o).
- Perceived risk R_p, which is related to an individual's feeling of fear in the face of an undesirable possible event, is not a property of the physical systems but is related to fear of the unknown. It may be a function of R_o, p_o, R_s, and p_s.

Because of the confusion between the concepts of objective and subjective risk, many characteristics of subjective risk are believed to be valid also for objective risk. Therefore, it is almost universally assumed that the imprecision of human judgment is equally prominent and destructive for all water resources risk evaluations and all risk assessments. This is perhaps the most important misconception that blocks the way to more effective societal disaster risk management. The ways society manages disaster risks appear to be dominated by considerations of perceived and subjective risks, while it is objective risks that kill people, damage the environment, and create property loss.

4.5.3 Probabilistic Approach

Probability is a concept widely accepted and practiced in integrated disaster management. To perform operations associated with probability, it is necessary to use sets—collections of elements, each with some specific characteristics. Boolean algebra provides a means for evaluating sets. In probability theory, the elements that comprise a set are outcomes of an experiment. Thus, the universal set Ω represents the mutually exclusive listing of all possible outcomes of the experiment and is referred to as the *sample space* of the experiment. In examining the outcomes of rolling a dice, the sample space is $S = (1, 2, 3, 4, 5, 6)$. This sample space consists of six items (elements) or sample points. In probability concepts, a combination of several sample points is called an *event*. An event is, therefore, a subset of the sample space. For example, the event of "an odd outcome when rolling a dice" represents a subset containing sample points 1, 3, and 5.

Associated with any event E of a sample space S is a probability, shown by $\Pr(E)$ and obtained from the following equation:

$$\Pr(E) = \frac{m(E)}{m(S)} \tag{4.4}$$

where $m(.)$ denotes the number of elements in the set (.).

The probability of getting an odd number when tossing a dice is determined by using m (odd outcomes) = 3 and m (sample space) = 6. In this case, Pr (odd outcomes) = 3/6 = 0.5. Note that Equation (4.4) represents a comparison of the relative size of the subset represented by the event E and the sample space S. This is true when all sample points are equally likely to be the outcome. When all sample points are not equally likely to be the outcome, the sample points may be weighted according to their relative frequency of occurrence over many trials or according to expert judgment. In disaster management practice, we use three major conceptual interpretations of probability.

Classical Interpretation of Probability (Equally Likely Concept) In this interpretation, the probability of an event E can be obtained from Equation (4.4), provided that the sample space contains N equally likely and different outcomes, that is, $m(S) = N$, n of which have an outcome (event) E, that is, $m(E) = n$. Thus, $\Pr(E) = n/N$. This definition is often inadequate for disaster management applications. For example, if failures of a pump to start in a flood-affected area are observed, it is unknown whether all failures are equally likely to occur. Nor is it clear whether the whole spectrum of possible events is observed. That case is not similar to rolling a perfect dice, with each side having an equal probability of 1/6 at any time in the future.

Frequency Interpretation of Probability In this interpretation, the limitation on knowledge about the overall sample space is remedied by defining the probability as the limit of n/N as N becomes large. Therefore, $\Pr(E) = \lim_{N=\infty} (n/N)$. Thus, if we have observed 2000 starts of a pump in which 20 failed, and if we assume that 2000 is a large number, then the probability of the pump failure to start is 20/2000 = 0.01. The frequency interpretation is the most widely used classical definition in disaster management today. However, some argue that because it does not cover cases in which little or no experience (or evidence) is available, or cases where estimates concerning the observations are intuitive, a broader definition is required. This has led to the third interpretation of probability.

Subjective Interpretation of Probability In this interpretation, $\Pr(E)$ is a measure of the degree of belief one holds in a specified event E. To better understand this interpretation, consider the probability of improving an evacuation system by making a plan change. The manager believes that such a change will result in a performance improvement in one out of three evacuation missions in which the plan is used. It would be difficult to describe this problem through the first two interpretations. That is, the classical interpretation is inadequate since there is no reason to believe that performance is as likely to improve as to not improve. The frequency interpretation is not applicable because no historical data exist to show how often a plan change resulted in improving the evacuation. Thus, the subjective interpretation provides a broad definition of the probability concept.

4.5.4 A Fuzzy Set Approach

One of the main goals of integrated disaster management is to ensure that disaster protection performs satisfactorily under a wide range of possible future disaster conditions. This premise is particularly true of large and complex disaster management systems. Disaster management systems include people, infrastructure, and environment. These elements are interconnected in complicated networks across broad geographical regions. Each element is vulnerable to natural hazards or human error, whether unintentional as in the case of operational errors and mistakes, or from intentional causes, such as a terrorist act.

The sources of uncertainty are many and diverse, as was discussed earlier, and as a result they provide a great challenge to integrated disaster management. The goal to ensure failsafe protection system performance may be unattainable. Adopting high safety factors is one way to avoid the uncertainty of potential failures. However, making safety the first priority may render the system solution infeasible. Therefore, known uncertainty sources must be quantified.

The problem of disaster system reliability has received considerable attention from statisticians and probability scientists. Probabilistic (stochastic) risk analysis has been used extensively to deal with the problem of uncertainty in disaster management. A prior knowledge of the probability density functions of both resistance and load, and their joint probability distribution function, is a prerequisite of the probabilistic approach. In practice, data on previous disasters is usually insufficient to provide such information. Even if data are available to estimate these distributions, approximations are almost always necessary to calculate disaster risk. Subjective judgment of the disaster decision maker in estimating the probability distribution of a random event—the subjective probability approach—is another approach to deal with a lack of data. The third approach is Bayes's theory, where an expert judgment is integrated with observed information. The choice of a Bayesian approach or any subjective probability distribution presents real challenges. For instance, it is difficult to translate prior knowledge into a meaningful probability distribution, especially in the case of multiparameter problems. In both subjective probability and Bayesian approaches, the degree of accuracy is strongly dependent on a realistic estimation of the decision-maker's judgment.

Until recently the probabilistic approach was the only approach for disaster systems reliability analysis. However, it fails to address the problem of uncertainty that goes along with human input, subjectivity, a lack of history and records. There is a real need to convert to new approaches that can compensate for the ambiguity or uncertainty of human perception.

The fuzzy set theory was intentionally developed to try to capture judgmental belief, or the uncertainty that is caused by the lack of knowledge. Relative to the probability theory, it has some degree of freedom with respect to aggregation operators, types of fuzzy sets (membership functions), and so on, which enables it to be adapted to different contexts. During the last 40 years, the fuzzy set theory and fuzzy logic have contributed successfully to technological development in different application areas such as mathematics, algorithms, standard models, and real-world problems of different kinds (Zadeh, 1965; Zimmermann, 1996). More recently disaster literature is showing a slow introduction of the fuzzy set theory in disaster management. Altay and Green (2006) report about 5% of research effort using fuzzy sets in their review. Among many examples, Chongfu (1996) provides an excellent introduction of fuzzy risk definition of natural hazards with emphasis on urban hazards. More recently Karimi and Hullermeier (2007) present a system for assessing the risk of natural disasters, particularly under highly uncertain conditions, that is, where neither the statistical data nor the physical knowledge required for a purely probabilistic risk analysis are sufficient. The theoretical foundation of this study is based on employing the fuzzy set theory. The likelihood of natural hazards is expressed by fuzzy probability. Moreover, uncertainties about the correlation of the parameters of hazard intensity, damage, and loss, that is, vulnerability relations, have been considered by means of fuzzy relations. The composition of fuzzy probability of hazard and fuzzy vulnerability relation yields the fuzzy probability of damage (or loss). The system has been applied for assessing the earthquake risk in Istanbul metropolitan area.

Shortly after the fuzzy set theory was first developed in the late 1960s, there were a number of claims that fuzziness was nothing but probability in disguise. Probability and fuzziness are related, but they are different concepts. Fuzziness is a type of deterministic uncertainty. It describes the *event class ambiguity*. Fuzziness measures the *degree to which* an event occurs, not whether it occurs. At issue is whether the event class can be unambiguously distinguished from its opposite. Probability, in contrast, arises from the question of *whether or not* an event occurs. Moreover, it assumes that the event class is crisply defined and that the law of noncontradiction holds—that is, for any property and for any definite subject, it is not the case that both the subject possesses that property and that the subject does not possess that property. Fuzziness occurs when the law of noncontradiction (and equivalently the law of excluded middle—for any property and for any individual, either that individual possesses that property or that individual does not possess that property) is violated. However, it seems more appropriate to investigate fuzzy probability for the latter case than to completely dismiss probability as a special case of fuzziness. In essence, whenever there is an experiment for which we are not capable of "computing" the outcome, a probabilistic approach may be used to estimate the likelihood of a possible outcome belonging to an event class. A fuzzy theory extends the traditional notion of a probability when there are outcomes that belong to several event classes at the same time, but to different degrees. Fuzziness and probability are orthogonal concepts that characterize different aspects of human experience. Hence, it is important to note that neither fuzziness nor probability governs physical processes in nature. These concepts were introduced by humans to compensate for our own limitations.

Let us review two examples that show a difference between fuzziness and probability.

Russell's paradox: That the laws of noncontradiction and excluded middle can be violated was pointed out by Bertrand Russell with the tale of the barber. Russell's barber is a bewhiskered man who lives in a town and shaves a man if and only if the man does not shave himself. The question is, who shaves the barber? If he shaves himself, then by definition he does not. But if he does not shave himself, then by definition he does. So he does and he does not. This is a contradiction or *paradox*. It has been shown that this paradoxical situation can be numerically resolved as follows. Let *S* be the proposition that the barber shaves himself and *not-S* the proposition that he does not. Since S implies *not-S* and vice versa, the two propositions are logically equivalent, that is, $S = not\text{-}S$. Fuzzy set theory allows for an event class to coexist with its opposite at the same time, but to different degrees, or in the case of paradox to the same degree, which is different from 0 or 1.

Misleading similarities: There are many similarities between fuzziness and probability. The largest, but superficial and misleading, similarity is that both systems quantify uncertainty using numbers in the unit interval [0, 1]. This means that both systems describe and quantify the uncertainty numerically. The structural similarity arising from lattice theory is that both systems algebraically manipulate sets and propositions associatively, commutatively, and distributively. These similarities are misleading because a key distinction comes from what the two systems are trying to model. Another distinction is in the idea of observation. Clearly, the two models possess different kinds of information: fuzzy memberships, which quantify similarities of objects to imprecisely defined properties; and probabilities, which provide information on expectations over a large number of experiments.

4.6 COMPUTER SUPPORT: DECISION SUPPORT SYSTEMS

The current trend in the application of systems approach to disaster management includes development of decision support systems (DSS) as a way of integrating data, models, and user-machine interface (Thompson et al., 2006). Various application examples include Simard and Eenigenburg (1990), Brown and Vassiliou (1993), Hobeika et al. (1994), Jianshe et al. (1994), Gheorghe and Vamanu (1995), Simonović (1999), de Silva and Eglese (2000), Ahmad and Simonović (2002), and Shim et al. (2002), among others.

Development of DSS is closely related to computers. The computer has moved out of primarily data processing, through the user's office, into knowledge processing. Whether the work takes the form of a laptop PC or a desktop multiprocessing workstation is not important. It is important that the computer is a "silent partner" for more effective decision making in a decision support system environment. The main factor responsible for involving computers in decision making is treatment of information as the sixth resource (besides people, machines, money, materials, and management).

A decision support system allows decision makers to combine personal judgment with computer output, in a user-machine interface, to produce meaningful information for support in a decision-making process. Such systems are capable of assisting in solution of all problems (structured, semistructured, and unstructured) using all information available on request. They use quantitative models and database elements for problem solving. They are an integral part of the decision-maker's approach to problem identification and solution (Simonović, 1996). This definition of DSS emphasizes the decision maker's insight and judgment at all stages of problem identification and/or problem solving. Important characteristics of DSS for integrated disaster management include accessibility, flexibility, facilitation, learning, interaction, and ease of use.

DSS is primarily concerned with supporting decision making in terms of problem identification and problem solving at all decision-making levels. The most important role of DSS is identifying the necessary steps of the decision-making process or decisions to help the decision makers in fulfilling their organizational duties and responsibilities.

DSS provides support to the user and does not replace the individual. The emphasis is on the enhancement of a decision-making process by allowing use of quantitative models appropriate to the problem. In this way objective (quantitative), measurement introduced by models is combined with the subjective (qualitative) factors introduced by the user. The interaction of the two is the most effective way in reaching a decision.

The term system includes both the user and the machine. The machine is a computer that, for now, operates in interactive mode through an input/output terminal. The system also implies availability of quantitative models and some type of database. In the framework of this definition of DSS, these elements are more providing a service to the decision maker than directly delivering a decision.

The complexity of disaster management tasks suggests a new set of requirements on the tools to be used in integrated disaster management.

Problem identification: Integrated disaster management contains both semistructured and nonstructured problems. The management problem that can be formulated in an algorithmic way (a computer program) is called well structured. Decisions in this case are relatively straightforward because alternative solutions are known. If the management problem involves lack of data or knowledge, nonquantifiable variables, and a very complex description, then it is semi- or nonstructured. In this case, structuring of the problem must be done by the human in the human–machine system.

Because judgment and intuition are critical in examining and resolving many disaster management problems, an effective DSS involves problem identification. This process includes searching the decision-making domain for future problems that need to be anticipated and solved. Future opportunities can be identified and implemented to address the long-term consequences of current decisions.

Problem formulation (learning): Disaster management DSS have been used in situations in which there is a clear problem definition. DSS serve to solve such problems. However, the concept of a "problem" as it relates to disaster management may be expanded to include two perspectives: (i) problem as objective reality; or (ii) problem as mental construct. In the first case, a problem is viewed as unsatisfactory objective reality discovered by observations and facts. The decision maker or expert has to define the problem. As a problem exists objectively, all participants in the decision-making process see it in the same way. Here, problem formulation is a preliminary step to DSS design. The second situation presents an alternative view, considering a problem to be a subjective presentation conceived by a participant confronted with the reality perceived as unsatisfactory. Here, common threshold values have to be defined by the different participants in the decision-making process before another procedure can take place. This approach requires integration of the problem formulation process into the context of a DSS. It is important to note that problem formulation for disaster management is more a social process than a technical one.

"What If" capability (*adaptability*): The DSS environment allows a number of "what if" questions to be asked and answered. The main benefit of DSS is that a number of decisions can be tried without having to deal with the consequences. In this way, DSS can guide decision makers through most optimistic, most pessimistic, and in-between scenarios.

Many issues related to the implementation of integrated disaster management can be examined using the "what if" approach. The ability to ask "what if" questions, to quantify uncertainties, and to recognize the sensitivity of results to varying assumptions stimulates the creative and analytical process of decision making. The process provides a common ground for communication. Since the decision maker can use the tool directly, higher quality decisions may be made on a timelier basis.

Use of analytical models (*facilitation*): The integration and administration of mathematical models within a general framework is a specific feature of the DSS concept. Since disaster management is principally concerned with the present and the immediate implications of management decisions, modeling capability is very important to grasp and use to manage disasters. For problem identification and problem solving, decision makers should deal with analysis. This belief underlines the need for DSS modeling capabilities for:

- retrieval of data,
- execution of ad hoc analysis,
- evaluation of consequences of proposed actions, and
- proposal of solutions.

Typical models, that include database management system functions, as data queries and data manipulation, range from simple arithmetic functions and statistical operations to the ability to call up optimization and simulation models. The scope of a DSS development is in the integration of such different facilities. The idea of DSS integrates different fields of science and puts weight on social circumstances that may decide or influence problem definitions and solution approaches.

User-machine interface (*interaction*): Whether the user is using a microcomputer or a powerful workstation is not the important issue anymore. What is important is an access to an interactive processing mode incorporating a user-machine interface that provides answers to identified problems or "what if" questions. The user-machine interface provides answers that decision makers can understand, when such information is needed, under their direct control. Therefore, DSS are intended to help decision makers throughout the process of identifying and solving their problems. The merging of the computer output with the judgment of the disaster manager provides a better basis for making efficient decisions.

Computers are more than number crunchers or storage devices. With progress in the field of Artificial Intelligence (AI), computers are more capable to support humans in creative and analytical thinking. It is important to note that DSS are not general problem solvers. They are a part of a complex user-machine system with the emphasis being placed on the "user" rather than on the "machine." Therefore, DSS are only possible tools to manage the complexity of disaster decision making.

Use of graphics (fast response): Closely related to the previous two characteristics is the use of graphics and more recently GIS. In a DSS environment, graphic/map displays of results allow users to grasp quickly the essence of large amounts of data and to reduce considerably the printout into a few readily understandable graphs, charts, and maps. It is the best way to select the important information in a user-machine interface, so that the user retains control during the decision-making process.

Some of the examples that will be presented in this book will illustrate DSS concepts presented here.

REFERENCES

Ahmad, S. and S.P. Simonović (2002), "A decision support tool for evaluation of impacts of flood management policies," *Hydrological Sciences and Technology*, 17(1–4):11–23.

AIR Worldwide Corporation (2009), http://www.air-worldwide.com, last accessed September 2009.

Altay, N. and W.G. Green III (2006), "OR/MS research in disaster operations management," *European Journal of Operational Research*, 175:475–493.

Brown, G.G. and A.L., Vassiliou (1993), "Optimizing disaster relief—real-time operational and tactical decision support," *Naval Research Logistics*, 40(1):1–23.

Chen, A., J. Zhao, and N. Chen (2009), "A recoverability assessment model and application in emergency management," in *Proceedings NATO and Research and Technology Organization Symposium: C3I for Crisis, Emergency and Consequence Management*, Bucharest, Rumania, May, available online, http://ftp.rta.nato.int/public//PubFullText/RTO/MP/RTO-MP-IST-086///MP-IST-086–09.doc, last accessed September 2009.

Chongfu, H. (1996), "Fuzzy risk assessment of urban natural hazards," *Fuzzy Sets and Systems*, 83:271–282.

Chow, W.K. and C.H. Lui (2002), "Numerical studies on evacuation design in a karaoke," *Building and Environment*, 37:285–294.

de Silva, F.N. and R.W. Eglese (2000), "Integrating simulation modelling and GIS: Spatial decision support systems for evacuation planning," *Journal of the Operational Research Society*, 51(4):423–430.

EQECAT (2009), http://www.eqecat.com, last accessed September 2009.

Fawcett, W. and C.S. Oliveira (2000), "Casualty treatment after earthquake disasters: Development of a regional simulation model," *Disasters*, 24(3):271–287.

Filippoupolitis, A. and E. Gelenbe (2009), "A distributed decision support system for building evacuation," in *Proceedings of the 2nd IEEE International Conference on Human System Interaction*, Catania, Italy, May, IEEE, New York, pp. 323–330.

Forrester, J.W. (1990), *Principles of Systems*, Productivity Press, Portland, OR, first published in 1968.

Gheorghe, A.V., and D. Vamanu (1995), "A pilot decision-support system for nuclear-power emergency management," *Safety Science*, 20(1):13–26.

Georgiadou, P.S., I.A. Papazoglou, C.T. Kiranoudis, and N.C. Markatos (2007), "Modeling emergency evacuation for major hazard industrial sites," *Reliability Engineering and System Safety*, 92:1388–1402.

High Performance Systems (1992), *Stella II: An Introduction to Systems Thinking*, High Performance Systems, Inc., Hanover, NH.

Hobeika, A.G., S. Kim, and R.E. Beckwith (1994), "A decision support system for developing evacuation plans around nuclear-power stations," *Interfaces*, 24(5):22–35.

Jianshe, D., W. Shuning, Y. Xiaoyin (1994), "Computerized support systems for emergency decision making," *Annals of Operations Research*, 51:315–325.

Kaplan, S. and B.J. Garrick (1981), "On the quantitative definition of risk," *Risk Analysis*, 1(1):165–188.

Karimi, I. and E. Hullermeier (2007), "Risk assessment system of natural hazards: A new approach based on fuzzy probability," *Fuzzy Sets and Systems*, 158:987–999.

Kohler, A., S. Julich, and L. Bloemertz (2004), "Risk analysis – a basis for disaster risk management," *Guidelines*, German Federal Ministry for Economic Cooperation and Development, Section 42, Governance and Democracy, Eschborn, Germany, p. 71.

Lane, D.C., C. Monefeldt, and J.V. Rosenhead (2000), "Looking in the wrong place for healthcare improvements: A system dynamics study of an accident and emergency department," *Journal of the Operational Research Society*, 51:518–531.

Lane, D.C., C. Monefeldt, and E. Husemann (2003), "Client Involvement in simulation model building: Hints and insights from a case study in a London hospital," *Health Care Management Science*, 6:105–116.

Ling, C.W. (1993), *Characterising Uncertainty: A Taxonomy and an Analysis of Extreme Events*, MSc Thesis, School of Engineering and Applied Science, University of Virginia, VA.

Lowrance, W.W. (1976), *Of Acceptable Risk*, William Kaufman, Inc., Los Altos, CA.

Lyneis, J., R. Kimberly, and S. Todd (1994), "Professional dynamo: Simulation software to facilitate management learning and decision making," in Morecroft, J. and J. Sterman (eds) *Modelling for Learning Organizations*, Pegasus Communications, Waltham, MA.

Palmer, R. N., W. J. Werick, A. MacEwan and A. W. Woods (1999), "Modelling water resources opportunities, challenges and trade-offs: The use of shared vision modelling for negotiation and conflict resolution," in Erin M. Wilson (ed.), *Proceedings ASCE Water Resources Planning and Management Conference*, 6–9 June, Tempe, AZ.

Powersim Corporation (1996), *Powersim 2.5 Reference Manual*, Powersim Corporation, Inc., Herndon, VI.

Risk Management Solutions (2009), http://www.rms.com, last accessed September, 2009.

Saadatseresht, M., A. Mansourian, and M. Taleai (2009), "Evacuation planning using multiobjective evolutionary optimization approach," *European Journal of Operational Research*, 198:305–314.

Shim, K.C., D.G. Fontane, and J.W. Labadie (2002), "Spatial decision support system for integrated river basin flood control," *Journal of Water Resources Planning and Management*, 128(3):190–201.

Slovic, P. (2000), *The Perception of Risk*, Earthscan, London, UK.

Smith, S. et al. (2009), "A novel approach to multihazard modeling and simulation," *Disaster Medicine and Public Health Preparedness*, 3:75–87.

Simard, A.J. and J.E. Eenigenburg (1990), "An executive information system to support wildfire disaster declarations," *Interfaces*, 20(6):53–66.

Simonović, S.P. (1996), "Decision support systems for sustainable management of water resources 1. General principles," *Water International*, 21(4):223–232 (the best paper award).

Simonović, S.P. (1999), "Decision support system for flood management in the red river basin," *Canadian Water Resources Journal*, 24(3):203–223.

Sterman, J.D. (2000), *Business Dynamics: Systems Thinking and Modelling for a Complex World*, McGraw Hill, New York.

Steward, D, and T.T.H. Wan (2007), "The role of simulation and modeling in disaster management," *J Med Syst*, 31:125–130.

Thompson, S., N. Altay, W.G.G. Green III, and J. Lapetina (2006), "Improving disaster response efforts with decision support systems," *International Journal of Emergency Management*, 4(4):250–263.

UNDP (2004), "Reducing disaster risk: A challenge for development," *A Global Report*, UNDP Bureau for Crisis Prevention and Recovery, New York, p. 146.

Ventana Systems (1996), *Vensim User's Guide*, Ventana Systems, Inc., Belmont, MA.

Wu, S. et al. (2008), "Agent-based discrete event simulation modeling for disaster responses," in Fowler, J.W. and Mason, S.J. (eds), *Proceedings of the 2008 Industrial Engineering Research Conference*, Vancouver, British Columbia, pp. 1908–1913.

Zadeh, L.A. (1965), "Fuzzy sets," *Information Control*, 8:338–353.

Zimmermann, H.J. (1996), *Fuzzy Set Theory–and Its Applications*, 2nd revised edn, Kluwer Academic Publishers, Boston, MA.

Zeleny, M. (1973), "Compromise Programming", in Cochrane J. and M. Zeleny (eds) *Multiple Criteria Decision Making*, University of South Carolina Press, Columbia, SC.

EXERCISES

4.1 What is the difference between simulation and optimization?

4.2 What is the difference between system dynamics simulation and classical simulation?

4.3 Define system structure. What does the term "dynamics" refer to?

4.4 What is the outcome of optimization analysis? Are there multiple optimal solutions to a single optimization problem?

4.5 What is a feasible solution? Is an optimal solution always feasible?

4.6 Is there an optimal solution for a problem with multiple objectives? Why?

4.7 What is a nondominated solution?

4.8 Describe, using words and a flow diagram, how you might simulate the operation of a hurricane disaster relief program over time. List all assumptions. To simulate a disaster relief program, what data do you need to have or know? Identify a feedback relationship(s) in your model.

4.9 You are hired to determine the allocation of water X_j to four municipalities hit by an earthquake (j–municipality A, B, C, and D). Each of the municipalities has a demand $D_j(X_j)$. The total water available is Q. Produce a flow chart showing how you can find the allocation to each municipality that results in the maximum demand satisfaction.

4.10 Consider the following seven alternatives for the combined use of storage space for disaster prevention purposes (permanent occupation of space 10^3 m^3/day) and nondisaster use that can be sold (10^6 \$/year):

Alternative	Disaster Space Use	Nondisaster Space Sale
1	25	78
2	32	56
3	18	100
4	7	112
5	27	71
6	20	92
7	11	105

(a) Which alternative would be the best in your opinion, and why?

(b) Which alternative would be the worst in your opinion, and why?

(c) Provide an argument for selecting alternative 6, even though other alternatives exist that can provide more space for disaster storage and higher levels space sale for other uses.

(d) What relative weight would you assign to these two objectives, and why?

PART III
Implementation of Systems Analysis to Management of Disasters

5 Simulation

Simulation models *describe* how a system operates, and are used to determine changes resulting from a specific course of action. Such models are referred to as *cause-and-effect* models in the introductory discussion in Section 4.1. They describe the state of the system in response to various inputs but give no direct measure of what decisions should be taken to improve the performance of the system. Therefore, simulation is a problem-solving technique that contains the following phases:

1. Development of a model of the system;
2. Operation of the model (i.e., generation of outputs resulting from the application of inputs); and
3. Observation and interpretation of the resulting outputs. The essence of simulation is an iterative process of modeling and experimentation.

5.1 DEFINITIONS

The *traditional simulation* procedure involves decomposition of the problem in order to aid the system description. When the main elements of the system are identified, the proper mathematical description is provided for each of them. The procedure continues with computer coding of the mathematical description of the model. Each model parameter is then calibrated and the model performance is verified using data that has not been seen during the calibration process. The completed model is then operated using a set of input data. Detailed analysis of the resulting output is the final step in the simulation procedure.

A completed model can be reused many times with alternative input data. If there is a need for modification of the system description or model structure, the whole process starts again and the model has to be recoded, calibrated, and verified again before its use.

The major components of a simulation model are:

- *Inputs*: quantities that "drive" the model (in flood disaster management, for example, a principal input is the set of streamflows, rainfall sequences, etc.).

Systems Approach to Management of Disasters: Methods and Applications, By Slobodan P. Simonović

- *Physical relationships*: mathematical expressions of the relationships among the physical variables of the system being modeled (continuity, energy conservation, etc.).
- *Nonphysical relationships*: those that define economic variables, political conflicts, public awareness, etc.
- *Operation rules*: the rules that govern operational control.
- *Outputs*: the final product of operations on inputs by the physical and nonphysical relations in accordance with operating rules.

System dynamics simulation is introduced in Section 4.2 as a rigorous method of system description, which facilitates feedback analysis via a simulation model of the effects of alternative system structures and control policies on system behavior. The advantages of system dynamics simulation over classical simulation are presented in Section 4.2. Briefly, they include:

- the simplicity of use of system dynamics simulation applications;
- the applicability of system dynamics general principles to social, natural, and physical systems;
- the ability to address how structural changes in one part of a system might affect the behavior of the system as a whole;
- combined predictive (determining the behavior of a system under particular input conditions) and learning (discovery of unexpected system behavior under particular input conditions) functionality; and
- active involvement of stakeholders in the modeling process.

The rest of this chapter focuses on system dynamics simulation as a method for management of disasters.

5.2 SYSTEM DYNAMICS SIMULATION

5.2.1 Introduction

System dynamics is an academic discipline introduced in the 1960s that has gradually developed into a tool useful in the analysis of social, economic, physical, chemical, biological, and ecological systems (Forrester, 1990; Sterman, 2000). In the context of this book, in Chapter 3, a *system* was defined as a collection of elements that continually interact over time to form a unified whole. The underlying pattern of interactions between the elements of a system is called the *structure* of the system. The term *dynamics* refers to change over time. If something is dynamic, it is constantly changing in response to the stimuli influencing it. A dynamic system is thus a system in which the variables interact to stimulate changes over time. The way in which the elements, or variables, composing a system vary over time is referred to as the *behavior* of the system. One feature that is common to all systems is that a system's structure determines its behavior. System dynamics links the behavior of a system to its underlying structure. It can be used to analyze how the structure of a physical, biological, or any other system can lead to the behavior the system exhibits.

The system dynamics simulation approach relies on understanding complex interrelationships existing between different elements within a system, by developing a model that can simulate and quantify the behavior of the system. The major steps that are carried out in the development of a system dynamics simulation model include:

1. understanding the system and its boundaries;
2. identifying the key variables;
3. describing the physical processes or variables through mathematical relationships;
4. mapping the structure of the model; and
5. simulating the model for understanding its behavior.

Advances made during the last decade in computer software have brought about considerable simplification in the development of system dynamics simulation models. The accompanying CD-ROM includes all the system dynamics models developed in the text, using the state-of-the-art simulation software Vensim PLE (Ventana Systems, 1995), which is available from the Ventana Systems, Inc., Web site (http://www.vensim.com, last accessed January 2010). This software is free for educational use. In the SYSTEMDYNAMICS directory on the CD-ROM, there is a read.me file that contains program installation instructions. The Tutorial subdirectory contains a short tutorial for Vensim PLE developed by Professor Craig Kirkwood at Arizona State University. The Examples subdirectory contains all the examples from this chapter.

Vensim PLE is an ideal tool for personal learning of system dynamics. Like similar programs, it uses the principles of object-oriented programming. It provides a set of graphical objects with their mathematical functions for easy representation of the system structure and the development of computer code. Simulation models can be easily and quickly developed using this type of software tool. Such models are easy to modify, easy to understand, and present results clearly to a wide audience of users.

5.2.2 System Structure and Patterns of Behavior

In starting to consider system structure, we first generalize from the specific events associated with the problem to considering patterns of behavior that characterize the situation. Usually this requires investigation of how one or more variables of interest change over time (e.g., flow of aid, traffic on a road, or number of people in a refugee camp). That is, we ask, what patterns of behavior do these variables display? The system dynamics simulation approach gains much of its power as a problem-solving method from the fact that similar patterns of behavior show up in a variety of different situations, and the underlying system structures that cause these characteristic patterns are known. Thus, once we have identified a pattern of behavior for a problem, we can look for the system structure that is known to cause that pattern. If we can find and modify this system structure, there is the possibility of permanently eliminating the problem.

Feedback Relationships The difference between the two basic types of feedback is important in understanding dynamic behavior. Section 3.1.6 introduced the concept of feedback. A positive, or reinforcing, feedback loop reinforces change with even more change (see Section 3.1.6). This can lead to rapid growth at an ever-increasing rate. This type of growth pattern is often referred to as *exponential growth*. Note that in the early stages of the growth, it seems to be slow, but then it speeds up. Thus, the nature of growth in a disaster management system that has a positive feedback loop can be deceptive. Examples that fit this category are after disaster water demand and disaster affected population growth. Positive feedback is quite common in managed systems, and may be valuable as an engine of growth. In a disaster management system, however, positive feedback is undesirable and should be designed out. Figure 5.1a shows an example of a generic positive feedback loop, and Figure 5.1b shows the corresponding behavior.

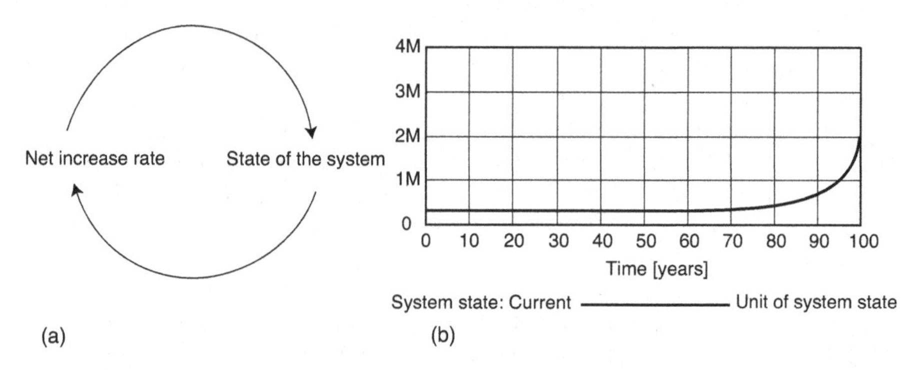

Figure 5.1 (a) Positive feedback loop; (b) system behavior.

A negative, or balancing, feedback loop seeks a goal. If the current level of the variable of interest is above the goal, then the loop structure pushes its value down, while if the current level is below the goal, the loop structure pushes its value up. Many disaster management processes contain negative feedback loops that provide useful stability, but which can also resist needed changes (see Section 3.1.6). The essential idea of negative feedback is that, when there is a difference between the desired and actual states of the system, action is generated according to the system's policy in an attempt to eliminate the difference. Figure 5.2a shows a negative feedback loop, and Figure 5.2b shows a typical pattern of behavior.

Example 1

Let us consider an evacuation planning example shown in Figure 5.3. The population of a small community under the threat of flooding is being evacuated. There are few

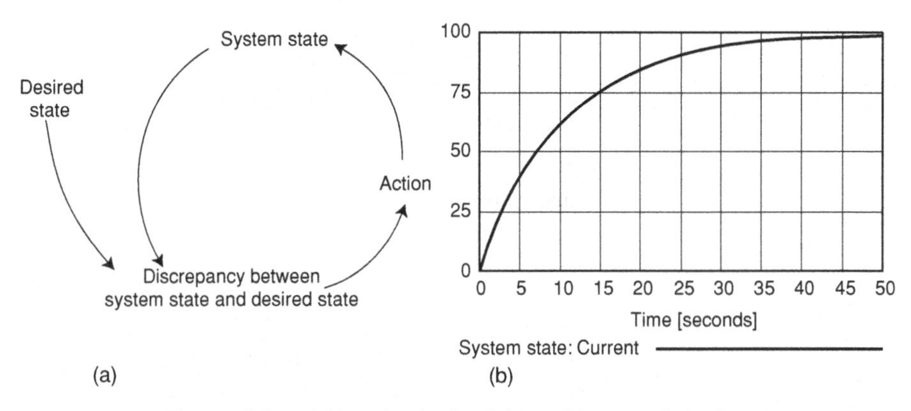

Figure 5.2 (a) Negative feedback loop; (b) system behavior.

shelters available. Emergency management team (EMT) receives the information from all shelters. For the purpose of simplicity we will focus only on one shelter, shelter A. The number of available spaces in shelter A is used by the EMT to issue an order to move the people to shelter A. The rising number of people occupying the shelter A forwarded to EMT causes gradual reduction in the number of people being allocated to this shelter.

In dealing with dynamic systems, it is important to see the system from several viewpoints: as a physical arrangement (Figure 5.3), as a flow diagram (Figure 5.4), and as a set of actions and consequences that can be shown graphically through time.

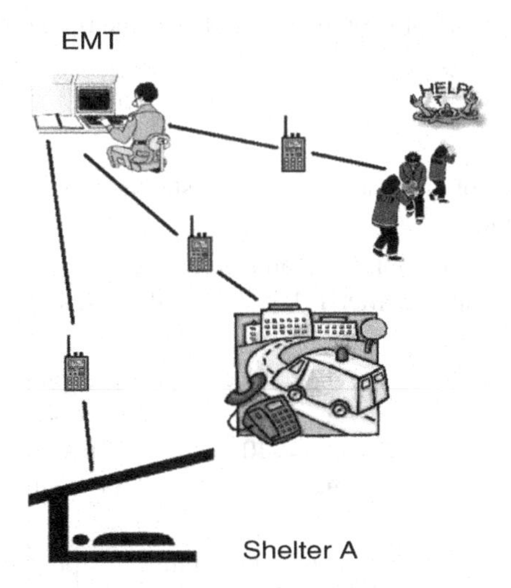

Figure 5.3 An evacuation planning example.

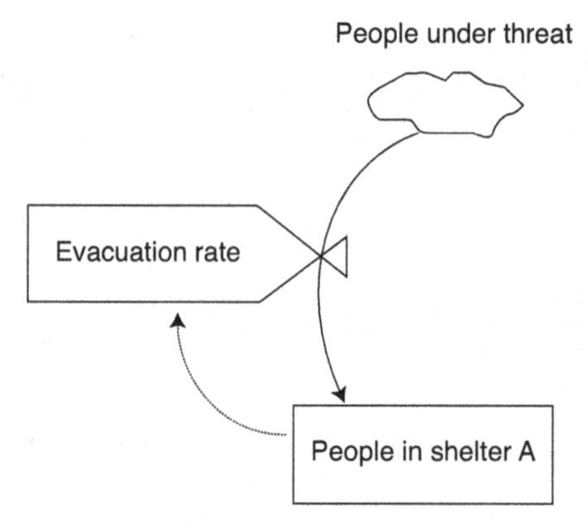

Figure 5.4 Evacuation flow diagram.

(a) If evacuation rate from the affected community *ER* is set at 10 people per day, how many people will enter the shelter A in 10 days?

$$\begin{aligned} ER &= 10[\text{people/day}] \\ PA &= 10[\text{people/day}] \times 10[\text{days}] \\ PA &= 100[\text{people}] \end{aligned}$$

(b) The units of measure must always accompany any numerical value to define the quantity. The units of measure of evacuation rate are:

$$ER[\text{people/day}]$$

Then the number of people in shelter A is measured in:

$$PA[\text{people}]$$

(c) Let us now plot the evacuation rate to shelter A of 10 [people/day]. Figure 5.5 shows the plot.

(d) If the $ER = 40$ [people/day] and shelter A is empty, calculate the number of people in the shelter after 1 day, 2 days, 4 days, and 6 days.

$$\begin{aligned} PA_1 &= 40 \times 1 = 40[\text{people}] \\ PA_2 &= 40 \times 2 = 80[\text{people}] \\ PA_4 &= 40 \times 4 = 160[\text{people}] \\ PA_6 &= 40 \times 6 = 240[\text{people}] \end{aligned}$$

(e) The plot in Figure 5.6a shows the number of people in shelter A for every point in time.

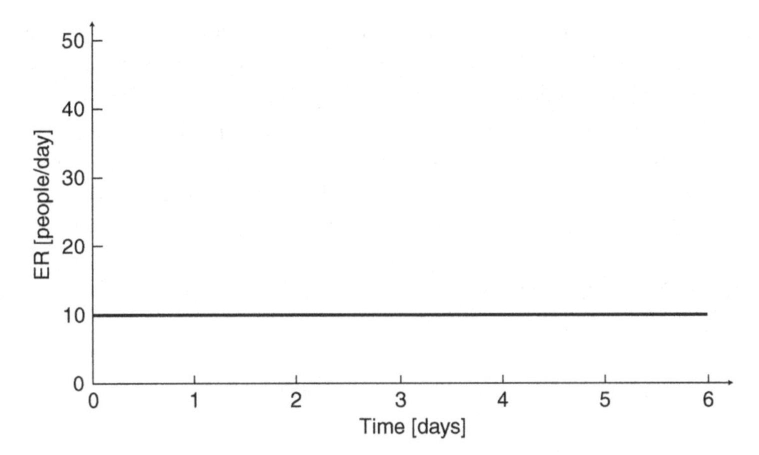

Figure 5.5 Constant evacuation flow rate.

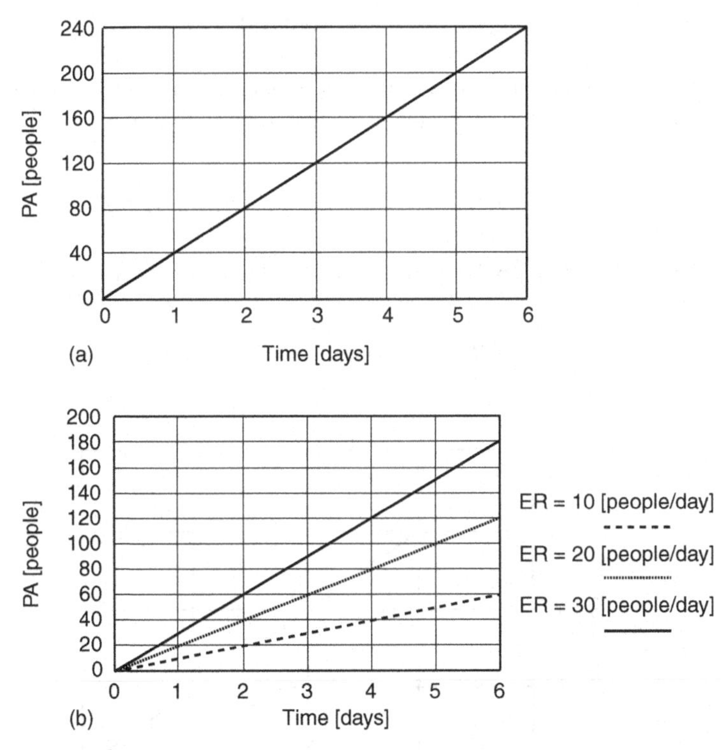

Figure 5.6 Quantity time graph: (a) the number of people in shelter A for every point in time and (b) the number of people for an evacuation rate of 10, 20, and 30 [people/day].

(f) Using a similar graph (Figure 5.6b), plot the people in shelter A and the time relationship showing the number of people for an evacuation rate of 10, 20, and 30 [people/day].

(g) The EMT is using the following information: the evacuation rate $ER =$ 20[people/day] when the shelter A is empty and declines proportionally to zero when the number of people in the shelter $PA = 200$[people]. Show the evacuation rate versus the number of people (volume) in shelter A on the graph.

Express ER as an equation.

$$ER = \frac{1}{T}(200 - PA) \tag{5.1}$$

Find T when $PA = 0$.

$$20 = \frac{1}{T}(200 - 0)$$
$$T = 10[\text{days}]$$

Replace T from the previous problem in Equation (5.1) and find ER when $PA =$ 150[people]. Check the value using the graph in Figure 5.7.

$$ER = \frac{1}{10}(200 - 150) = 5[\text{people/day}]$$

(h) Using $PA = 80$[people], find the ER from the equation $ER = 1/10(200 - PA)$?

$$ER = \frac{1}{10}(200 - 80) = 12[\text{people/day}]$$

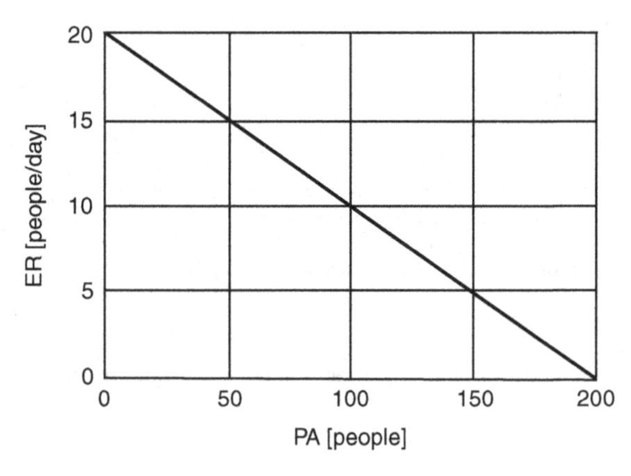

Figure 5.7 Evacuation rate volume graph.

(i) How many people will be added to the shelter during the next 4 days?

$$PA = ER \times T = 12 \times 4 = 48[\text{people}]$$

(j) Using Table 5.1 calculate the evacuation rate and the number of people in shelter A every day. Complete the table for 10 days and plot the curves of the number of people and evacuation rate (Figure 5.8) as your calculation progresses. (It is important for learning that you actually do these calculations and the plotting. Pay attention to the way in which the variables are changing and why.)

TABLE 5.1 Input Data for the Evacuation Problem

Days	Change in Number of People [people]	Number of People [people]	Evacuation Rate [people/day]
0		0.00	20.00
1	20.00	20.00	18.00
2	36.00	56.00	14.40
3	43.20	99.20	10.08
4	40.32	139.52	6.05
5	30.24	169.76	3.02
6	18.14	187.90	1.21
7	8.47	196.37	0.36
8	2.90	199.27	0.07
9	0.65	199.93	0.01
10	0.07	200.00	0.00

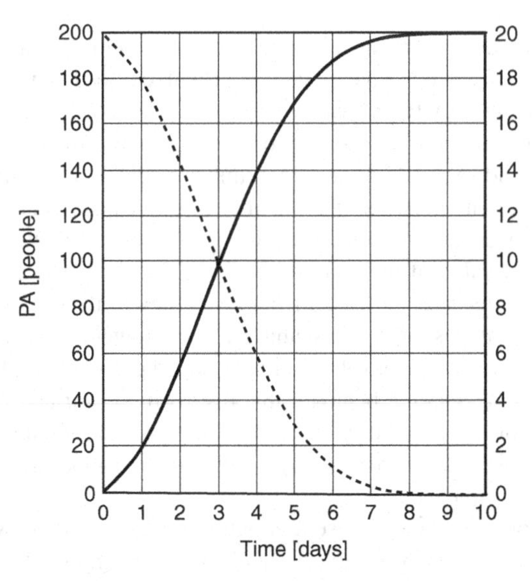

Figure 5.8 Evacuation problem data.

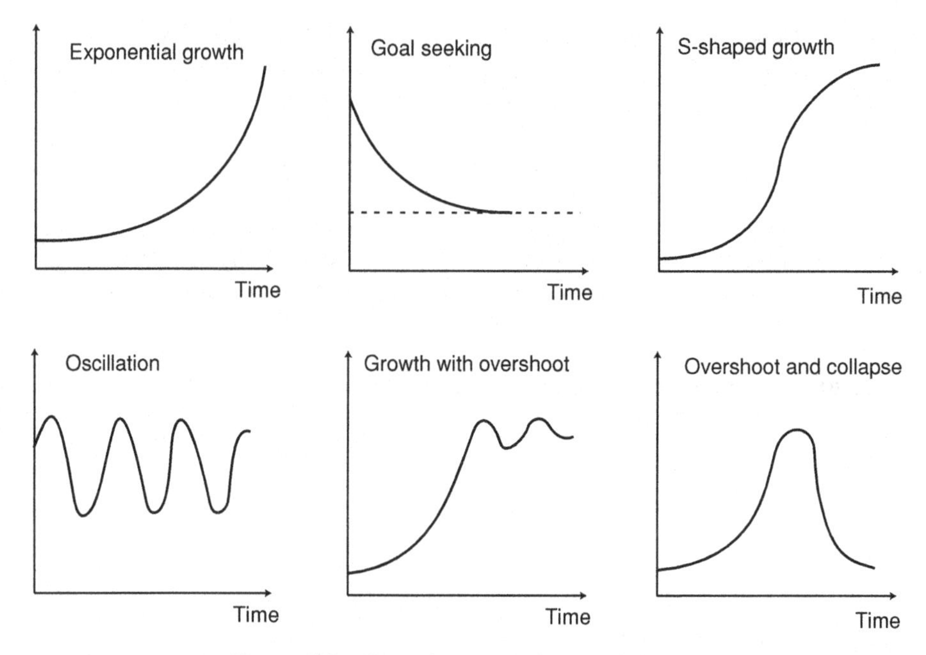

Figure 5.9 General patterns of system behavior.

Generic Patterns of Behavior The six patterns of behavior shown in Figure 5.9 often show up, either individually or in combinations, in complex systems (such as disaster management) simulations. In this figure, the vertical axis shows a variable of interest.

With exponential growth, an initial quantity of something starts to grow and the rate of growth increases. The term exponential growth comes from a mathematical model for this increasing growth process where the growth follows a particular functional form called the exponential (see Section 3.1.6). Such growth is seen in increases in pollution, demand for postdisaster aid, number of disaster victims, and so on.

With goal-seeking behavior, the quantity of interest starts either above or below a goal level and over time moves toward the goal. Figure 5.9 shows one possible case where the initial value of the quantity is above the goal. The curve might represent the way the disaster aid is released from an agency to the disaster area. The change toward the final value is rapid at first and becomes slower as the discrepancy decreases between the present and final value. With S-shaped growth, initial exponential growth is followed by goal-seeking behavior that results in the variable leveling off.

With oscillation, the quantity of interest fluctuates around a level. Note that oscillation initially resembles exponential growth, and then appears to be S-shaped growth before reversing direction.

Common combinations of these four patterns include growth with overshoot. With this behavior, the quantity of interest will overshoot the goal first on one side and then the other. The amplitude of these overshoots declines until the quantity finally stabilizes at the goal. Such behavior can result from excessive time delays in the feedback loop or from too violent an effort to correct a discrepancy between the current state of the system and the system goal (as with adjusting the temperature of water in a shower).

Another combination is overshoot and collapse. In this behavior, when resources are initially ample, the positive growth loop dominates and the state of the system grows exponentially. As it grows, resource adequacy drops. The negative loop gradually gains in strength, and unlike with S-shaped growth, the state of the system starts to decline.

Example 2
Which curve in Figure 5.9 would best describe:

(a) how the population of a small town changes with time after the evacuation process due to wildfire starts? Answer: the goal-seeking curve.
(b) the position of a pendulum, which is displaced and allowed to swing? Answer: the oscillation curve (note that these oscillations will gradually reduce with time).
(c) the learning process? Answer: the growth and overshoot curve.
(d) the amount of capital equipment over time in industrialization, where capital equipment is used to produce more capital equipment? Answer: exponential growth curve.

Delays Delays are a critical source of dynamics in nearly all systems. They are sometimes a source of instability and oscillatory system behavior. They are omnipresent in management of disasters. It takes time to measure and report extreme precipitation or flood flow. It takes time to make decisions on how to operate a pump. It also takes time for decisions to affect the state of a system. The simplest definition of a delay is *a process whose output lags behind its input in some fashion* (Sterman, 2000; Simonović, 2009).

There are two types of delays. *Material delay* captures the physical flow of material. Consider, for example, the repair of road destroyed by an earthquake. It takes time for construction material to be provided and then to be transported to the required location so that the repair can be initialized. Other examples of material delay include the delivery of water and medications as well as the progression of various emergency management tasks. In each, there are physical units (cubic meters of soil, square meters of space, boxes of medications, or engineering drawings) moving through the process.

Other delays represent the gradual adjustment of perceptions or beliefs—these are *information delays*. The delay between a change in the temporary flood protection level and our belief in the flood forecast is an example of information delay. There is a delay between the receipt of new information and the updating of our perception. For example, flood risk perception is directly related to the extent of the flood forecast. If the forecast changes, there will be a delay in the change of our risk perception.

5.3 SYSTEM DYNAMICS SIMULATION MODELING PROCESS

Although there is no universally accepted process for developing and using good-quality system dynamics models, there are some basic practices that are quite commonly used (Ford and Flynn, 2005). The following steps are a useful guideline (Ventana Systems, 2003b).

Issue statement: The issue statement is simply a statement of the problem that makes it clear what the purpose of the model will be. Clarity of purpose is essential to effective model development. It is difficult to develop a model of a system or process without specifying how the system needs to be improved or what specific behavior is problematic. Having a clear problem in mind makes it easier to develop models with good practical applicability.

Variable identification: Identify some key quantities that will need to be included in the model for the model to be able to address the issues at hand. Usually a number of these are very obvious. It can sometimes be useful just to write down all of the variables that might be important and try to rank them in order to identify the most important ones.

Reference modes: A reference mode is a pattern of behavior over time. Reference modes are drawn as graphs over time for key variables, but are not necessarily graphs of observed behavior. Rather, they are cartoons that show a particular characteristic of behavior that is interesting. For example, a history of volunteering during an emergency may be growing but bumpy, and the reference mode may be the up and down movement around the growth trend. Reference modes can refer to either past behavior or future behavior. They can represent what you expect to have happen, what you fear will happen, and what you hope will happen. They should be drawn with an explicitly labeled time axis to help refine, clarify, and bound a problem statement.

Reality check: Define some reality check statements about how things must interrelate. These include a basic understanding of what actors are involved and how they interact, along with the consequences for some variables of significant changes in other variables. Reality check information is often simply recorded as notes about what connections need to exist. It is based on knowledge of the system being modeled.

Dynamic hypotheses: A dynamic hypothesis is a theory about what structure exists that generates the reference modes. A dynamic hypothesis can be stated verbally, as a causal loop diagram or as a stock and flow diagram (detailed presentation follows). The dynamic hypotheses you generate can be used to determine what will be kept in models and what will be excluded. Like all hypotheses, dynamic hypotheses are not always right. Refinement and revision is an important part of developing good models.

Simulation model: A simulation model is the refinement and closure of a set of dynamic hypotheses to an explicit set of mathematical relationships. Simulation models generate behavior through simulation. A simulation model provides a laboratory in which you can experiment to understand how different elements of structure determine behavior. This process is iterative and flexible. As you continue to work with a problem you will gain understanding that changes the way you need to think about the things you have done before. Various computer-based tools available provide explicit support for naming variables, writing reality check information, developing dynamic hypotheses, and building simulation models. Creating good issue statements and developing reference modes can easily be done with pencil and paper or using other technologies. Dynamic hypotheses can be developed as visual models in one of the computer-based tools, or simply sketched out with pencil and paper. Simulation is one stage where it is necessary to use the computer for at least part of the process.

Three steps of the modeling process introduced need a more detailed explanation: (i) development of a causal loop diagram; (ii) development of the stock and flow diagram; and (iii) model simulation. Causal loop diagrams are an important tool for representing the feedback structure of systems being modeled. They quickly capture dynamic hypotheses, elicit and capture the mental models, and communicate the important feedbacks that are responsible for the problem. They are used effectively at the start of modeling process. However, causal diagrams have a number of limitations and can be misused. One of the most important limitations is their inability to capture the stock and flow structure of systems.

Stocks and flows, as it will be seen later, are the two central concepts of dynamic system theory. They generate dynamics. Stocks describe the process of accumulation that is equivalent to integration in calculus. The slope of the trajectory of a stock at any time is its derivative, the net rate of change. The ability to relate stocks and flows intuitively is essential for all modeling efforts.

Formalizing a conceptual model through causal and stock and flow diagramming often provides important insight even before the model is ready to be simulated. Simulation is a phase where the real test of the model understanding occurs. Equations and parameter estimation are a way to resolve ambiguity and test initial hypotheses. System dynamics simulation practice includes a large variety of tests one can apply during the model testing and implementation. When the confidence is developed in the structure and behavior of the model, it can be used to design and evaluate policies for improvement.

5.3.1 Causal Loop Diagram

To better understand the system structures that cause the patterns of behavior discussed in Section 5.2.2, let us introduce a notation for representing system structures. When an element of a system indirectly influences itself, the portion of the system involved is called a *feedback loop*. A map of the feedback structure—an *annotated causal loop diagram*—of a simple system, such as that shown in Figure 5.10, is a starting point for analyzing what is causing a particular pattern of behavior. This figure

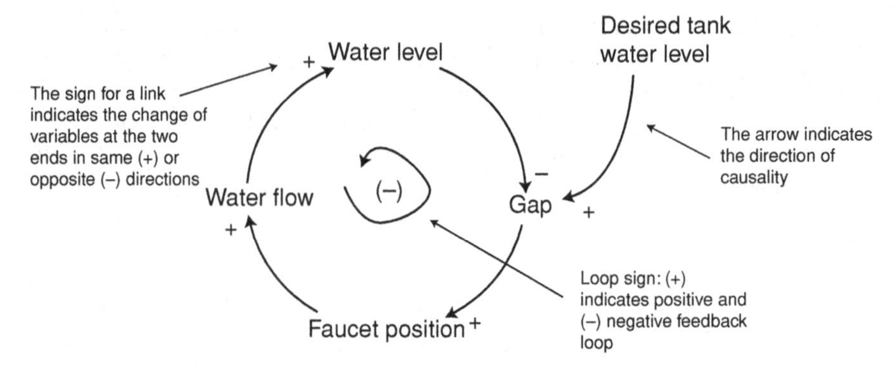

Figure 5.10 Causal loop diagram for "filling a temporary storage tank with water."

considers a simple process, filling a temporary storage tank with water. It includes elements, and arrows (which are called *causal links*) linking them, and also includes a sign (either + or −) on each link. These signs have the following meanings:

- A causal link from one element A to another element B is negative (i.e., −) if either (a) A subtracts from B or (b) a change in A produces a change in B in the opposite direction.
- A causal link from one element A to another element B is positive (i.e., +) if either (a) A adds to B or (b) a change in A produces a change in B in the same direction.

Let us start from the element *faucet position* at the bottom of the diagram. If it is increased (i.e., the faucet (tap) is opened further) then the water flow increases. Therefore, the sign on the link from *faucet position* to *water flow* is positive. Similarly, if the *water flow* increases, then the *water level* in the tank will increase. Therefore, the sign on the link between these two elements is positive. The next element along the chain of causal influences is the *gap*, which is the difference between the *desired tank water level* and the (actual) *water level* (i.e., *gap* = *desired tank water level* − *water level*). From this definition, it follows that an increase in *water level* decreases *gap*, and therefore the sign on the link between these two elements is negative. Finally, to close the causal loop back to *faucet position*, a greater value for *gap* presumably leads to an increase in *faucet position* (as you attempt to fill the tank), and therefore the sign on the link between these two elements is positive. There is one additional link in this diagram, from *desired tank water level* to *gap*. From the definition of *gap* given above, the influence is in the same direction along this link, and therefore the sign on the link is positive.

In addition to the signs on each link, a complete loop is also given a sign. The sign for a particular loop is determined by counting the number of minus (−) signs on all the links that make up the loop. Specifically, a feedback loop is called positive, indicated by a (+), if it contains an even number of negative causal links, and it is called negative, indicated by a (−), if it contains an odd number of negative causal links.

Thus, the sign of a loop is the algebraic product of the signs of its links. Often a small looping arrow is drawn around the feedback loop sign to more clearly indicate that the sign refers to the loop, as is done in Figure 5.10. Note that in this diagram there is a single feedback loop, and that this loop has one negative sign on its links. Since 1 is an odd number, the entire loop is negative.

To start drawing a causal loop diagram, decide which events are of interest in developing a better understanding of system structure. From these events, move to showing (perhaps only qualitatively) the pattern of behavior over time for the quantities of interest. Finally, once the pattern of behavior is determined, use the concepts of positive and negative feedback loops, with their associated generic patterns of behavior, to begin constructing a causal loop diagram that will explain the observed pattern of behavior.

The following tutorial for drawing causal loop diagrams is based on guidelines by Forrester (1990) and Senge (1990):

1. Think of the elements in a causal loop diagram as variables that can go up or down, but do not worry if you cannot readily think of existing measuring scales for these variables.
   - Use nouns or noun phrases to represent the elements, rather than verbs. That is, the actions in a causal loop diagram are represented by the links (arrows) and not by the elements.
   - Be sure that the definition of an element makes it clear which direction is positive and which is negative.
   - Generally, it is clearer if you use an element name for which the positive sense is preferable.
   - Causal links should imply a direction of causation, and not simply a time sequence. That is, a positive link from element A to element B does not mean first A occurs and then B occurs. Rather it means, when A increases then B increases.
2. As you construct links in your diagram, think about possible unexpected side effects that might occur in addition to the influences you are drawing. As you identify these, decide whether links should be added to represent them.
3. For negative feedback loops, there is a goal. It is usually clearer if this goal is explicitly shown along with the gap that is driving the loop toward the goal.
4. A difference between actual and perceived states of a process can often be important in explaining patterns of behavior. Thus, it may be important to include causal loop elements for both the actual value of a variable and the perceived value. In many cases, there is a lag (delay) before the actual state is perceived. For example, when there is a change in drinking water quality, it usually takes a while before we perceive this change.
5. There are often differences between short-term and long-term consequences of actions, and these may need to be distinguished with different loops.

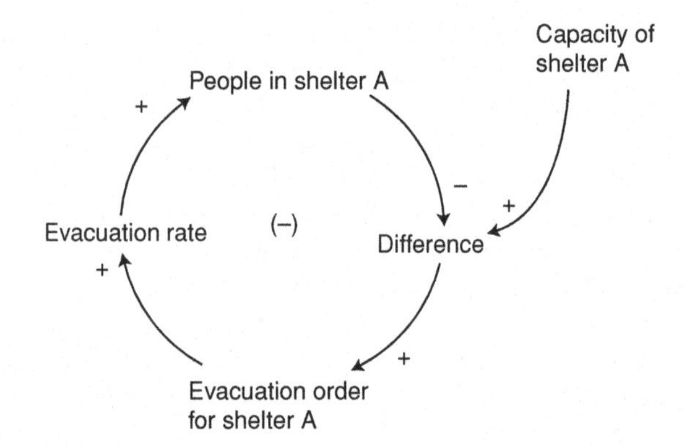

Figure 5.11 A causal diagram of the evacuation planning system.

6. If a link between two elements needs a lot of explaining, you probably need to add intermediate elements between the two existing elements that will more clearly specify what is happening.
7. Keep the diagram as simple as possible, subject to the earlier suggestions. The purpose of the diagram is not to describe every detail of the process, but to show those aspects of the feedback structure that lead to the observed pattern of behavior.

Example 3
Develop a causal loop diagram of the evacuation planning problem from Example 1 in Section 5.2.2 and identify the character of the feedback relationship.

Figure 5.11 shows a causal diagram for the evacuation planning example. Start from the element *evacuation order for shelter A* at the bottom of the diagram. If the *evacuation order for shelter A* is increased (i.e., the order is for more people to be directed to shelter A), the *evacuation rate* increases. Therefore, the sign on the link from *evacuation order for shelter A* to *evacuation rate* is positive.

Similarly, if the *evacuation rate* increases, then the number of *people in shelter A* will increase. Therefore, the sign on the link between these two elements is positive. The next element along the chain of causal influences is the *difference*, which is the difference between the *capacity of shelter A* and the (actual) *people in shelter A* (i.e., *difference* = *capacity of shelter A* − *people in shelter A*). From this definition, it follows that an increase in *people in shelter A* decreases *difference*, and therefore, the sign on the link between these two elements is negative. Finally, to close the causal loop back to *evacuation order for shelter A*, a greater value for *difference* leads to an increase in *evacuation order for shelter A* (as you attempt to fill the shelter A), and therefore the sign on the link between these two elements is positive. There is one additional link in this diagram, from *capacity of shelter A* to *difference*. From the definition of *difference* given above, the influence is in the same direction along this link, and therefore the sign on the link is positive. In this case we have one negative sign, and therefore this feedback loop is negative or balancing, indicated by a minus sign in parentheses. The causal diagram of the evacuation planning example is on the CD-ROM, in the directory SYSTEM DYNAMICS, subdirectory Examples, Example 1.

5.3.2 Stock and Flow Diagram

Causal loop diagrams are an important tool for representing the feedback structure of systems being modeled. They are able to quickly capture dynamic hypotheses, elicit and capture the mental models, and communicate the important feedbacks that are responsible for the problem. They are used effectively at the start of the modeling process. One of the main limitations of causal diagrams is their inability to capture the stock and flow structure of systems.

Stocks and flows, together with feedbacks, are the two main concepts of dynamic systems theory. Stock and flow notation provides a general way to graphically characterize any system process. *Any* process? This ambitious statement by Forrester (1990) is based on the characteristics that are generally shared by all management processes and the components that make up these processes. It is a remarkable fact that all such processes can be characterized in terms of variables of two types, stocks (levels, accumulations) and flows (rates).

Stocks are accumulations. They characterize the state of the system and generate the information upon which decisions and actions are based. Stocks give systems inertia and provide them with memory. Stocks create delays by accumulating the difference between the inflow to a process and its outflow. By decoupling rates of flow, stocks are the source of disequilibrium dynamics in systems.

Stocks and flows are familiar to all of us. The number of boxes holding supplies is a stock. The amount of water in a tank is a stock. The number of people in a shelter is a stock. Stocks are altered by inflows and outflows. Tank storage is increased by the inflow of water and decreased by the release provided to users. Despite everyday experience of stocks and flows, quite often we fail to distinguish clearly between them. Is a water shortage during a disaster a stock or a flow?

Diagramming Notation System dynamics simulation and the computer tools for its implementation use a particular diagramming notation for stocks and flows. The Vensim notation is shown in Figure 5.12.

Stocks are represented by rectangles (suggesting a container holding the contents of the stock). Stocks can be used to depict both material and nonmaterial accumulations. The magnitudes of stocks within a system persist even if the magnitudes of all the activities fall to zero. When you take a snapshot of a system, only the accumulations that the activities had filled and drained would appear in the picture. The picture would show the state of the system at that point in time. Because they accumulate, stocks often act as "buffers" within a system. In this role, stocks enable inflows and outflows to be out of balance with each other—that is, out of equilibrium.

Flows are used to depict activities (i.e., things in motion). They are represented by a pipe (arrow) pointing into or out of the stock. Valves (flow regulators) control the

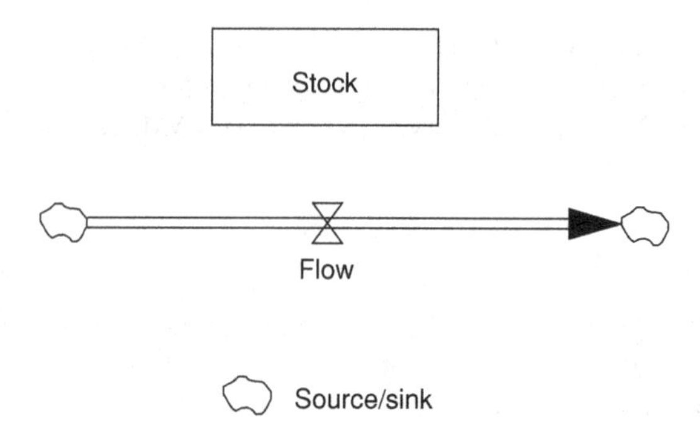

Figure 5.12 Stock and flow diagramming notation.

flows. Clouds represent the sources and sinks for the flows. A source represents the stock from which a flow originating outside the boundary of the model arises; sinks represent the stocks into which flows leaving the model boundary drain. Sources and sinks are assumed to have infinite capacity and can never constrain the flows they support. If there is an accumulation of something, that accumulation must result from some activity, a flow of something. And if there is a flow of something, there must be an associated buildup or depletion. "Stuff" flows through the pipe, in the direction indicated by the arrowhead. The flow volume is calculated by the algebraic expression, or number, that you enter into the flow regulator. You can imagine that large volumes cause the valve to be open wide, while small volumes cause it to be shut down. Flows can have several attributes. They can be conserved or nonconserved, unidirectional or bidirectional, and not unit converted or unit converted. A conserved flow draws down one stock as it fills another. The "stuff" that is flowing is conserved, in the sense that it only changes its location within the system (not its magnitude). In most cases, flows are expressed in the same units of "stuff" as the stocks to which they are attached. For example, a storage holds boxes. The flow would also be expressed in terms of the number of boxes, with the suffix "per [time]". When the units are not converted, the only difference between stock and flow expression is in this suffix. Stocks and flows are inseparable. Both are necessary for generating change over time, or dynamics. If we want only a static snapshot of reality, stocks alone would be sufficient. But without flows, no change in the magnitude of the stocks could occur. In order to move from snapshots to continuous presentations, we need flows. Figure 5.12 is on the CD-ROM, in the directory SYSTEM DYNAMICS, subdirectory Examples, Example 2. The structure of all stock and flow diagrams is composed of these elements.

Vensim notation offers two more graphical objects that complete the system dynamics syntax. They are auxiliary variables and arrows. *Auxiliary variables* often modify the activities within the system. They transform inputs into outputs. They can represent either information or material quantities. They are often used to break out the detail of the logic which otherwise would be buried within a flow regulator. Unlike stocks, auxiliary variables do *not* accumulate. The value for an auxiliary variable is recalculated from scratch in each time step. Auxiliary variables thus have no "memory." Auxiliary variables play one of four roles: stock related, flow related, stock/flow related, and external input related. In their stock-related role, they can provide an alternative way to measure the magnitude of a stock and are sometimes used to substitute for a stock. In their flow-related role, they can be used to "roll up" the net of several flow processes, or to break out the components of the logic of a flow, so as to avoid diagram clutter. Finally, auxiliary variables can be used as external inputs including some time series inputs (often implemented via the graphical function), as well as various built-in functions. Vensim notation for auxiliary variables is a text box with the name of the variable.

Arrows link stocks to auxiliary variables, stocks to flow regulators, flow regulators to flow regulators, auxiliary variables to flow regulators, and auxiliary variables to other auxiliary variables. Arrows represent inputs and outputs, *not* inflows and outflows! Arrows do not take on numerical values. They only transmit values taken on by other building blocks. Vensim notation for an arrow is a simple line with arrowhead at the end indicating the direction of relationship between variables connected by the arrow.

Identifying Stocks and Flows System dynamics simulation modeling process starts with the development of a causal diagram. The next step in the process is the development of a stock and flow diagram. Conversion of causal diagrams into stock and flow diagrams is not a straightforward task. Causal diagrams cannot be directly converted into stock and flow diagrams. Conversion requires full understanding of the difference between stocks and flows and implementation of basic principles how they can be connected to represent feedbacks.

The distinction between stocks and flows is very important and sometimes not obvious. In mathematics, system dynamics, control theory, and system dynamics originating scientific disciplines, stocks are also known as *integrals* or *state variables*. Flows are also known as *rates* or *derivatives*.

The units of measure can help you distinguish stocks from flows. Stocks are usually a quantity such as the number of boxes in storage, amount of water in storage, people affected by a disaster, or dollars in relief account. The associated flows must be measured in the same units *per time*; for example, the rate at which boxes are delivered or distributed from storage per day, the rate at which water is added per second to the storage, the rate in person per day, or the rate of expenditure from an account in dollars per day. Note that the choice of time period is arbitrary. You are free to select any measurement system you like as long as you remain consistent.

5.3.3 Generic Principles of System Dynamics Simulation Modeling

Forrester (1990) presents a set of principles that are of help in understanding and implementing system dynamics simulations. They are reproduced here.

1. *A feedback system is a closed system.* Its dynamic behavior arises within its internal structure. Any interaction that is essential to the behavior mode being investigated must be included inside the system boundary.

2. *Every decision is made within a feedback loop.* The decision controls action that alters the system state that influences the decision. A decision process can be part of more than one feedback loop.
3. *The feedback loop is the basic structural element of the system.* Dynamic behavior is generated by feedback. The more complex systems are aggregations of interacting feedback loops.
4. *A feedback loop consists of stocks and flows.* Except for constants these two are sufficient to represent a feedback loop. Both are necessary.
5. *Stocks are integrations.* The stocks integrate the results of action in a system. The stock variables cannot change instantaneously. The stocks create system continuity between points in time.
6. *Stocks are changed only by the flows.* A stock variable is computed by the change, due to flow variables, that alters the previous value of the stock. The earlier value of the stock is carried forward from the previous period. It is altered by flows over the intervening time interval. The present value of a stock variable can be computed without the present values of any other stock variables.
7. *Stocks and flows are not distinguished by units of measure.* The units of measure of a variable do not distinguish between a stock and a flow. The identification must recognize the difference between a variable created by integration and one that is a policy statement in the system.
8. *No flow can be measured except as an average over a period of time.* No flow can control another flow without an intervening stock variable.
9. *Flows depend only on stocks and constants.* No flow variable depends directly on any other flow variable. The flow equations of a system are of simple algebraic form. They do not involve time or the solution interval. They are not dependent on their own past values.
10. *Stock and flow variables must alternate.* Any path through the structure of a system encounters alternating stock and flow variables.
11. *Stocks completely describe the system condition.* Only the values of the stock variables are needed to fully describe the condition of a system. Flow variables are not needed because they can be computed from stocks.
12. *A policy or flow equation recognizes a local goal toward which that decision strives.* It compares the goal with the current system condition to detect a discrepancy, and uses the discrepancy to guide the action.

Example 4

Identify a few accumulations (stocks) that exist within an earthquake-stricken place.

An example could be "homeless people"—people without safe home. Another example could be "injured people"—people with injuries that require medical attention. One more example could be "residential damage"—number of residential homes destroyed by the earthquake.

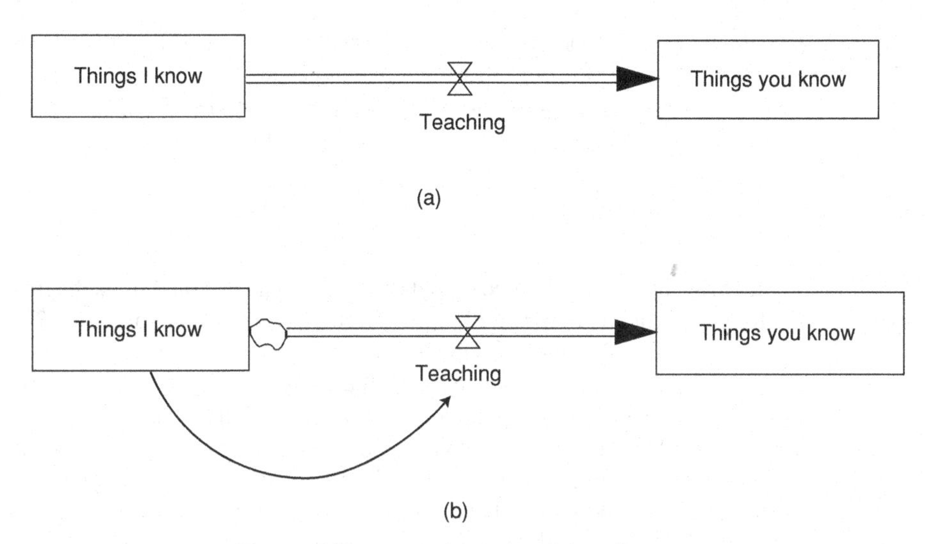

Figure 5.13 Example stock and flow diagrams.

Example 5
Consider the following situation: I am teaching you what I know. Figure 5.13 provides two possible stock and flow diagrams of that process. For each one, explain what you like and do not like about the diagram as a representation of the process.

Figure 5.13 is on the CD-ROM, in the directory SYSTEM DYNAMICS, subdirectory Examples, Example 3. Careful analysis of the two sketches in Figure 5.13 will show a difference in the interpretation of teaching activity. In both graphs, *teaching* is shown as flow. By definition, flow changes the attached stocks. The interpretation of Figure 5.13a is as follows. Through *teaching*, the stock of *things I know* is being depleted and the stock of *things you know* is increasing. The last part of this statement is correct but the initial part is not. Teaching as an activity does not drain the stock of the instructor's knowledge. Therefore, Figure 5.13a is not correct. A more appropriate flow diagram of teaching activity is Figure 5.13b, where the stock of *things I know* provides the source of information that goes into the *teaching* flow, which helps the stock of *things you know* to increase.

Example 6
Create an annotated causal diagram and a stock and flow diagram to represent each of the key feedback loop processes described below:

(a) The large-scale disaster causes a temporary refugee camp (tent city) to grow. As the camp grows and more refugees come, addition of new tents also increases since there are more refugees.

(b) City operates the flood-protection infrastructure (pumps, dikes, diversions, etc). With time the value of infrastructure depreciates.

(c) When the volunteering team performs well in the field, its confidence grows. With this increasing self-confidence comes even better performance.

Let us look at each of these in turn.

(a) Figure 5.14 shows an annotated causal diagram (a) and flow and stock diagram (b) for our problem. Two variables are selected to describe the problem: *Tent city*—a number of refugees living in the camp [people]; and *Refugees*—a number of people moving to the camp [people/time interval]. A stock and flow diagram is enhanced by the addition of an auxiliary variable—*Refugee growth rate*. This variable will assist in the calculation of refugee inflow rate (growth rate × people in the tent city). The loop shown in the diagram is a positive feedback loop. Note that construction is represented here as an activity that is implicitly part of the two variables representing stock and flow.

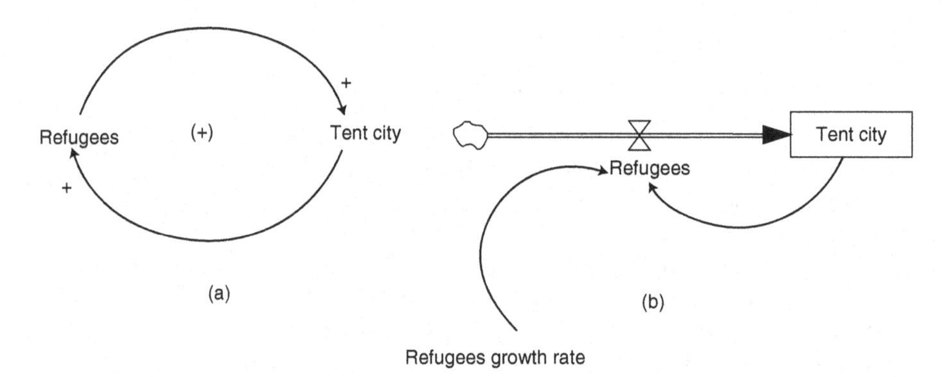

Figure 5.14 Tent city growth: (a) causal diagram and (b) stock and flow diagram.

(b) Figure 5.15 shows an annotated causal diagram (a) and flow and stock diagram (b) for our problem. Two variables are selected to describe the problem: *Flood-protection infrastructure value*—monetary value of the flood-protection infrastructure operated by the city [$]; and *Depreciation*—a change rate of the value of flood-protection infrastructure [$/time interval]. A stock and flow diagram is enhanced by the addition of an auxiliary variable—*Asset lifetime*. This variable will assist in the calculation of the depreciation rate (value/lifetime). The loop shown in the diagram is a negative feedback loop. Obviously this system will reach a stable equilibrium at 0 value—when the value of asset declines to zero, no further depreciation can be taken.

(c) Volunteering team's performance is shown in Figure 5.16. The loop representing team's performance is again a positive feedback loop. In this example both team's performance and team's self-confidence are shown as stocks, and the

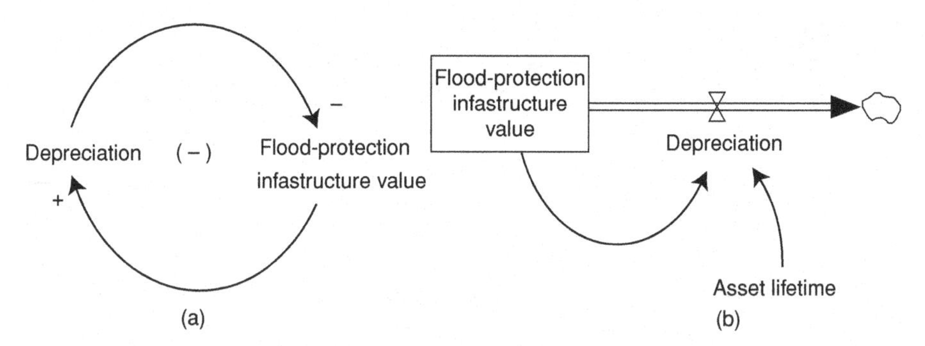

Figure 5.15 Flood-protection infrastructure depreciation: (a) causal diagram and (b) stock and flow diagram.

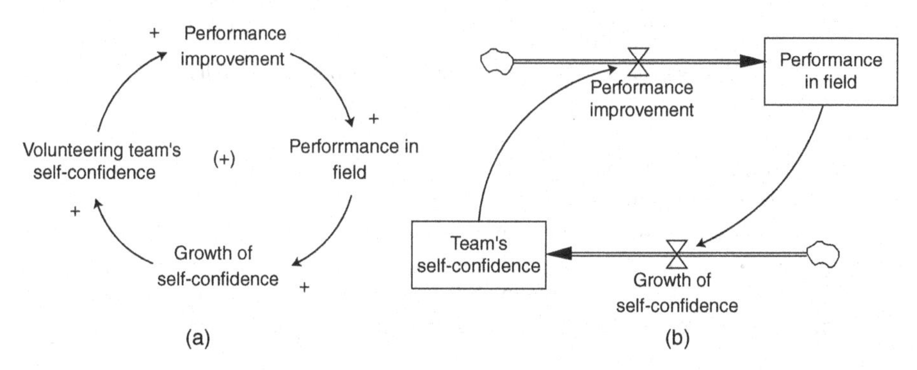

Figure 5.16 Volunteering team's performance in the field: (a) causal diagram and (b) stock and flow diagram.

increase in both is shown using flows. Note the reinforcing character of this loop, and consider what units can be used to represent the variables.

5.3.4 Numerical Simulation

The stocks and flows are building blocks of system dynamics simulation models. They allow the user to focus on assembling a model structure and assist in converting a structure into mathematical equations. The equations describe how the system changes. These changes, accumulated step-by-step unfold the behavior pattern of the system. The process of step-by-step solution of model equations is called simulation. The equations (the instructions) for how to compute the next time step are called a "simulation model." Simulation is an effective replacement of analytical solution to the equations that would express the system condition in terms of any future time. Most dynamic behavior in social systems (including management of disasters) can only be represented by models that are nonlinear and so complex that analytical

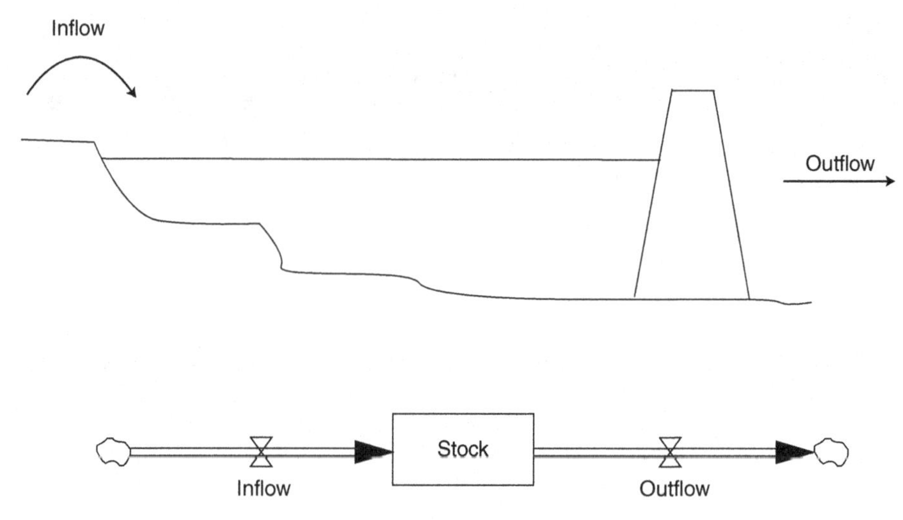

Figure 5.17 Hydraulic metaphor for stock and flow diagram.

mathematical solutions are impossible. For such systems, only the simulation process using step-by-step numerical solution is available.

The stock and flow diagramming notation from the previous section is based on a hydraulic metaphor (Sterman, 2000)—the flow of water into and out of a reservoir (Figure 5.17). Despite the simple metaphor, the stock and flow diagram has a precise mathematical meaning. Stocks accumulate or integrate their flows:

$$Stock(t) = \int_{t_0}^{t_c} [Inflow(t) - Outflow(t)\mathrm{d}t + Stock(t_0)], \tag{5.2}$$

where *Inflow*(t) is the value of the inflow at any time t_c between the initial time t_0 and the current time t_c. Equivalently, the net rate of change of any stock can be represented with its derivative:

$$\frac{\mathrm{d}(Stock)}{\mathrm{d}t} = Inflow(t) - Outflow(t) \tag{5.3}$$

The differential Equation (5.3) is the basis of system dynamics simulation. System dynamics simulation software tools such as Vensim use the principles of object-oriented programming to help users develop the model structure using objects that represent stocks, flows, auxiliary variables, and arrows. The mathematical equations corresponding to a particular structure are written by the tool itself.

System dynamics models are systems of nonlinear ordinary differential equations such as Equation (5.3). With simpler notation—replacement of the integral sign with the *INTEGRAL*() function, replacement of *Inflow* with *I*, *Outflow* with *O*, and *Stock* with S—we can rewrite Equation (5.2) as follows:

$$S_t = INTEGRAL(I_t - O_t), S_{t_0} \tag{5.4}$$

and express inflows and outflows as:

$$\begin{aligned} I_t &= f(S_t, U_t, C) \\ O_t &= g(S_t, U_t, C) \end{aligned} \tag{5.5}$$

where U is any exogenous variable and C is a constant.

For any realistic system dynamics model, analytic solutions cannot be found and the behavior of the model must be computed numerically. Most system dynamics tools such as Vensim provide a basic tool for numerical integration known as the Euler method, and some more sophisticated tools such as Runge–Kutta. Details of numerical integration methods can be found, for example, in Atkinson (1985). We shall concentrate here on some basic principles and practical considerations.

Denoting the time interval between periods as $\mathrm{d}t$, the assumption of constant flows during the time interval implies:

$$S_{t+\mathrm{d}t} = S_t + \mathrm{d}t \times (I_t - O_t) \tag{5.6}$$

Equation (5.6) is the most basic technique, known as *Euler integration*. The assumption that the flows remain constant throughout the time interval $\mathrm{d}t$ is reasonable if the dynamics of the system are slow enough and $\mathrm{d}t$ is small enough. The definitions of "reasonable" and "slow enough" depend on the required accuracy, which depends on the purpose of the model. As the time step gets smaller, the accuracy of Euler's approximation improves. At the limit, when $\mathrm{d}t$ becomes an infinitesimal moment of time, Equation (5.6) reduces to the exact continuous-time differential equation governing the dynamics of the system:

$$S_t = \lim_{\mathrm{d}t \to 0} \frac{S_{t+\mathrm{d}t} - S_t}{\mathrm{d}t} = \frac{\mathrm{d}S}{\mathrm{d}t} = (I_t - O_t) \tag{5.7}$$

Vensim uses Euler integration as its default simulation method. The only difference between the numerical and analytic solution of the underlying differential equation system is the size of $\mathrm{d}t$. The differential equation uses an infinitesimal, a true instant. Digital computers use discrete steps and a finite time step. The use of a finite time step and resulting approximations of flows over the interval introduce error, known as integration error or $\mathrm{d}t$ error. This error depends on how quickly the flows change relative to the time step. The faster the dynamics of the system, or the longer the $\mathrm{d}t$, the larger the integration error. That points us to one of the most common questions about using system dynamics simulation: how should we select the time step? Here are some practical recommendations:

- Select a time step for your model that is a power of 2, such as 2, 1, 0.5, and 0.25.
- Make sure your time step is evenly divisible into the interval between data points.
- Select a time step one-fourth to one-tenth as large as the smallest time constant in your model.
- Test for integration error by cutting the time step in half and running the model again. If there are no significant differences, then the original value is fine. If the behavior changes significantly, continue to cut the time step in half until the differences in behavior no longer matter.
- Note that Euler integration is almost always fine in models where there are large errors in parameters, initial conditions, historical data, and especially model structure. Test the robustness of your results to Euler by running the model with a higher-order method such as fourth-order Runge–Kutta. If there are no significant differences, Euler is fine.

Euler integration is simple and adequate for many disaster management applications. It is advisable to spend more time on improving the model rather than fine-tuning the numerical integration method. However, there are some systems and some model purposes where Euler is not appropriate, because either the errors it generates are too large or the time step required to gain the needed accuracy slows model execution too much.

There are many more advanced techniques for numerical integration of differential equations. The most popular of those available in Vensim are the Runge–Kutta methods. Euler's method assumes the flows at time t remain constant over the entire interval of time $t + \mathrm{d}t$, that is, that the average flow over the interval equals the flow at the start of the interval. The Runge–Kutta method finds a better approximation of the average rate between t and $t + \mathrm{d}t$. First, provisional estimates of the stocks at $t + \mathrm{d}t$ are calculated by Euler's method. Next, the flows at time $t + \mathrm{d}t$ are calculated from the Euler estimate of the stocks at time $t + \mathrm{d}t$. The estimated flows at time t and $t + \mathrm{d}t$ are averaged and used to calculate the value of the stocks at $t + \mathrm{d}t$. This method is known as second-order Runge–Kutta.

Higher-order Runge–Kutta methods work in essentially the same way, but estimate the average flow over subintervals within $[t, t + \mathrm{d}t]$ to yield a still better approximation. Vensim offers the fourth-order Runge–Kutta. While Runge–Kutta requires more computation per time step, the accuracy of the approximation is much greater than with Euler's method. Integration errors for a comparable choice of $\mathrm{d}t$ are much smaller and propagate at much smaller rates, allowing the modeler to use a larger time step or gain additional accuracy. For the details of Runge–Kutta methods available in Vensim, consult the *Vensim Reference Manual* (Ventana Systems, 2003a) available for download from the Ventana Web site (http://www.vensim.com, last accessed January 2010).

Later in this chapter (Section 5.4), two detailed examples are presented that illustrate the model simulation process with derivation of model equations.

5.3.5 Policy Design and Evaluation—Model Use

Once the confidence has been developed in the structure and behavior of the system dynamics simulation model, it can be used to design and evaluate policies for improvement. Policy design involves various approaches such as (a) change of model parameters and/or (b) creation of new strategies, structures, and decision rules. Regardless of the approach used, the aim of any system dynamics simulation model experimentation is the exploration of model variable behaviors between different simulation runs. The desire is to see how the modeled system behaves normally, and then how changes in policies or physical parameters alter that behavior.

It has been shown that the feedback structure of a system determines its dynamics—system behavior over time. Most of the time high leverage policies involve changing the dominant feedback loops by redesigning the stock and flow structure, eliminating time delays, changing the flow, and quality of information available at key decision points, or fundamentally recreating the decision processes in the system.

The robustness of policies and their sensitivity to uncertainties in model parameters and structure must be assessed, including their performance under a wide range of alternative scenarios. The interactions of different policies must also be considered; because real systems are highly nonlinear, the impact of combination policies is usually not the sum of their impacts alone. Often policies interfere with one another; sometimes, they reinforce one another and generate substantial synergies.

From a policy perspective, model sensitivity to a parameter change or model structure change means that a "high-leverage" point has been discovered—such changes may represent useful intervention points in the real world. From a scientific perspective, less sensitivity in uncertain parameters is preferable, since this lower sensitivity means that the process or physical characteristic associated with that parameter does not affect model behavior strongly. A lack of understanding of the physical system involved therefore changes neither the model results nor the conclusions drawn from model behavior.

5.4 SYSTEM DYNAMICS SIMULATION MODELING EXAMPLES

This section illustrates the development of a simple system dynamics simulation model (i.e., the last step of the modeling process elaborated above). Specifically, we develop and investigate a simple model of infectious disease. The purpose of this example is to familiarize you with what is required to build a system dynamics simulation model, and how such a model can be used. The second section provides a description of more complex system dynamics model of acute infection and immunization that can completely eradicate a disease. Both models are modified after Sterman (2000) and belong to the study of dynamics of disease.

5.4.1 A Simple Flu Epidemic Model

Epidemics of infectious diseases (such as flu) are characterized with the cumulative number of new cases rising exponentially until it peaks and then falls as the epidemic ends. The epidemic usually begins with a single infected person (patient zero). The flu spreads through contact and by inhalation of virus-laden aerosols released when infected individuals cough and sneeze. The flu spreads slowly at first, but as more and more people get ill and become infectious, the number they infect grows exponentially. The epidemic ends because of the depletion of the pool of susceptible people.

Example 7
Figure 5.18 shows the causal diagram of a flu epidemic. The total population of the community or region represented in the model is divided into two categories: those susceptible to the flu, *SFP*, and those who are infectious, *IFP*. As people are infected, they move from the susceptible category to the infectious category. The causal diagram shows presence of two feedback loops that drive the dynamics of the flu infection. The main simplifying assumptions used in this model are (i) the population of the region affected by the flu is constant (births, deaths, migrations, and other important variables are ignored); (ii) the population is homogeneous—all members of the community are interacting at the same average rate; (iii) the flu is not altering people's lifestyles—infected people interact at the same average rate as susceptible; and (iv) there is no recovery, quarantine, or immunization—once infected, people remain infectious indefinitely (this assumption applies to chronic infections, not acute illnesses).

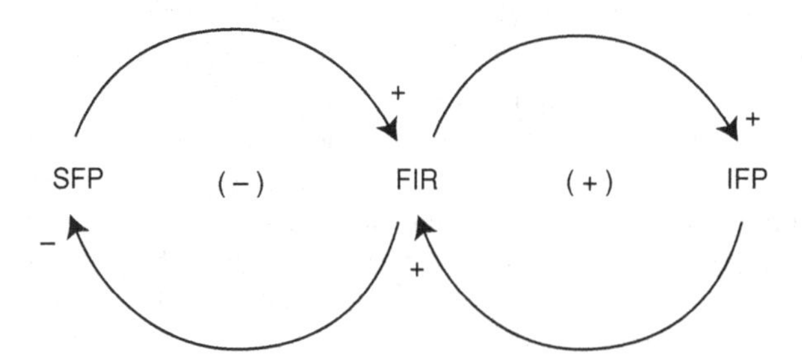

Figure 5.18 Causal diagram of a simple flu epidemic model.

The model in Figure 5.18 contains two loops: one positive and one negative. Flu spreads as members of the *IFP* come into contact with and pass the disease to the members of *SFP*, increasing the *IFP* still further (the positive loop) while at the same time depleting the *SFP* (the negative loop). The flu infection rate, *FIR*, is the rate of increase in *IFP* and decrease in *SFP*.

Figure 5.19 shows a stock and flow diagram of the flu epidemic model. Both populations, *SFP* and *IFP*, are modeled as stocks (in number of people) and the flu infection rate, *FIR*, as a flow (number of people/time interval).

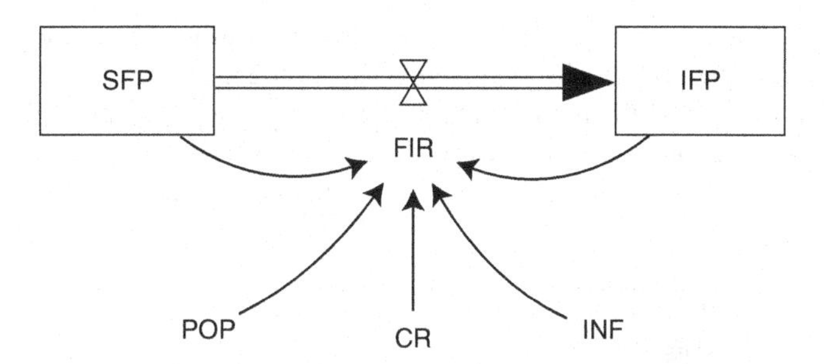

Figure 5.19 Stock and flow diagram of a simple flu epidemic model.

The stock and flow diagram in Figure 5.19 shows three auxiliary variables included in the model to assist the computation: *POP*—the total population of the community or a region (number of people), *CR*—contact rate (people contacted per person per time period, or 1/time period), and *INF*—infectivity (defined as the probability that a person will become infected after exposure to someone with the flu).

Let us now prepare the model equations and data. Starting with the stock variables, we shall use as initial value for $SFP = 1000$ and as initial value for *INF* a small number or even a single individual (1). The $POP = 10{,}000$, $CR = 2$, and $INF = 0.5$. The time step is assumed to be 1 day and simulation time period is assumed to be 120 days.

The flow equation is next. People in the community interact at a certain rate (the contact rate, *CR*. Thus, the susceptible population generates $SFP \times CR$ encounters per time period. Some of these encounters are with infectious people. If infectious people interact at the same rate as susceptible people (they are confined to bed), then there is the probability that any randomly selected encounter is an encounter with an infectious individual. Not every encounter with an infectious person results in infection. The infectivity, *INF*, of the flu is the probability that a person becomes infected after contact with an infectious person. The infection rate is, therefore, the total number of encounters $SFP \times CR$ multiplied by the probability that any of those encounters is with an infectious individual *INF/POP* multiplied by the probability that an encounter with an infectious person results in infection:

$$FIR = CR \times INF \times SFP \times \frac{IFP}{POP} \tag{5.8}$$

Since the total population is fixed:

$$POP = SFP + IFP \tag{5.9}$$

the flow Equation (5.8) gets the final form:

$$FIR = CR \times INF \times IFP \times \left(1 - \frac{IFP}{POP}\right) \tag{5.10}$$

Our model is now complete and ready for simulation. Vensim equations of the complete flu epidemic model are given in Figure 5.20.

The simulation of the model should answer one major question: How is the total infected population going to change over time? This calculation will offer more insights: when we can expect the peak of infection; how long will take for the total

```
(01)   CR = 2
       Units: 1/Day
       People in the community interact at a certain rate (CR, measured
       in people contacted per person per time period, or 1/time
       periods).
(02)   FINAL TIME = 120
       Units: Day
       The final time for the simulation.
(03)   FIR = (SFP × INF × CR) × IFP/POP
       Units: People/Day
       The infection rate is modeled using equation (5.10) in the text.
(04)   IFP = INTEG (FIR, 1)
       Units: People
       The total population of the community or region considered is
       divided into two categories: those susceptible to the flu,
       SFP, and those who are infectious, IFP
(05)   INF = 0.5
       Units: fraction
       The infectivity INF of the disease is the probability that a
       person will become infected after exposure to someone with the
       flu.
(06)   INITIAL TIME = 0
       Units: Day
       The initial time for the simulation.
(07)   POP = 10,000
       Units: People
       POP is the total population in the community
(08)   SAVEPER = TIME STEP
       Units: Day
       The frequency with which output is stored.
(09)   SFP = INTEG (−FIR, 1000)
       Units: People
       The total population of the community or region considered is
       divided into two categories: those susceptible to the flu,
       SFP, and those who are infectious, IFP
(10)   TIME STEP = 0.125
       Units: Day
       The time step for the simulation.
```

Figure 5.20 Vensim equations of the flu epidemic model.

population to get infected (due to the assumption that once infected each person remains infectious indefinitely). Model simulation results are shown in Figure 5.21 for low value of $INF = 0.5$ and high value of $INF = 0.7$.

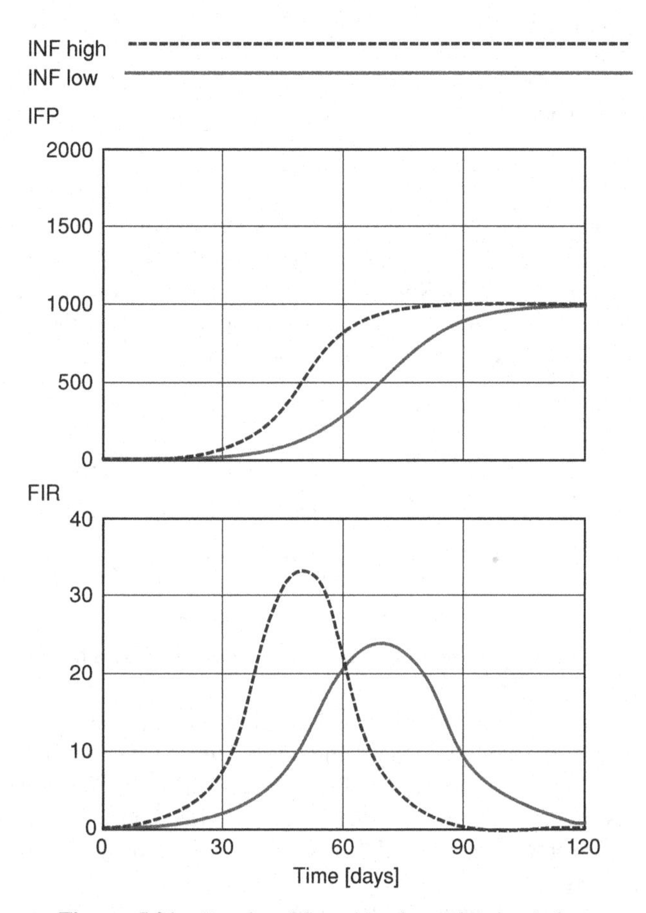

Figure 5.21 Results of flu epidemic model simulations.

Once an infectious person arrives in the community, every susceptible person eventually becomes infected. Infection rate follows a bell-shaped curve and the total infected population follows the classic S-shaped pattern. S-shaped growth is the consequence of two feedbacks acting together to determine the system behavior. The positive feedback, which controls increase in infectious population, dominates the first part of the diagram (exponential growth). The second part of the diagram (goal-seeking pattern) reflects the dominance of negative feedback that controls the decline in susceptible population.

The complete flu epidemic model is available on the CD-ROM, in directory SYSTEMDYNAMICS, subdirectory Examples, Example 4.

5.4.2 A More Complex Flu Epidemic Model with Recovery

The most restrictive and unrealistic assumption of the model in Section 5.4.1 is the assumption that flu is chronic—infected people remain infectious indefinitely. This assumption may be reasonable for some diseases. However, many infectious diseases produce a period of acute infectiousness and illness that is followed by recovery. The following is the description of such a model that is based on the original work of Kermack and McKendrick (1927, reprinted in 1991) and adopted from Sterman (2000).

Example 8
The model with recovery contains three stocks: the susceptible population, *SFP*, the infectious population, *IFP*, and the recovered population, *RFP* (long known as the SIR model, the Kermack–McKendrick formulation is widely used in epidemiology). Figure 5.22 shows the causal diagram of the model which is the extension of the causal diagram of model from Example 7 shown in Figure 5.18.

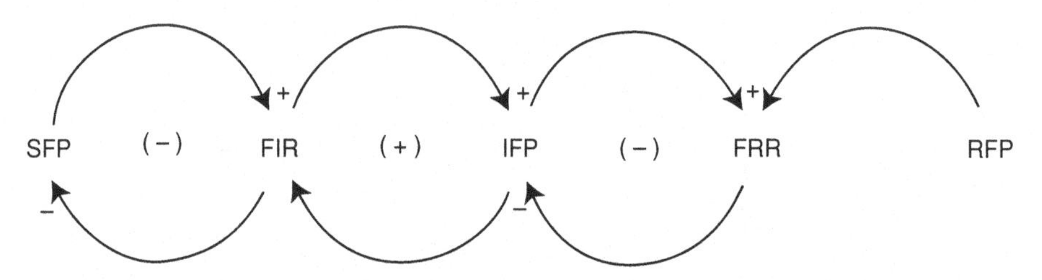

Figure 5.22 Causal diagram of an epidemic model with recovery.

People contracting the flu become infectious for a certain period of time but then recover and develop permanent immunity. The assumption that people recover creates one additional feedback—the negative recovery loop. The greater the number of infectious individuals, the greater the recovery rate and the smaller the number of infectious people remaining. All other assumptions of the simple model from Example 7 are retained.

Figure 5.23 shows a stock and flow diagram of the flu epidemic model with recovery. All three populations, *SFP*, *IFP*, and *RFP*, are modeled as stocks (in the

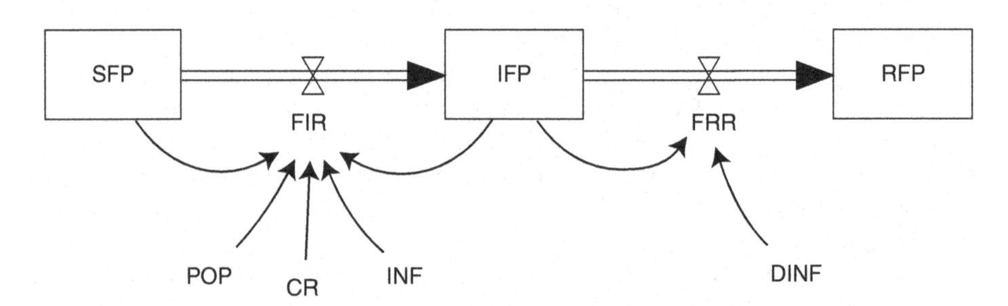

Figure 5.23 Stock and flow diagram of an epidemic model with recovery.

number of people). The flu infection rate, *FIR* and the flu recovery rate, *FRR*, are modeled as flows (the number of people/time interval).

The stock and flow diagram in Figure 5.23 shows three auxiliary variables included in the model presented in Example 7: *POP*—the total population of the community or a region (number of people), *CR*—contact rate (people contacted per person per time period, or 1/time period), *INF*—infectivity (defined as the probability that a person will become infected after exposure to someone with the flu), and a newly added auxiliary variable *DINF*—average duration of infectivity (number of days a person is infectious).

Let us now prepare the model equations and data. Starting with the stock variables, we shall use as initial value for $SFP = 1000$, as initial value for *INF* a single individual (1) and as initial value of $RFP = 0$. The $POP = 10{,}000$, $CR = 6$, $INF = 0.25$, and $DINF = 2$ days. The time step is assumed to be 1 day and simulation time period is assumed to be 30 days.

The new flow equation can be presented in several ways. The average duration of infectivity, *DINF*, is assumed to be constant and the recovery process is assumed to follow a first-order negative feedback process:

$$FRR = \frac{IFP}{DINF} \tag{5.11}$$

The average duration of infectivity, *DINF*, represents the average length of time people are infectious (measured in days). The assumption that the recovery rate is a first-order feedback process means that people do not all recover after exactly the same time. A given population of infectious individuals will decline exponentially, with some people recovering rapidly and others more slowly. The infection rate is formulated exactly as in Example 7, Equation (5.8).

Our model with recovery is now complete and ready for simulation. Vensim equations of the complete flu epidemic model with recovery are given in Figure 5.24.

The simulation of the more complex model can provide quantitative insights into model behavior. It is now possible for the disease to die out without causing an epidemic. If the infection rate is less than the recovery rate, the infectious population will fall. As it falls, so too will the infection rate. The infectious population can therefore fall to zero before everyone contracts the disease.

Running the simulations we can now answer various questions. Let us first concentrate on the following question: Under what circumstances will the introduction of an infectious individual to the population cause an epidemic? Intuitively, for an epidemic to occur the infection rate must exceed the recovery rate ($FIR > FRR$). If this is the case, the infectious population, *IFP*, will grow, leading to more new cases. Therefore, for an epidemic to occur, each infective must, on average, pass the disease on to more than one other person prior to recovering.

The question of whether an epidemic will occur can be addressed through the analysis of feedback loop dominance. If the positive loop (see Figure 5.22) dominates the other two, then the introduction of even a single infective individual to a community triggers an epidemic. The infection rate, *FIR*, will exceed the recovery rate, *FRR*,

(01)	CR = 6 Units: 1/Day Contact Rate, CR, measured in people contacted per person per time period, or 1/time periods.
(02)	DINF = 2 Units: Day The average length of time that a person is infectious.
(03)	FINAL TIME = 30 Units: Day The final time for the simulation.
(04)	FIR = CR × INF × SFP × IFP/POP Units: People/Day The infection rate is the total number of encounters multiplied by the probability that any of those encounters is with an infectious individual, and finally multiplied by the probability that an encounter with an infectious person results in infection.
(05)	FRR = IFP/DINF Units: People/Day The rate at which the infected population recover and become immune.
(06)	IFP = INTEG (FIR − FRR,1) Units: People The infectious population accumulates the infection rate less the recovery rate.
(07)	INF = 0.25 Units: Dimensionless The infectivity, INF, is the probability that a person will become infected after exposure to someone with the disease.
(08)	INITIAL TIME = 0 Units: Day The initial time for the simulation.
(09)	POP = 10,000 Units: People The total population is constant
(10)	RFP = INTEG (FRR, 0) Units: People The recovered population RFP accumulates the recovery rate
(11)	SAVEPER = TIME STEP Units: Day The frequency with which output is stored.
(12)	SFP = INTEG (−FIR, POP − IFP − RFP) Units: People The susceptible population is reduced by the infection rate. The initial susceptible population is the total population less the initial number of infectives and any initially recovered individuals.
(13)	TIME STEP = 0.125 Units: Day The time step for the simulation.

Figure 5.24 Vensim equations of the flu epidemic model with recovery.

causing the infection rate to grow further, until depletion of the susceptible population, *SFP*, finally limits the epidemic. If, however, the positive loop is weaker than the negative loops, an epidemic will not occur since infectious people will recover faster than new cases arise. The number of new cases created by each infective prior to their recovery, and therefore the strength of the different loops, depends on the average duration of infection and the number of new cases each infective generates per time period. The higher the contact rate, *CR*, or the greater the infectivity of the disease, *INF*, the stronger the positive loop. Likewise, the larger the fraction of the total population susceptible to infection, the weaker the negative loop that depletes the susceptible population, *SFP*. Finally, the longer the average duration of infection, *DINF*, the weaker the negative loop that increases the recovery and the more likely an epidemic will occur.

Fine interaction between the three feedback loops may help us answer the question: What is the threshold of the positive feedback dominance? For any given susceptible population size, *SFP*, there is some critical combination of contact rate, *CR*, infectivity, *INF*, and disease duration, *DINF*, just great enough for the positive loop to dominate the negative loops. That threshold is known as the *tipping point*. Below the tipping point, the system is stable, and if the disease is introduced into the community, there may be a few new cases, but on average, people will recover faster than new cases are generated. Negative feedback dominates and the population is resistant to an epidemic. Past the tipping point, the positive loop dominates. The system is unstable and once a disease arrives, it can spread very fast—by positive feedback—limited only by the depletion of the susceptible population, *SFP*.

Figures 5.25 and 5.26 show a simulation of the model where the system is well past the tipping point. The population of the community is $POP = 10{,}000$ and initially everyone is susceptible to the disease. At the beginning of simulation, a single

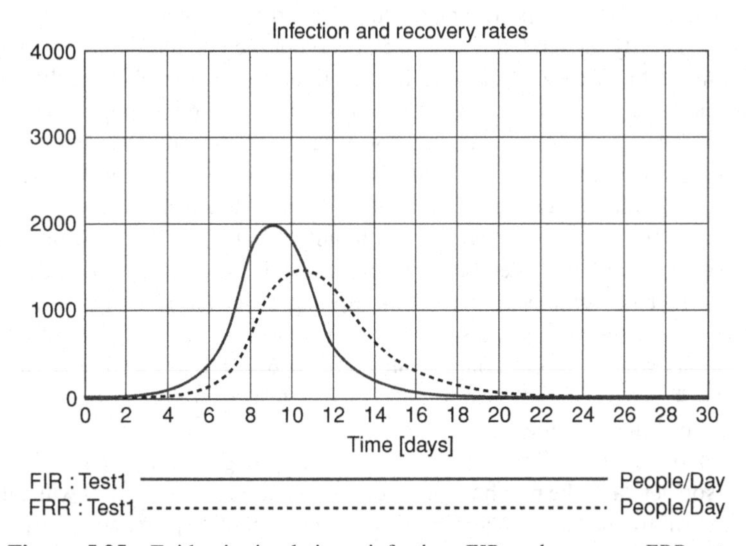

Figure 5.25 Epidemic simulation—infection, *FIR*, and recovery, *FRR*, rates.

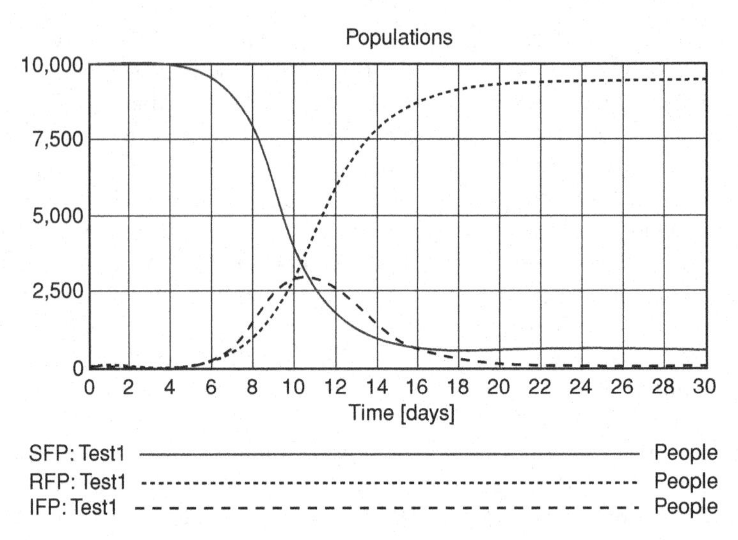

Figure 5.26 Epidemic simulation—susceptible, *SFP*, infectious, *IFP*, and recovered, *RFP*, populations.

infective individual arrives in the community. The average duration of infection is *DINF* = 2 days, and infectivity is *INF* = 0.25. The average contact rate is *CR* = 6 people per person per day.

Each infective therefore generates 1.5 new cases per day and an average of three new cases before they recover. The positive loop therefore dominates and the epidemic spreads quickly. From the Figure 5.25, we see that the infection rate, *FIR*, peaks at more than 2000 people per day around day nine, and from Figure 5.26, we see that at its peak more than one-quarter of the population is infectious. The susceptible population, *SFP*, falls rapidly, and it is this depletion of potential new cases that halts the epidemic. By the tenth day, the size of susceptible population, *SFP*, is so low that the number of new cases declines. The infectious population, *SFP*, peaks and falls as people now recover faster than new cases arise. The susceptible population, *SFP*, continues to fall, though at a slower and slower rate, until the epidemic ends. In less than 3 weeks, a single infectious individual led to a massive epidemic involving nearly the entire community. Note the epidemic ends before the susceptible population, *SFP*, falls to zero (not every person in the community gets infected). The stronger the positive loop, however, the fewer the people remain uninfected at the end of the epidemic.

To further our understanding of the tipping point let us focus on Figure 5.27. It shows the dynamics of susceptible population, *SFP*, of results of five model simulations with different contact rates (*CR* = 6 test 1 in the figure, *CR* = 5 test 2, *CR* = 4 test 3, *CR* = 3 test 4, and *CR* = 2 test 5) in several simulations of the model. The other parameters are as above.

From the Figure 5.27 (test 5) we see that the tipping point is at *CR* = 2 (2 contacts per person per day), the number of new cases each infective generates while infectious is just equal to one (2 contacts per person per day × 0.25 probability of infection × 2 days of infectivity). Contacts at a rate less than two per person per day do not

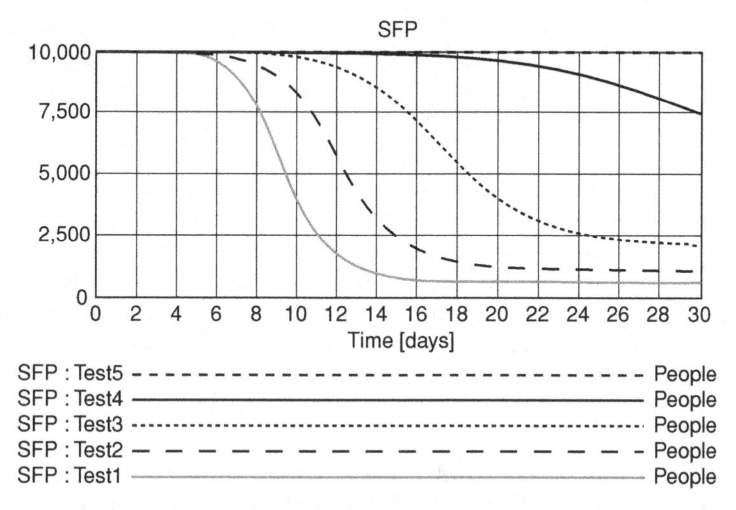

Figure 5.27 Epidemic dynamics.

cause an epidemic. When the contact rate rises above the critical threshold of two ($CR > 2$) the system becomes unstable, and an epidemic occurs. The higher the contact rate, the stronger the positive loop relative to the negative loop that controls recovery, and the faster the epidemic progresses. Further, the stronger the positive loop the greater the population contracting the disease.

Any change that increases the strength of the positive loops will yield similar results. An increase in infectivity strengthens the positive loop and is identical in impact to an increase in contact rate, *CR*. An increase in the duration of the infectious period, *DINF*, weakens the negative recovery loop and also pushes the system farther from the tipping point.

Based on the system dynamics epidemic model developed here, we can calculate the exact location of tipping point. For an epidemic to occur, the infection rate, *INF*, must exceed the recovery rate, *FRR*:

$$INF > FRR \tag{5.12}$$

Replacing Equations (5.10) and (5.11) in Equation (5.12) leads to:

$$CR \times INF \times SFP \times \frac{IFP}{POP} > \frac{IFP}{DINF} \tag{5.13}$$

or equivalently,

$$CR \times INF \times DINF \times \frac{SFP}{POP} > 1 \tag{5.14}$$

In Equation (5.14), the product of the contact rate, infectivity, and the average duration of infection yields the dimensionless ratio, $CR \times INF \times DINF$, known as the *contact number*. However, not all the contacts will result in a new case. The number of infectious contacts that actually result in the infection of a susceptible person depends on the probability that the infectives encounter susceptibles. Assuming the population is homogeneous, the probability of encountering a susceptible is given by the prevalence of susceptibles in the population, SFP/POP. The expression $CR \times INF \times DINF \times (SFP/POP)$ is also known as the *reproduction rate* for the epidemic. Equation (5.14) therefore defines the tipping point or threshold at which an epidemic occurs in a population and is known as the threshold theorem in epidemiology.

The complete flu epidemic model with recovery is available on the CD-ROM, in directory SYSTEMDYNAMICS, subdirectory Examples, Example 5.

5.5 AN EXAMPLE OF DISASTER MANAGEMENT SIMULATION—FLOOD EVACUATION SIMULATION MODEL

Simulation models play an important role in management of disasters. Sections 4.1 and 4.2 provide examples from the literature. Here, we shall look into one system dynamics model selected from my experience in the application of system dynamics—a flood evacuation simulation for the Red River basin in Manitoba, Canada (Simonović and Ahmad, 2005). The model presents an interesting application of system dynamics simulation for capturing human behavior during emergency flood evacuation. The model simulates the acceptance of evacuation orders by the residents of the area under threat, the number of families in the process of evacuation, and the time required for all evacuees to reach safety. The model is conceptualized around the flooding conditions (both physical and management) and a core set of social and psychological factors that determine human behavior before and during the flood evacuation. The main purpose of the model is to assess the effectiveness of different flood emergency management procedures. Each procedure consists of the choice of a flood warning method, warning consistency, timing of evacuation order, coherence of the community, upstream flooding conditions, and a set of weights assigned to different warning distribution methods. The model use and effectiveness were tested through the evaluation of different flood evacuation options in the Red River basin, Canada.

5.5.1 Introduction

Preparation for emergency action must be taken before a crisis for several reasons. Conditions in a disaster-affected region tend to be chaotic. Communication is difficult and command structures can break down because of logistical or communications failure. Human behavior during the emergency is hard to control and predict. Complaints cannot normally be addressed during the emergency. Experience with emergency evacuation in the Red River basin (Manitoba, Canada) during a major flood in 1997 unveiled an abundance of problems that the population affected by the disaster had with policies and their implementation. They were not happy with the timing of the evacuation orders, evacuation process implementation, order of command, and many related issues. The details of the Red River flood of 1997 are in Section 1.1 of this book. The literature confirms that there is a very similar situation in other kinds of disasters. There is an obvious need to improve:

- our understanding of the social side of emergency management processes,
- our understanding of human behavior during emergencies,
- the communication between the population affected by the disaster and emergency management authorities, and
- preparedness through simulation, or investigation of "what-if" scenarios.

The proper understanding of human behavior in response to a disaster, and the ability to capture it in a dynamic model, are valuable additions to emergency management policy analysis. This example develops a theoretical framework for studying flood evacuation emergency planning in a more holistic way, integrating a broad range of social and cultural responses to the evacuation process. It also provides new insights by developing a dynamic model of the process, which is converted into a gaming format for policy analysis, training and for other practical applications. The model integrates empirical survey data to fit the characteristics of specific communities.

5.5.2 Human Behavior During Disasters

The modeling process required a very detailed consideration of major factors that affect human behavior during disasters (such as individual risk perception, disaster recognition, and acceptance of the evacuation order). It was found that core factors determining how people cope with floods are their economic status and previous experience with flooding. Stress indicators were measured using fear, desperation, action, depression, and family health indexes. Issues covered in modeling flood knowledge were the flood warning system, contributing topographical factors, contributing effects of urbanization, and political trends. Both economic status and previous experience with flooding are incorporated in the model and are discussed below.

The amount of human effort involved in coping with natural hazards varies greatly, and there are several different levels and thresholds. A social group moves from one level to another in a cumulative fashion as it acquires experience. Factors that influence this movement are the severity of a hazard, recency of a hazard, intensity and extensiveness of human activities in the area, and the wealth of the society. We can model this by proposing an awareness threshold that precedes an acceptance level. This is followed by an action threshold, which leads people to modify and prevent events, and then an intolerance threshold, marked by "change use" or "change location" steps. The Red River basin evacuation model uses a structure that divides the process into three phases: concern, danger recognition, and evacuation decision.

The relocation of residents after a natural disaster contributes to environmental, social, and psychological stress. Research has showed that people from the same neighborhood prefer to be evacuated together. Evacuation orders that direct people from close social environments to different temporary accommodation do not meet with ready acceptance, and can delay the general evacuation process.

5.5.3 A System Dynamics Simulation Model

Flood management is aimed at reducing the potential harmful impact of floods on people, the environment, and the economy of the region. In Canada, the flood management process can be divided into three major phases: planning, emergency management, and postflood recovery. During the *planning* phase, different alternative measures (structural and nonstructural) are analyzed and compared for possible implementation to reduce flood damage. *Emergency management* involves regular appraisal of the current flood situation and daily operation of flood control structures to minimize damage. Following appraisal of the situation, decisions may be made to evacuate areas. *Postflood recovery* involves decisions regarding the return to normal everyday life. The main concerns during this phase are provision of assistance to flood victims and rehabilitation of damaged properties. This example focuses on issues related to emergency management, provision of assistance, and conduct of the evacuation process. Human behavior during evacuation, in response to a disaster warning, is captured in a system dynamics model that allows emergency managers to develop the "best" possible response strategy in order to minimize the negative impacts of a flood disaster. Theoretical knowledge collected from the relevant literature was used to conceptualize the model. Model relationships and all other necessary data were obtained through interviews conducted in the Red River basin immediately after the flood of 1997.

The human decision-making process in response to a disaster warning can be divided into four psychological phases:

1. concern,
2. danger recognition,
3. acceptance, and
4. evacuation decision.

The factors that play an important role in the decision-making process can also be divided into four groups: initial conditions, social factors, external factors, and psychological factors (denoted as IF, SF, EF, and PF in Figure 5.28).

Figure 5.28 shows the conceptual framework of the behavioral flood evacuation model. The four groups of factors are identified with their acronyms. The vertical arrow alongside each of the variables indicates the direction of causal relationship between the variable and the psychological phase under consideration. For example, if a family has had previous flood experience, its concern rate will be lower than that of a family without this experience. Therefore, an arrow pointing down is shown along the variable *flood experience*. Variables in italics are the policy variables, and

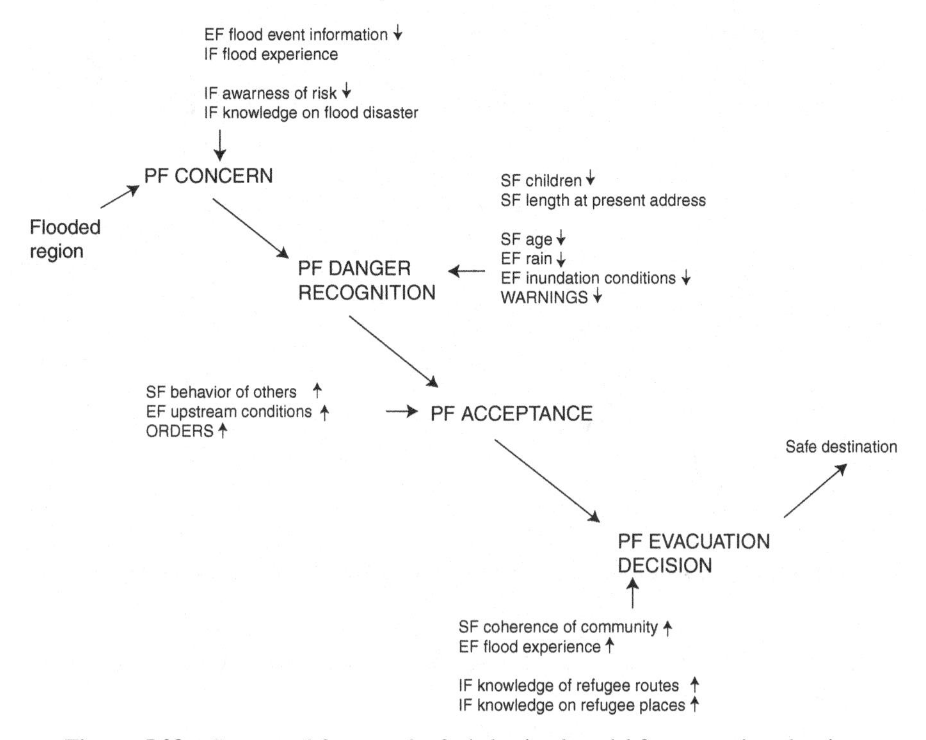

Figure 5.28 Conceptual framework of a behavioral model for evacuation planning.

can be changed by emergency managers. A detailed discussion of the links between the phases and the main groups of factors follows.

Model Variables: Initial Conditions and Social Factors This concerns social and demographic aspects of the population, such as income, age group, and daily life pattern, and includes attributes such as an inhabitant's experience of natural disasters, awareness of being at risk, and knowledge about disasters. Each family living in a disaster-prone area has a certain degree of disaster awareness and a life pattern of its own. This disaster and risk awareness forms a set of initial conditions for that family's behavior when the disaster hits the area. The behavioral patterns of the household are further affected by the information provided about the disaster and by physical parameters such as the intensity of the disaster and the size of the area affected. Initial conditions trigger a *concern*. On the basis of the data collected in the Red River basin, concern is higher if experience with flooding is missing, the sense of risk is high, and the event (precipitation, flood peak, water levels, etc.) is large. In the model, *concern* is defined as the first phase of the decision-making process, when an individual or family is aware of risk, and has basic information on the type of disaster and its impacts (Figure 5.28). This concern is always present, even when

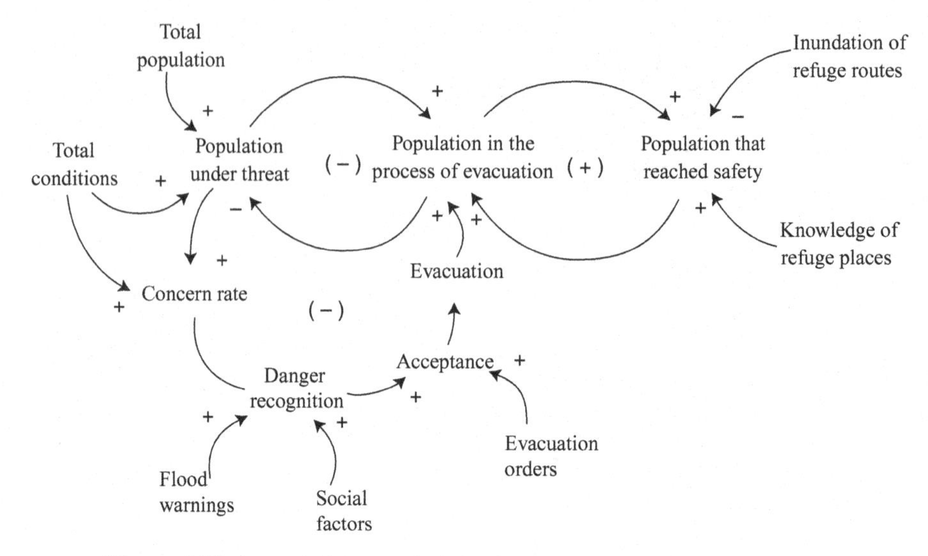

Figure 5.29 Causal diagram of a behavioral model for evacuation planning.

there is no imminent threat of a disaster. Initial conditions provide a background for an individual's perception of danger.

The process through which initial conditions affect an individual's behavior is complicated, and involves a chain of complex psychological reactions. The model considers two categories of households: those that have experienced disaster before and those that have not. For households with previous experience, it considers the extent of damage they experienced, whether they have evacuated in the past, whether they evacuated in the most recent disaster, the reasons for any decision not to evacuate, and damage to property. Information required from households with no disaster experience includes their knowledge about disasters in the area, the criteria they would use in order to decide whether evacuation is necessary, and awareness of the risk to their property. In the Red River basin, this information was obtained through personal interviews.

Depending on the severity of the situation *concern* may develop into *danger recognition*, defined as a new variable and calculated in the model using a different equation. There is a positive causal relationship between these two variables, which is shown in Figure 5.29. In this second phase, an individual or family is aware of imminent threat and is on alert.

Model Variables: External Factors The external factors that play a vital role in forming responses to disaster situations are information provided by the media, responsible emergency management authorities, and personal experience of the physical conditions. An evacuation order directly affects the acceptance and therefore the evacuation decision, as it initiates the action of evacuation itself (Figure 5.28).

Experience of weather conditions affects the decision more indirectly through *danger recognition*. Each household's *danger recognition* level changes over time, and affects not only the evacuation decision but also broader attitudes to the disaster. People with a high danger recognition rate try to get as much information as possible about a disaster event from all possible sources.

Model Variables: Psychological Factors Psychological factors are used in this model to represent all phases of the evacuation decision-making process. They cover the phases of *concern*, *danger recognition*, *acceptance*, and *evacuation decision*. An inhabitant's decision to evacuate is a result of external factors influencing the initial conditions. Social factors such as age and presence of dependants in the family (children or elderly people), in combination with external factors such as awareness of heavy rain and inundation conditions, give rise to *danger recognition*. Once the *danger recognition* rate reaches a certain threshold level, evacuation orders and the behavior of others can precipitate an *evacuation decision*. The reaction of each inhabitant to external factors differs. For example, there are households that do not evacuate in spite of receipt of an evacuation order, and there are those who evacuate even before an evacuation order is issued. To incorporate these behaviors in the model, a variable called *acceptance level* is introduced. This measures the extent to which a household accepts the danger. The *evacuation decision* results from the interaction between the *acceptance level* and the trigger information. An evacuation order and the behavior of other households are considered the trigger information in the model.

After a household decides to evacuate, it has to determine a place of refuge and a route to it. Its ability to reach safety depends on household members' knowledge of these things. This knowledge affects the behavior of the inhabitants after they decide to evacuate. Lack of knowledge will lose them valuable time in the evacuation process. The behavior of a family with little knowledge of a route may be affected by the behavior of other families with better knowledge.

Policy/Decision Variables The flood evacuation system dynamics model was developed to investigate different emergency policy/decision options. The two main sets of policy/decision variables concern flood warnings (media used for dissemination and consistency) and evacuation orders (media used for dissemination and timing). Warnings can be disseminated using the television, radio, mail, Internet and visits. For dissemination of evacuation orders only two options are considered, mail and visits to the household.

Model Structure The variables discussed above are interrelated through the model structure. The basic causal diagram (not including all the variables for simplicity of presentation) is shown as Figure 5.29. This identifies the main feedback forces that determine the behavior of the system captured with our model. There are two feedback loops in the center of the model connecting three stocks: population under threat, population in process of evacuation, and population that reached safety. The loop on the left-hand side represents negative feedback, and the loop on the right-hand side represents positive feedback. The reference behavior mode is S-shaped growth (see Section 5.2.2) (this was confirmed by the results discussed later). The growth of population that reaches safety is exponential at first, but then gradually slows with the decrease in the population under threat. The upper boundary value of the system state is also the goal, and equals the total population under threat that can be evacuated. Figure 5.29 also shows a negative feedback loop that links the psychological variables (*concern*, *danger recognition*, *acceptance*, and *evacuation decision*) with the main stocks.

Mathematical Relationships The data set used to develop this model is derived from a field survey of families that were evacuated during the 1997 flood. A questionnaire was administered to 52 households involving more than 200 respondents in 6 different community types. These communities represented a broad range of people affected by the 1997 flood, including:

- an urban community (Kingston Row and Crescent in Winnipeg),
- a rural community protected by a ring dike (St. Adolphe),
- a rural community without structural protection (Ste. Agathe),
- a suburban community (St. Norbert),
- an urban fringe community (Grande Pointe), and
- rural estates/farmers.

The survey was conducted less than 1 month after the flood, when many families were still in the process of recovery and under considerable stress. Both closed and open-ended questions were used. The data collected directly by the survey were verified through the process of public hearings organized by the International Joint Commission at five locations, on two occasions: immediately after the flood (autumn 1997) and before submission of the final report (spring 2000). There were more than 2000 participants in these hearings. Note that the relationships developed and used in the model apply only to the communities in south Manitoba. It is expected that major value systems captured by the survey will not change with time since the population in the flooded regions of the Red River basin tends to be stable in both size and characteristics.

The data collected were processed to establish different relationships among the variables in the evacuation model. For example, the relationship in Equation (5.15) describes the relative importance (weight) of each variable used for representation of the *concern* rate.

$$\begin{aligned} Concern = {} & (Awareness\_of\_Flood\_Disaster) \times 0.1 \\ & + (Previous\_Flood\_Experience) \times 0.7 \\ & + (Awareness\_of\_Risk) \times 0.2 \end{aligned} \tag{5.15}$$

The relationship derived from *danger recognition* is:

$$\begin{aligned} Danger\_Recognition = {} & (Age\_Factor) \times 0.05 + (Impact\_of\_Warning) \times 0.3 \\ & + (Concern) \times 0.3 + (Rain\_Factor) \times 0.1 \\ & + (Inundation\_Factor) \times 0.15 + (Children\_Factor) \times 0.05 \\ & + (Stay\_Factor) \times 0.05 \end{aligned} \tag{5.16}$$

This equation describes different variables involved in the calculation of *acceptance level*:

$$\begin{aligned} Acceptance\_Level = {} & (Danger\_Recognition) \times 0.2 + (Behavior\_of\_Others) \times 0.3 \\ & + (Order\_Impacts) \times 0.2 + (Flooding\_Factor) \times 0.3 \end{aligned} \tag{5.17}$$

Finally, the *evacuation decision* is expressed as a function of the *acceptance level*, previous experience with evacuation and disaster claims, and support available from the community where the family lives:

$$\begin{aligned} Decision = {} & (Acceptance\_Level) \times 0.7 + (Experience\_Factor) \\ & \times 0.2 + (Support\_Factor) \times 0.1 \end{aligned} \tag{5.18}$$

All the variables in Equations (5.15) through (5.18) are restricted to values between 0 and 1. Quantification of weights is done through the model calibration procedure. Data collected from the Manitoba Emergency Management Organization (MEMO) provided details on the evacuation process (length, timing, and number of people) and were compared with the outcome of the model simulations. The weights that generated model output that matched the observed data the best were selected and used in the model.

Other relationships between model variables require graphical description. These relationships are developed from the data collected through the field survey. For example, a graphical relationship for the *Flooding_Factor*, which is a function of *Upstream_Community_Flooded*, is shown in Figure 5.30. Relative values of the *Flooding_Factor* are between 0 and 1. The value of 0 indicates no upstream flooding information. The value of 1 indicates the full knowledge of the upstream flooding situation. Relative values of the *Upstream_Community_Flooded* are also between 0 and 1. They are derived from the survey data by calculating the relative ratio of people aware of upstream flooding, if it existed. The shape of the graph reflects the notion that the more knowledge about upstream flooding was available to people, the higher the attention that was given to this information in their process of making a decision about personal and family evacuation.

Delays and random number generation functions are used for describing different processes in the model. For example, reaching a place of refuge after evacuation is conditioned on the individual's knowledge of its location and inundation of access

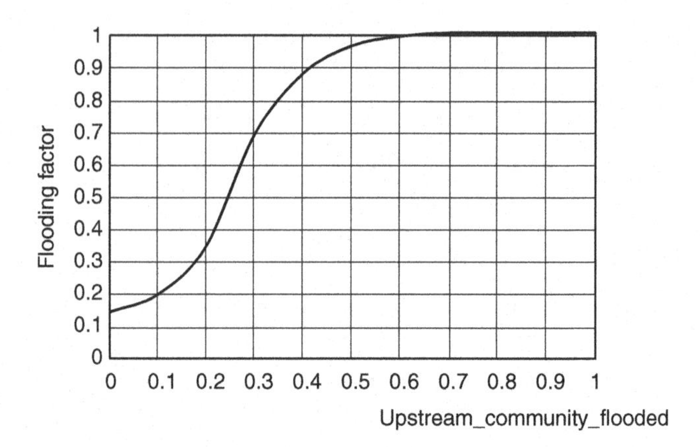

Figure 5.30 A graphical relationship between *Flooding_Factor* and *Upstream_Community_Flooded*.

routes. There are "information" and "material delays" involved in making a decision to evacuate and in reaching a place of refuge. An example of information delay is the time difference between the moment when a flood warning is issued and the time an individual takes to make an evacuation decision. An example of material delay shown in Equation (5.19) is the time required by people in the process of evacuation to reach the final destination. It is captured in the model in the following form:

$$\begin{aligned} \mathit{Reaching} = {} & \mathit{DELAY}((\mathit{Evacuation\_in\_Process}) \times (\mathit{Knowledge\_of\_Refuge\_Place}) \\ & \times 0.2 + (\mathit{Route\_Factor}) \times \mathit{0.8\ Random(1, 5, 5)}) \end{aligned} \tag{5.19}$$

There are three main stocks in the model: *Population*, *Evacuation_in_Process*, and *Reached_the_Destination*. The *Population* stock represents the total number of households in the area under threat (52 families with more than 200 individuals in seven different communities in the Red River basin). The outflow from this stock is the number of families that decide to evacuate. The population stock in mathematical form is expressed as:

$$\frac{\mathrm{d}(\mathit{Population}(t))}{\mathrm{d}t} = -\mathit{Evacuating}(t) \tag{5.20}$$

The second important stock in the model is *Evacuation_in_Process*. This stock represents the difference between the number of families that have decided to evacuate and the number of families that have reached a refuge:

$$\frac{\mathrm{d}(\mathrm{d}\mathit{Evacuation\_in\_Process}(t))}{\mathrm{d}t} = -\mathit{Evacuating}(t) - \mathit{Reaching} \tag{5.21}$$

The third stock accumulates the number of families that have reached safety:

$$\frac{\mathrm{d}(Reached\_the\_Destination(t))}{\mathrm{d}t} = -Reaching(t) \tag{5.22}$$

The flood evacuation system dynamics model was developed in four sectors that are linked together: initial conditions, social factors, psychological factors, and external factors. After mapping each sector and defining connections between different sectors, decision rules are developed and incorporated in the model using logical statements with an IF-THEN-ELSE structure. The following rule states that if the threshold value for an evacuation decision is less than or equal to, for example, 0.65, then there will be no evacuation; otherwise there will be an evacuation:

$$\text{IF Decision} \leq 0.65 \text{ THEN 0 ELSE1} \tag{5.23}$$

The evacuation model depends on the input from other models. For example, the information on dynamically varying water levels is provided by a hydrodynamic model, and a geographic information system (G1S) provides data on the spatial location of each community, its distance from the river, and its relative location to other communities (upstream or downstream).

Model Use The model interface shown in Figure 5.31 is the main control window for the use and navigation of the evacuation model. The model is developed using Stella system dynamics environment (http://www.iseesystems.com/, last accessed January 2010) that has similar characteristics as Vensim. The main use for this model is to develop different scenarios for assessing the impact of different emergency evacuation policy options. An appropriate interface is required to provide the user with an easy process of scenario development and assessment. The upper section of the interface window provides an introduction to the model and may be browsed by scrolling the text within the window. In order to use the model properly, policy selection is required. Switches and sliders are used to set the value for different variables. All sliders offer the choice of a value between 0 and 1. Zero always indicates the lowest level of importance with 1 always indicating the highest level. The model is ready for simulation when all values are selected. A simulation run is started by clicking on the "Run Model" button.

The graph window on the interface shows in real time the results of model calculations by redrawing the two lines (time series) shown (1 and 2). Completion of the graph indicates the end of the simulation. A line (numbered 2) shows the number of families (out of 50 stored in the model database) that have reached a refuge. Line number 1 shows the number of families on the way. Both variables are shown as functions of time. The total simulation horizon is 96 hours, or 4 days. The shape of these two lines is a function of the policy selected, and encompasses the warning distribution, the evacuation orders distribution, characteristics of the community, awareness of incoming flood, and the weights given by community members to different warning and evacuation order distribution modes.

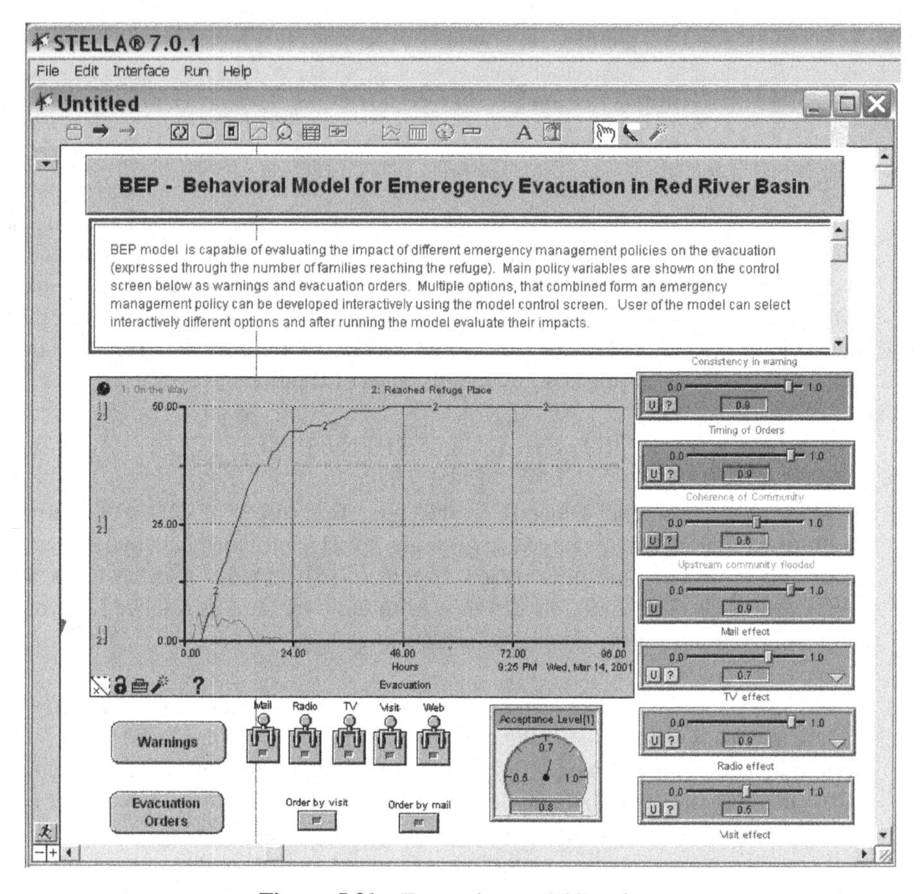

Figure 5.31 Evacuation model interface.

5.5.4 Application of the Evacuation Model to the Analyses of Flood Emergency Procedures in the Red River Basin, Manitoba, Canada

The main purpose of the model is to allow for the different policy options available to flood emergency managers to be evaluated before an emergency situation occurs. In this study, the different policy choices related to the evacuation warning dissemination in particular are investigated using the model. To demonstrate the utility of the model, a set of experiments is designed for testing the efficiency of evacuation procedures (measured in number of hours required for the evacuation of all 50 families) in the Red River Basin, Manitoba, Canada. Methodology used for testing the efficiency of flood evacuation procedures is sufficiently general to be applied to other types of disasters.

The following is the description of rigorous procedure used in testing the efficiency of evacuation procedures by running the flood evacuation system dynamics simulation model:

1. Identify policy variables.
2. Test the sensitivity of simulation model results by changing the value of different policy variables.
   (a) Determine the order of testing.
   (b) Identify range for each policy variable.
   (c) Make a record of each simulation.
3. Compare the sensitivity results.
   (a) Order all policy variables according to their impact on the simulation results.
   (b) Measure the impact of each policy variable.
4. Develop analysis scenarios.
   (a) Use the results of sensitivity analysis to develop extreme scenarios.
   (b) Develop a middle scenario.
5. Analyze simulation results for each scenario.
6. Present your results to decision makers.

Selection of Policy Variables Policy variables selected during the development of the model are divided into three different groups. The first group includes binary variables that describe two main activities preceding the flood evacuation: (a) *warning method* and (b) *mode of evacuation order dissemination*. Review of the MEMO procedures and a set of public meetings used to evaluate the data needs in the Red River Basin (Science Applications International Corp., 1999) prompted the following choice of policy variables that describe the flood warning method in the Red River Basin: *Mail, Radio, TV, Visit*, and *Web*. Two variables are selected to describe the possible mode of flood evacuation order dissemination: *Order by visit* and *Order by mail*. All these variables are of a binary nature. They can be used or not. Therefore, a switch representation is used in the model to allow user to select or deselect these variables.

The second group of policy variables is selected to describe local triggers of human behavior in the case of flood emergency in the Red River Basin. These variables are *Warning consistency*, *Timing of the order*, *Coherence of the community*, and *Flooding of upstream community*. These variables may take different values and for all of them a range from 0 to 1 is used. A value of 0 indicates a low level and a value of 1 indicates a high level. *Warning consistency* describes how much the flood warning information changes over time. The basic source of this information (time of peak, maximum water level, and duration of peak) is the Water Resources Branch of Manitoba Conservation. The value is determined from the comparison of warnings provided at different time, and the content of the warning information. *Timing of order* describes the moment when the evacuation orders are distributed to the public. The source of this information is the MEMO. Timing is measured in number of days before the flood peak and usually affects the individual effort invested in temporary protection of the property. Early orders have negative impact on the local protection effort (people do not have sufficient time to build, e.g., temporary dikes) and late orders may result in high risk to the residents under the threat. *Coherence of the community* describes the connections existing between individual members of the community. More coherent communities function more efficiently during the emergency (people helping each other). The final variable, *Flooding of upstream community*, describes the availability of information on the upstream conditions. It has been shown that this information plays an important role in individual decision making by providing the information to assess the personal level of risk and time available to make an appropriate decision. For the purpose of this study, all the data available for the flood of 1997 were collected from the MEMO and the Water Resources Branch of Manitoba Conservation.

The third group of policy variables describes the importance of different flood warning modes. Following variables are used: *Mail effect, TV effect, Radio effect, Visit effect*, and *Web effect*. Since these variables are used to indicate the weight given to each of flood warning policy choices, a scale from 0 to 1 is used again. A value of 0 indicates the lowest weight and a value of 1 the highest. Most of the data from this group of policy variables were collected for the flood of 1997 during the two public meetings used to evaluate the data needs in the Red River Basin (Science Applications International Corp., 1999).

Sensitivity Analyses of Flood Evacuation Strategies Six sensitivity experiments are performed to evaluate the impact of four main variables from the second group and two variables from the third group. Sensitivity is not performed on the variables from the first group since they describe real flood evacuation warning and order dissemination options. However, later in this section, they are used extensively in developing major scenarios for flood evacuation. Only two variables are selected from the third group after it was detected that the final simulation results are not very sensitive to the change of values in variables from this group.

Sensitivity to "Warning consistency" Results of the survey performed by Morris-Oswald and Simonović (1997) immediately after the flood of 1997 indicated that consistency in warning has been an issue of concern for most of the residents in the Red River valley affected by the flood. Therefore, this variable has been incorporated in the model structure playing an important role in the determination of danger recognition. A sensitivity test has been done with all of the warning and order dissemination modes used. Five simulation runs are performed simultaneously, and a comparative graph of final results is shown in Figure 5.32 where the values on the vertical axis represent the number of families. Line 1 corresponds to *Warning consistency* of 0 (inconsistent warnings) and line 5 corresponds to *Warning consistency* of 1 (very consistent warnings).

Two main observations are made from this analysis: (a) more consistent warnings increase considerably the efficiency of flood evacuation in the Red River Basin. Increase in efficiency, measured in the number of days necessary to reach the refuge, reaches 100%. In other words, the time necessary to move the residents to safety can be cut in half with the consistent warning system; and (b) when the *Warning consistency* increases above 0.5 (on the scale from 0 to 1), very minor improvement is observed in the flood evacuation efficiency.

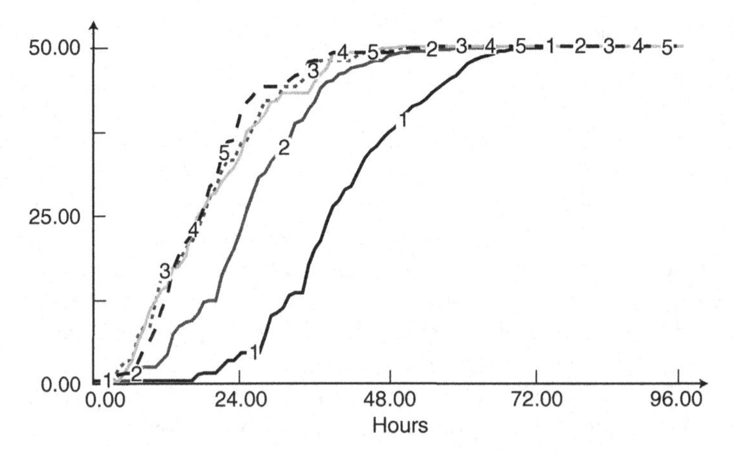

Figure 5.32 Sensitivity analysis of *Warning consistency*.

Sensitivity to "Timing of orders" Timing of evacuation orders is established to contribute to the evacuation order acceptance level. Sensitivity to this policy variable has been performed in the same way as for the *Warning consistency*. Five simulation experiments are performed simultaneously changing the values of this variable from 0 to 1. Figure 5.33 presents the result of sensitivity analysis (line 1 corresponds to late ordering of mandatory evacuation and line 5 corresponds to timely ordering of evacuation).

The following observations can be made from the sensitivity results: (a) *Timing of orders* is a very important policy variable. If the evacuation is not ordered on time some families will not be able to reach the safe place; (b) careful timing of the evacuation order may increase evacuation efficiency up to four times; and (c) when

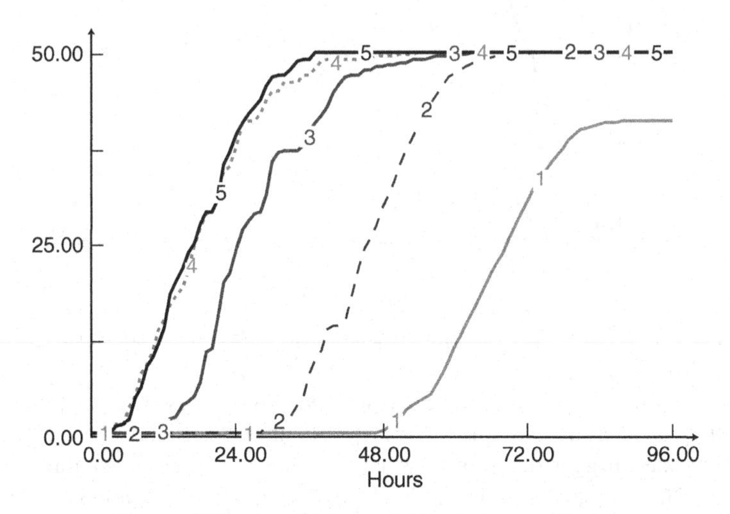

Figure 5.33 Sensitivity analysis of *Timing of order*.

the relative timing is above 0.75 (on the scale from 0 to 1) there is no improvement in evacuation efficiency.

Sensitivity to "Coherence of community" Study by Morris-Oswald and Simonović (1997) captures an important characteristic of human behavior in emergencies, possibly specific only for the Red River Basin. In more coherent communities, people do make decisions together and help each other much more efficiently than in less coherent communities. Example from the Red River Basin is observed with Mennonite communities that were very well organized during the flood emergency of 1997. Therefore, a sensitivity analysis of this variable is performed, having in mind that changing the coherence of the community will not be possible, but can be used into consideration when preparing for the emergency.

Five simultaneous simulation runs are performed as in the previous two cases. A final result is shown in Figure 5.34. Simple observations from this analysis are (a) community coherence affects evacuation efficiency very strongly. Incoherent communities (example for the value of 0 represented with line 1) may not succeed in the evacuation of all families to safety on time; (b) more coherent communities can be evacuated two to three times more efficiently than incoherent communities; and (c) when the coherence of the community reaches 0.75 (on the relative scale from 0 to 1) maximum efficiency of evacuation is already achieved.

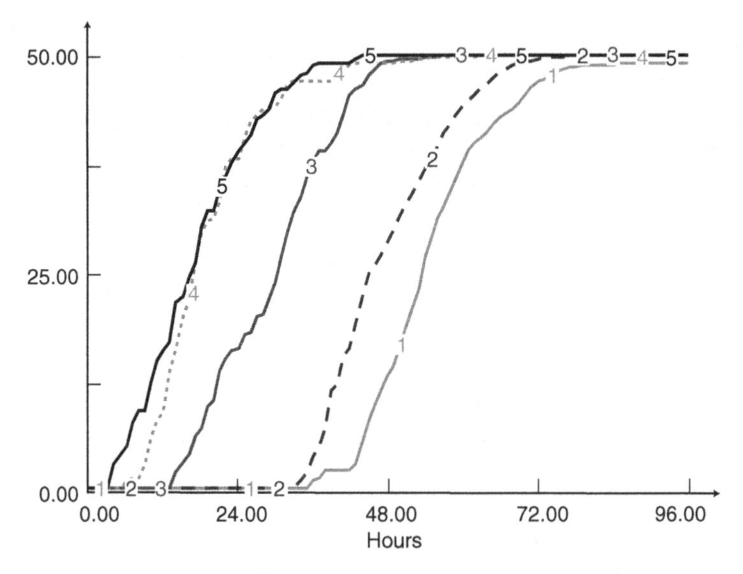

Figure 5.34 Sensitivity analysis of *Coherence of community*.

Sensitivity to "Upstream community flooding" Awareness of upstream community flooding is identified as one of the factors that determine evacuation order acceptance and trigger decision to evacuate from the place under threat. In today's world of fast communications, it can be expected that information on what is going on upstream will be available and used in making personal evacuation decisions.

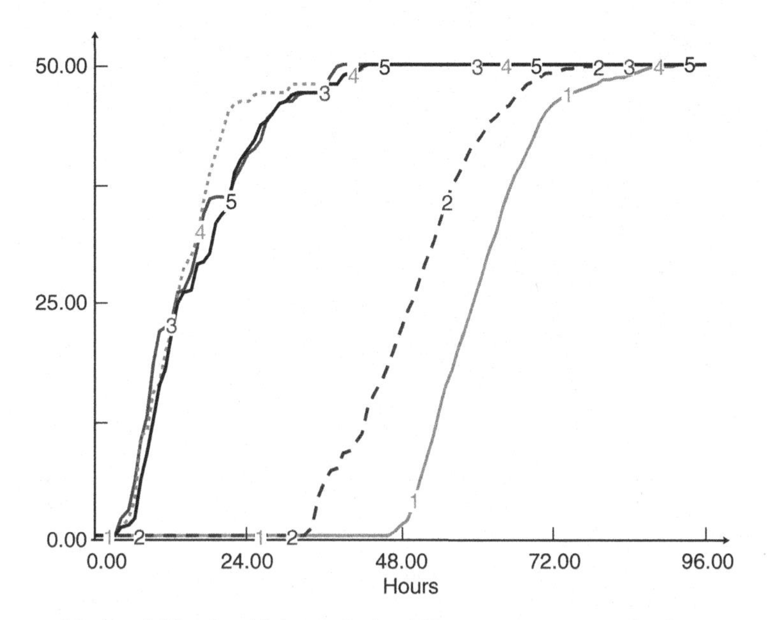

Figure 5.35 Sensitivity analysis of *Upstream community flooding*.

Impact of this variable is tested through the sensitivity analysis similar to the one above. Five simulations are performed simultaneously and the final result is shown in Figure 5.35.

The following are the observations that can be made from this analysis: (a) there is obvious grouping of results. If the value of *Upstream community flooding* exceeds 0.25 (on the relative scale between 0 and 1), efficiency of flood evacuation is improved three to four times; and (b) for the values above 0.25, the efficiency of evacuation cannot be increased any more.

Sensitivity to "Mail effect" A third group of policy variables includes weights that can be associated with different modes of flood warning or evacuation order distribution. These relative weights combined with impacts of warning affect danger recognition. During the development of the model it has been identified that all policy variables from the third group play less important role in determining human behavior during emergency.

Sensitivity analysis has been carried out in the same way as in the previous cases. Five simultaneous simulation runs are performed for the value of *Mail effect* between 0 and 1. Results of the sensitivity analysis are shown in Figure 5.36.

The main observation inferred from this result is variation in weight associated with *Mail effect* is not affecting the evacuation process to the great extent. Efficiency of evacuation can be improved by associating higher weight with a particular flood-warning mode. The largest observed increase in the efficiency is in the range of 30%.

Sensitivity to "Visit effect" In a similar way, the *Visit effect* is participating in the process of determining the dynamics of human behavior during an emergency. Testing

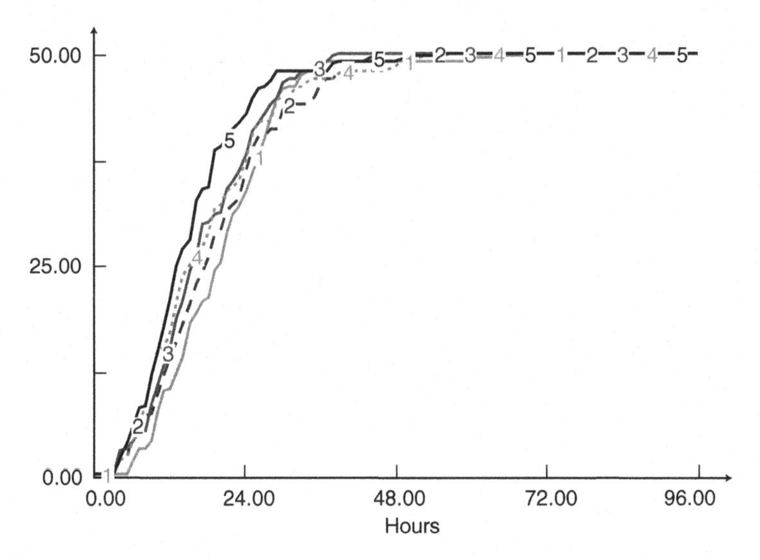

Figure 5.36 Sensitivity analysis of *Mail effect*.

of the sensitivity of the model performance to this variable is done as above and the results are shown in Figure 5.37.

Similar observations can be made from this figure: (a) variation in weight associated with *Visit effect* is not affecting the evacuation process to the great extent. The efficiency of evacuation can be improved by associating higher weight with the *Visit effect*; (b) the increase in efficiency is more prominent in the later stages of the evacuation process; and (c) the maximum improvement in efficiency is between 40% and 50%.

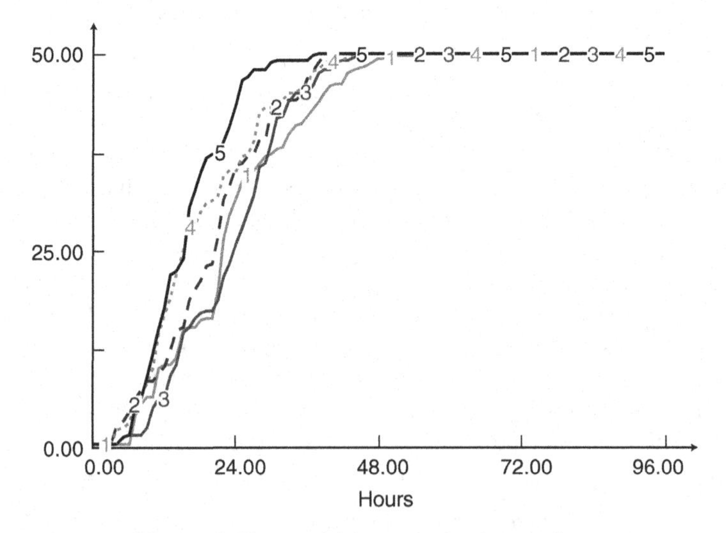

Figure 5.37 Sensitivity analysis of *Visit effect*.

Comparison of sensitivity results Sensitivity analyses conducted in this research revealed a number of important issues for emergency managers in the Red River Basin.

(i) *Timing of evacuation orders* is the most important variable that affects human behavior during the flood emergency in the Red River Basin. Therefore, extreme care should be given to proper forecasting of flood peak and the establishment of time when the evacuation order is issued. The second most important factor is the *Warning consistency*. Evacuation efficiency can be significantly improved by maintaining a high level of warning consistency.

(ii) Emergency managers can benefit from prior knowledge of *community coherence*. It is shown that the more coherent communities are much more efficient in dealing with the emergency.

(iii) *Awareness of upstream flooding* seems to be a motivating force for making a personal decision about evacuation. Insuring that the residents of the community under threat are informed about upstream emergency situation on time may make the evacuation process more efficient and the job of emergency managers much easier.

(iv) Determining the weight that residents are associating with different modes of flood warning can be used in order to make the evacuation process more efficient.

Flood Evacuation Scenarios The results of sensitivity analyses and survey conducted by Morris-Oswald and Simonović (1997) together with private communications with MEMO are used to demonstrate the utility of the model by developing four different scenarios and then comparing the results of model simulations. Scenario-based analysis is the main mode of utilization of the model. Therefore, this exercise should be considered more as a demonstration than the real use of the model. It is expected that this model will be used by emergency managers with some knowledge of the emergency type and region under the threat. Their experience and intuition should be used in developing scenarios and interpreting the results of model simulations.

Four scenarios created for model use demonstration are:

1. *MEMO scenario*: In this scenario, a set of policy variables is selected based on the MEMO operation during the 1997 flood. *Mail, Radio*, and *TV* flood warning modes are used and *Order by visit* mode of dissemination. Assumptions of high *Warning consistency* (0.9) and effective *Timing of order* (0.9) are made. Realistic values of *Coherence of the community* of 0.6 and *Flooding of upstream community* of 0.5 are used in this scenario. From the consultation with residents of the Red River Basin, a choice of weights is selected as *Mail effect* of 0.7; *TV effect* of 0.9; and *Radio effect* of 0.5.
2. *RESIDENTS scenario*: Flood consultations in the basin revealed that residents of the region had a different prospective of the evacuation process. This scenario is attempting to capture the view of residents. The main difference from the

MEMO scenario was identified in *Warning consistency* (0.5) and *Timing of the order* (0.4). The rest of the policy variables in this scenario are the same as in the MEMO scenario.

3. *GOOD scenario*: In this scenario, an attempt is made to demonstrate the value of using all modes of flood warning (*Mail, Radio, TV, Web*, and *Visit*) and both ways of disseminating evacuation orders (*Order by visit* and *Order by mail*). High level of *Warning consistency* (0.9) and *Timing of order* (0.9) are introduced, and realistic values of *Coherence of the community* of 0.6 and *Flooding of upstream community* of 0.5 are used. From the consultation with residents of the Red River Basin, a choice of weights is selected as *Mail effect* of 0.7; *TV effect* of 0.9; *Radio effect* of 0.5; *Visit effect* of 0.9; and *Web effect* of 0.9.
4. *BEST scenario*: In this scenario, all the variables are selected at the same level as in the GOOD scenario except for the *Flooding of upstream community*. A value of 0.9 is used in this scenario to demonstrate an opportunity for improving efficiency by providing timely information that, according to the sensitivity analysis, plays an important role in determining the danger recognition rate.

Simulation of four selected scenarios has been performed using the model, and final results are shown in Figure 5.38. The same graph is generated for all four scenarios. It presents a number of families that reached the refuge (line 2) and a number of families in the process of evacuation (line 1).

Visual comparison of four graphs is showing a considerable difference between scenarios in the evacuation starting time, evacuation efficiency (difference between the starting and ending time), and evacuation speed (slope of the line 2).

BEST scenario is obviously the most effective one and the RESIDENTS scenario is the most inefficient. Evacuation starting time is between twenty-eighth hour (RESIDENTS scenario) and fifth hour (BEST and GOOD scenarios). In the case of the least effective scenario all 50 families are in safety after 84 hours (RESIDENTS scenario) and in the case of the most efficient scenario after 47 hours (BEST scenario). Conditions created in the BEST and GOOD scenarios are conductive to the higher acceptance level and danger recognition rate. Therefore, the reaction is fast and the evacuation process is very efficient. Acceptance level calculated by the model for these two scenarios is at 0.8 level (at relative scale between 0 and 1). On the other side, MEMO and RESIDENTS scenarios offer the possibility for improvement. It seems that the acceptance level, calculated to be at 0.6 on the relative scale, causes the late reactions of the population. The late start and the slow process did not affect the final outcome. All of the families are evacuated to safety.

This evacuation result was expected since the data from Red River Basin were for the mandatory evacuation. However, the insight provided by the model simulation offers assistance to emergency managers. Key policy variables are identified and their impact is evaluated. Future emergency situations can be simulated and their impacts easily evaluated by using the model.

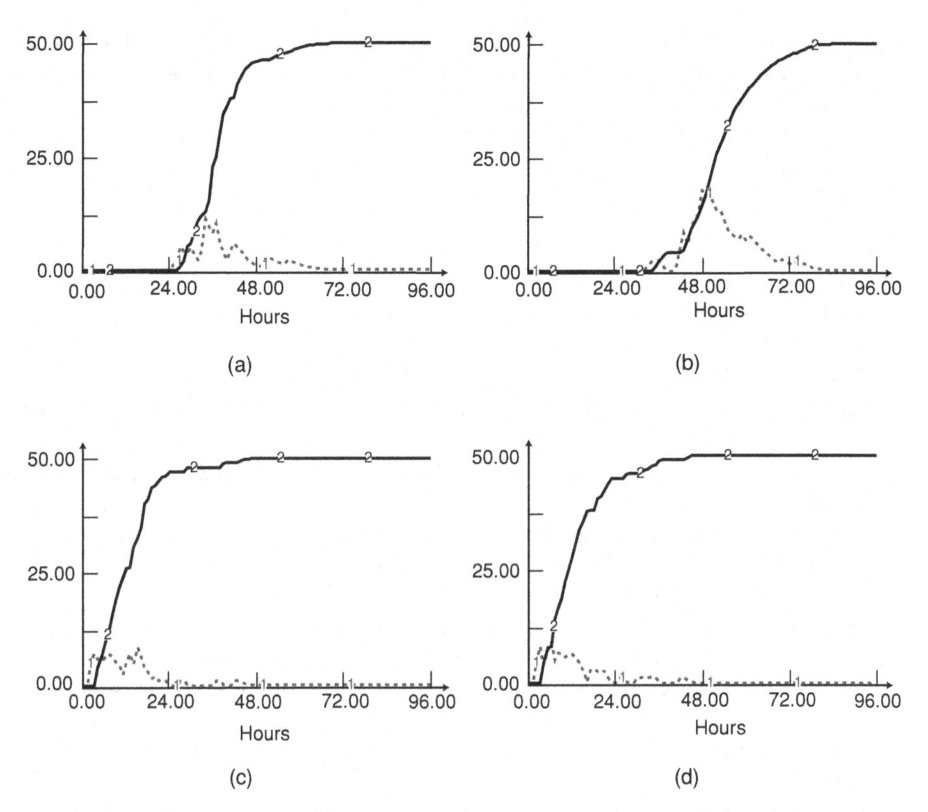

Figure 5.38 Model simulation results for different scenarios: (a) MEMO scenario, (b) RESIDENTS scenario, (c) GOOD scenario, and (d) BEST scenario.

5.5.5 Conclusions

The system dynamics model is shown to be capable of simulating the effect of different flood evacuation policies. The main advantage of using the system dynamics approach for modeling human behavior during emergency situations is that by understanding how a particular structure of feedback loops is capable of generating the observed behavior we get insights into potential solutions. Through the use of the model a number of "what if" questions can be asked and answers to them provided. Sensitivity analyses performed in this research and documented in this section provide for better understanding of the importance of different factors affecting the evacuation process. A numerous alternative management options were tested by developing simulation scenarios. In this way the model can guide emergency managers through most optimistic, most pessimistic, and in-between scenarios. The flood evacuation model is available for use by emergency managers directly, and it is expected that it can lead to higher quality of decisions and a higher level of emergency preparedness. The ability to capture specific characteristics of the evacuation process during the flood emergency and to answer questions makes this model a powerful planning and analysis tool aimed at preventing loss of life and the minimization of material damage. The model can be fine-tuned easily in the light of experience, or with the help of insight provided by an expert.

The original model has been converted into the Vensim software (without the user interface) and it is included on the CD-ROM, in directory SYSTEMDYNAMICS, subdirectory Examples, Example 6.

REFERENCES

Atkinson, K. (1985), *Elementary Numerical Analysis*, John Wiley, New York.

Ford, A. and H. Flynn (2005), "Statistical screening of system dynamics models", *System Dynamics Review*, 21(4):273–303.

Forrester, J.W. (1990), *Principles of Systems*, Productivity Press, Portland, OR, first published 1968.

Kermack, W. and A. McKendrick (1927), "Contributions to the mathematical theory of epidemics", *Proceedings of the Royal Society*, 115A:700–721. Reprinted, *Bulletin of Mathematical Biology* (1991), 53(1–2):33–55.

Morris-Oswald, M. and S.P. Simonović (1997), "Assessment of the social impact of flooding for use in flood management in the Red River Basin", *Report Prepared for the International Joint Commission*, University of Manitoba, Winnipeg, p. 73 (available from the IJC, www.ijc.org).

Science Applications International Corp. (1999), *Information/data Needs for Floodplain Management: The Red River Basin Workshop*, McLean, Virginia, p. 39 (available from the IJC, www.ijc.org).

Senge, P.M. (1990), *Fifth Discipline—The Art and Practice of the Learning Organization*, Doubleday, New York.

Simonović, S.P. (2009), *Managing Water Resources: Methods and Tools for a Systems Approach*, UNESCO Publishing, Paris and Earthscan Publishing, London.

Simonović, S.P. and S. Ahmad (2005), "Computer-based model for flood evacuation emergency planning," *Natural Hazards*, 34(1):25–51.

Sterman, J.D. (2000), *Business Dynamics: Systems Thinking and Modelling for a complex World*, McGraw Hill, New York.

Ventana Systems (1995), *Vensim User's Guide*, Ventana Systems, Inc., Belmont, MA.

Ventana Systems (2003a), *Vensim 5 Reference Manual*, Ventana Systems, Inc., Belmont, MA.

Ventana Systems (2003b), *Vensim 5 Modelling Guide*, Ventana Systems, Inc., Belmont, MA.

EXERCISES

5.1 In the famous excerpts below from Thomas Malthus' *First Essay on Population* (1798), the author implicitly describes feedback loops that influence the dynamics of population. Draw a causal diagram to show his feedback thinking. The first paragraph sets the stage; it is the second and third paragraphs that should be diagrammed:

> Population, when unchecked, increases in a geometrical ratio. Subsistence increases only in an arithmetical ratio . . .
>
> By that law of our nature which makes food necessary to the life of man, the effects of these two unequal powers must be kept equal. This implies a strong and constantly operating check on population from the difficulty of subsistence. This difficulty must fall somewhere; and must necessarily be severely felt by a large portion of mankind . . .
>
> Population, could it be supplied with food, would go on with unexhausted vigour, and the increase of one period would furnish the power of a greater increase the next, and this without any limit . . .
>
> Foresight of the difficulties attending the rearing of a family acts as a preventative check [acting on the birth rate]; and the actual distress of some of the lower classes, by which they are disabled from giving the proper food and attention to their children, acts as a positive check [acting on the death rate], to the natural increase of population.

You might choose to include in your causal diagram population, births, deaths, preventative checks, positive checks, food adequacy, and food.

5.2 The volunteering team is losing motivation and you are asked to help find the best way to generate maximum effort from the people. Your starting point is the diagram below. One group within your team argues that the greater the performance shortfall, the greater the motivation of team members will be. They argue that the secret of motivation is to set aggressive, even impossible goals to elicit maximum motivation and effort. The second group argues that the link to team motivation should be negative—a big performance shortfall simply causes frustration as team members conclude there is no chance to accomplish the goal.

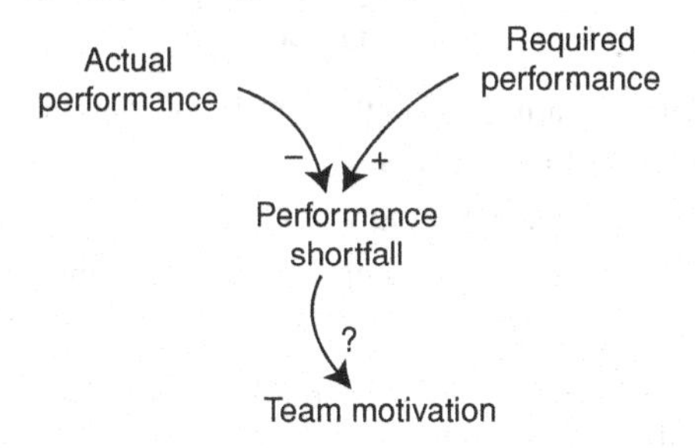

Expand the diagram to resolve the conflict between these two pathways? (Include both pathways in your diagram.)

(a) Which links dominate in each pathway?

(b) How can you tell which pathway is likely to dominate in any situation?

(c) How does team motivation feed back to performance, and how might actual performance affect the goal?

(d) Indicate these loops in your diagram and explain their importance?

5.3 Present a causal loop of your own (from your own thinking or from an article, or book of some sort). Explain it sufficiently in words so that your picture and the story it tries to tell are clear.

5.4 Identify at least one stock within the disaster management activity that your actions help to fill or deplete.

5.5 You have learned about the basic building blocks (objects) of the modeling tool Vensim that are used to develop system dynamics simulation models. With the knowledge you have gained so far, define the objects Auxiliary variable and Arrow in your own words. Comment on their use in Vensim models.

5.6 In each of the following groups, identify a stock and one or more related flows. Some of the words represent concepts that are not connected stocks or flows—they are just information in the system. Show in a causal diagram how you think those other concepts in the group are related to the stock and flow sets you identify. (Suggestion: do not add any more concepts to these lists.)

(a) Pipes, pipeline construction, water flow.

(b) Births, deaths, population, fecundity, life span.

(c) Knowledge, learning, forgetting, intelligence.

(d) Deficit, debt, income, spending, interest payments on debt, payments on debt principle.

5.7 Consider the causal diagram in Exercise 5.2. Convert the diagram into the stock and flow diagram. Identify the main stocks and flows.

5.8 Consider a stock with a single inflow rate R_1 and outflow rate R_2. Draw the behavior of the stock given the two sets of flow and inflow rates in diagram below. The initial volume of the stock is 100 units in both cases. Do not use a computer. This exercise should help you to develop intuition about stocks and flows, and the ability to relate their behavior.

5.9 Use the epidemic model in Example 5 on the CD-ROM. Simulate the model under various combinations of parameters.

(a) What determines whether an epidemic will occur?

(b) What determines the fraction of the population remaining uninfected in equilibrium? Why?

5.10 Use the flood evacuation model in Example 6 on the CD-ROM. Develop your own scenarios and simulate the model.

(a) What is the dominant loop in the model?

(b) What is the dominant parameter in your scenario?

(c) How does the performance of the model for your scenario compare to the four scenarios presented in Section 5.5.4? Why?

6 Optimization

The procedure of selecting the set of decision variables that maximizes/minimizes the objective function, subject to the systems constraints, is called the *optimization procedure*. Numerous optimization techniques are used in disaster management as reviewed in Section 4.3. One of the most significant contributions of optimization to the solution of wide variety of management problems was the introduction of linear programming (LP) in the late 1950s (Dantzig, 1963). LP is applied to problems that are formulated in terms of separable objective functions and linear constraints.

In this book, LP will be presented as one of the optimization methods. It is presented for its academic and practical significance. Various disaster management problems can be very efficiently addressed through the use of LP. They include the general transportation, allocation of resources, scheduling, among other problems. This Chapter will introduce the general LP problem and an algorithm for its solution. In addition, it will include a set of special LP applications and algorithms for their solution. The theoretical presentation in this chapter draws from Hillier and Lieberman (1990). All disaster management examples are carefully created to be sufficiently simple, demonstrate the optimization problem formulation, efficiency of various types of LP algorithms, and potential for implementation of LP in disaster management practice.

Disaster management literature offers a large scope for the application of LP. It is usually used with a fair level of complexity that is imposed by the application of the tool in real conditions. Some interesting examples from the literature are presented below.

Bryson et al. (2002) present an application to disaster recovery planning (DRP). DRP or disaster recovery strategy (DRS) is a system for internal control and security that focuses on quick restoration of service for critical organizational processes when there are operational failures due to natural or man-made disasters. A DRP aims to minimize potential loss by identifying, prioritizing, and safeguarding those organizational assets that are most valuable and that need the most protection. It is interesting that the field of DRP has attracted the interests of system analysis practitioners and vendors. However, very little formal management science research, education, and training has been done in this area (Sarker et al., 1996). It appears that

Systems Approach to Management of Disasters: Methods and Applications, By Slobodan P. Simonović

while much attention has been paid by vendors and organization to the development of hardware and software tools for addressing specific aspects of a disaster, little attention has been paid to formal training in systems modeling of the DRP process. The scarcity of system optimization modeling applications in practice shows quite clearly that the benefit of systems optimization methodology has not yet been brought to bear on this emerging discipline. Recently, Jenkins (2000) has explored the use of LP (integer) technique for selecting disaster scenarios. Bryson et al. (2002) present a model that provides support for decision making. They introduce the application of LP to DRP at the predisaster planning/development phase. They use a variation of LP known as mixed-integer mathematical decision model for selecting subplans for a DRP. The objective of the model is to maximize the total value of the coverage provided by the set of selected subplans subject to several constraints. The model that contains, real, binary, and general integer variables can be solved using a variety of methods, LP being one of them.

Ozdamar et al. (2004) present an application of optimization to logistics planning. Logistics planning in emergency situations involves dispatching commodities (e.g., medical materials and personnel, specialized rescue equipment and rescue teams, and food) to distribution centers in affected areas as soon as possible so that relief operations are accelerated. In this study, a planning model is developed that addresses the dynamic time-dependent transportation problem that needs to be solved repetitively at given time intervals during ongoing aid delivery. The model regenerates plans, incorporating new requests for aid materials, new supplies, and transportation means that become available during the optimization period. The plan indicates the optimal mixed pickup and delivery schedules for vehicles within the considered planning time horizon as well as the optimal quantities and types of loads picked up and delivered on these routes. In emergency logistics context, supply is available in limited quantities at the current time period and on specified future dates. Commodity demand is known with certainty at the current date, but can be forecasted for future dates. Unlike commercial environments, vehicles do not have to return to depots, because the next time the plan is regenerated, a node receiving commodities may become a depot or a former depot may have no supplies at all. As a result, there are no closed loop tours, and vehicles wait at their last stop until they receive the next order from the logistics coordination center. Hence, dispatch orders for vehicles consist of sets of "broken" routes that are generated in response to time-dependent supply/demand. The mathematical optimization model describes a setting that is considerably different than the conventional vehicle routing problem. In fact, the problem is a hybrid that integrates the multicommodity network flow problem and the vehicle routing problem. In this setting, vehicles are also treated as commodities. The model is readily decomposed into two multicommodity network flow problems, the first one being linear (for conventional commodities) and the second integer (for vehicle flows). The model and the solution methodology are implemented on a scenario based on the data for the Izmit earthquake in Turkey that took place on August 17, 1999.

Balcik and Beamon (2008) present an optimization model for facility location decisions for a humanitarian relief chain responding to quick-onset disasters. Their work was preceded by numerous studies in disaster relief logistics with focus on operational logistical activities in the relief chain and the objective of optimizing the flow of supplies through existing distribution networks. Knott (1987) considers the last mile delivery of food items from a distribution center to a number of refugee camps, assuming a single mode of transportation that makes direct deliveries to camps. The author develops an LP model to determine the number of trips to each camp to satisfy demand while minimizing the transportation cost or maximizing the amount of food delivered. Haghani and Oh (1996) determine detailed routing and scheduling plans for multiple transportation modes carrying various commodities from multiple supply points in a disaster relief operation. They assume that the commodity quantities are known and formulate a multicommodity, multimodal network flow problem with time windows as a large-scale LP model on a time-space network with the objective of minimizing the sum of the vehicular flow costs, commodity flow costs, supply/demand carry-over costs, and transfer costs over all time periods. Angelis et al. (2007) consider a multidepot, multivehicle routing, and scheduling problem for air delivery of emergency supply. Planes deliver full cargo to single clients from the warehouses in port cities. The authors set a service level for food distribution and develop a linear integer—programming model that maximizes the total satisfied demand. Based on the previous work, Balcik and Beamon (2008) present an optimization model that determines the number and locations of distribution centers in a relief network and the amount of relief supplies to be stocked at each distribution center to meet the needs of people affected by the disasters. Optimization model presented in the paper is a variation of the maximal covering location model and integrates (a) facility location and inventory decisions, (b) considers multiple item types, and (c) captures budgetary constraints and capacity restrictions. The experiments presented illustrate how the proposed model works on a realistic problem. Results show the effects of pre- and postdisaster relief funding on relief system's performance, specifically on response time and the proportion of demand satisfied.

Existence of highly sophisticated and complex optimization models for a variety of disaster management problems in the literature did not yet affect the practice to the expected level. To bridge the gap between research and practice requires full-scale introduction of management science to the disaster management community.

My personal experience is that the simpler models developed by the users are much more appreciated in practice than very complex models that require presence of an expert in order to be applied. This is why I am approaching optimization in this book from a very basic level. My intention is to reach disaster practitioners without experience in systems theory or operations research. I have selected an LP as the most powerful starting point for the development and application of more complex optimization models. I hope that theoretical explanations, the LP computer program that is provided on the accompanied CD-ROM, and practical examples presented in the following sections of this chapter will (a) help increase the understanding of the power of optimization in disaster management, (b) provide experience in optimization problem formulation and solution, and (c) help bridge the gap between theory and practice.

6.1 LINEAR PROGRAMMING

LP is one of the most widely used techniques in management of disasters. This section introduces the basic concepts of its optimization technique (Wagner, 1975; Hillier and Lieberman, 1990).

6.1.1 Formulation of Linear Optimization Models

Sections 3.3.2 and 3.3.3 provide two examples of linear optimization model formulations for a shortest supply route and resources allocation. Model formulation is the most difficult part of the process. Wagner (1975) offers the following guidelines for this stage of the optimization analysis:

- What are the key decisions to be made? What problem is being solved?
- What makes the real decision environment so complex as to require the use of a linear optimization model? What elements of complexity are incorporated in the model? What elements are ignored?
- What distinguishes a practical decision from an unusable one in this environment? What distinguishes a good decision from a poor one?
- As a decision maker, how would you employ the results of the analysis? What is your interpretation of results? In what ways might you want or need to temper the results because of factors not explicitly considered in the model?

In order to formulate the mathematical model in terms of linear relationships, four conditions must be satisfied:

Proportionality: For each activity, the total amounts of each input and the associated value of the objective are strictly proportional to the level of output—that is, to the activity level. Each activity is capable of continuous proportional expansion or reduction.

Additivity: Given the activity levels for each of the decision variables x_j, the total amounts of each input and the associated value of the objective are the sums of the inputs and objective values for each individual process.

Divisibility: Activity units can be divided into any fractional levels so that noninteger values for the decision variables are permissible.

Certainty: All the parameters of the linear model are known constants.

I emphasized in Section 3.1 that a mathematical model is only an idealized representation of the real problem. Approximations and simplifying assumptions generally are required in order for the model to be tractable. Adding too much detail and precision can make the model too complicated for useful analysis of the problem. All that is really needed is that there is a high correlation between the prediction of the model and the real problem. This applies certainly to LP. It is very common in real applications of LP that almost *none* of the four assumptions hold completely. Except perhaps for the *divisibility assumption*, minor differences are to be expected. This is especially true for the *certainty assumption*, so sensitivity analysis normally is a must to compensate for the violation of this assumption. However, it is important to examine how large the differences are. If any of the assumptions are violated in a major way then a number of alternative models are to be considered.

Example 1

The SIMON D&W is a producer of high-quality glass products, including windows and glass doors. It has three plants. Aluminum frames and hardware are made in plant 1, wood frames are made in plant 2, and plant 3 is used to produce the glass and assemble the products. Recent tornado disaster in the nearby community affected the demand for two products. One (product 1) is a 2.5 m glass door with aluminum framing. The other product (product 2) is a large (1.2 × 1.8 m) double-hung wood-framed window. The company can sell as much of either product as can be produced with the available capacity. However, because both products would be competing for the same production capacity in plant 3, it is not clear which *mix* between the two products would be the *most profitable*. Therefore, management collected the information—on (a) the percentage of each plant's production capacity that would be available for these products, (b) the percentages required by each product for each unit produced per day, and (c) the unit profit for each product—to help them make the decision. All the information is summarized in Table 6.1.

It was immediately recognized that this was an LP problem of the classic product mix type. Let us jointly proceed with the formulation and solution of the problem.

To formulate the mathematical (LP) model for this problem, let x_1 and x_2 represent the number of units produced per day of products 1 and 2, respectively, and let Z be the resulting contribution to profit per day. Thus, x_1 and x_2 are the *decision variables* for the model. Using the bottom row of Table 6.1, $Z = 3x_1 + 5x_2$. The objective is to choose the values of x_1 and x_2 so as to *maximize* $Z = 3x_1 + 5x_2$, subject to the restrictions imposed on their values by the limited plant capacities available. Table 6.1 implies that each unit of product 1 produced per day would use 1 percent of plant 1 capacity, whereas only 4 percent is available. This restriction is expressed mathematically by the inequality $x_1 \leq 4$. Similarly, plant 2 imposes the restriction that $2x_2 \leq 12$. The percentage of plant 3 capacity consumed by choosing x_1 and x_2 as the new products' production rates would be $3x_1 + 2x_2$. Therefore, the mathematical

TABLE 6.1 Input Data for Prototype Problem—Capacity Used per Unit Production Rate

Plant	Product 1	Product 2	Capacity Available
1	1	0	4
2	0	2	12
3	3	2	18
Unit profit	$3	$5	

statement of the plant 3 restriction is $3x_1 + 2x_2 \leq 18$. Finally, since production rates cannot be negative, it is necessary to restrict the decision variables to be nonnegative: $x_1 \geq 0$ and $x_2 \geq 0$.

To summarize, in the mathematical language of LP, the problem is to choose the values of x_1 and x_2 so as to:

$$
\begin{aligned}
&\text{Maximize } Z = 3x_1 + 5x_2 \\
&\text{subject to the restrictions} \\
&x_1 \leq 4 \\
&2x_2 \leq 12 \\
&3x_1 + 2x_2 \leq 18 \\
&x_1 \geq 0 \quad x_2 \geq 0
\end{aligned}
\tag{6.1}
$$

Equations (6.1) represent an LP formulation of our example. Let us now proceed with solving this problem. This is a very small problem with only two decision variables, two dimensions, so a graphical procedure can be used to solve it. This procedure involves constructing a two-dimensional graph with x_1 and x_2 as the axes. The first step is to identify the values of (x_1, x_2) that are permitted by the restrictions (constraints). This is done by drawing the lines that must border the range of permissible values. To begin, note that the nonnegativity restrictions, $x_1 \geq 0$ and $x_2 \geq 0$, require (x_1, x_2) to lie on the *positive* side of the axes. Next, observe that the restriction $x_1 \leq 4$ means that (x_1, x_2) cannot lie to the right of the line $x_1 = 4$. These results are shown in Figure 6.1, where the shaded area contains the only values of (x_1, x_2) that are still allowed.

In a similar fashion, the restriction $2x_2 \leq 12$ implies that the line $2x_2 = 12$ should be added to the boundary of the permissible region. The final restriction, $3x_1 + 2x_2 \leq 18$, requires plotting the points (x_1, x_2) such that $3x_1 + 2x_2 = 18$ (another line) to

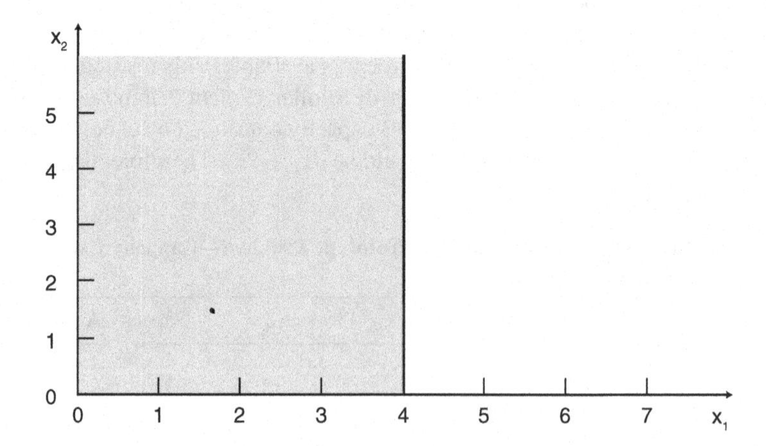

Figure 6.1 Possible values of (x_1, x_2) allowed by $x_1 \geq 0, x_2 \geq 0$ and $x_1 \leq 4$.

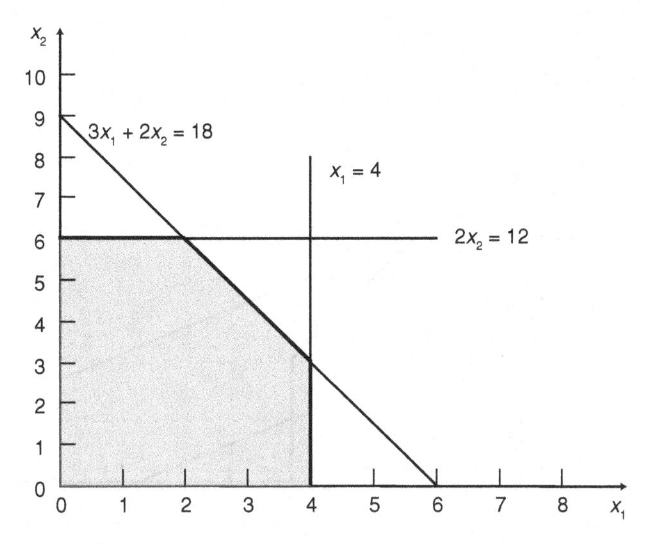

Figure 6.2 Possible values of (x_1, x_2).

complete the boundary (the points such that $3x_1 + 2x_2 \leq 18$ are those that lie either underneath or on the line $3x_1 + 2x_2 = 18$). The resulting region of possible values of (x_1, x_2) is shown in Figure 6.2.

The final step is to pick out the point in this region that maximizes the value of $Z = 3x_1 + 5x_2$. To illustrate how to perform this step, we will begin by trial and error. Try, for example, $Z = 10 = 3x_1 + 5x_2$ to see if there are in the permissible region any values of (x_1, x_2) that yield a value of Z as large as 10. By drawing the line $3x_1 + 5x_2 = 10$ (Figure 6.3), you can see that there are many points on this line that lie within the region. So let us try a larger value of Z, say, for example, $Z = 20 = 3x_1 + 5x_2$. Again, Figure 6.3 reveals that a segment of the line $3x_1 + 5x_2 = 20$ lies within the region, so that the maximum permissible value of Z must be at least 20. We can also see in Figure 6.3 that the two lines just constructed are parallel and that the line giving a larger value of Z ($Z = 20$) is further up and away from the origin than the other line ($Z = 10$). Thus, this trial-and-error procedure involves nothing more than drawing a family of parallel lines containing at least one point in the permissible region and selecting the line that is the greatest distance from the origin (in the direction of increasing values of Z). This line passes through the point (2, 6) as indicated in Figure 6.3, so that the equation is $3x_1 + 5x_2 = 3(2) + 5(6) = 36 = Z$. The point (2, 6) lies at the intersection of the two lines, $2x_2 = 12$ and $3x_1 + 2x_2 = 18$, shown in Figure 6.2, so that this point can be calculated algebraically as the simultaneous solution of these two equations.

The company used this approach to find that the optimal solution is $x_1 = 2$, $x_2 = 6$, with $Z = 36$. This solution indicates that the SIMON D&W should produce products 1 and 2 at the rate of 2 per day and 6 per day, respectively, with a resulting profitability of $36 per day. No other mix of the two products would be so profitable *according to the model.*

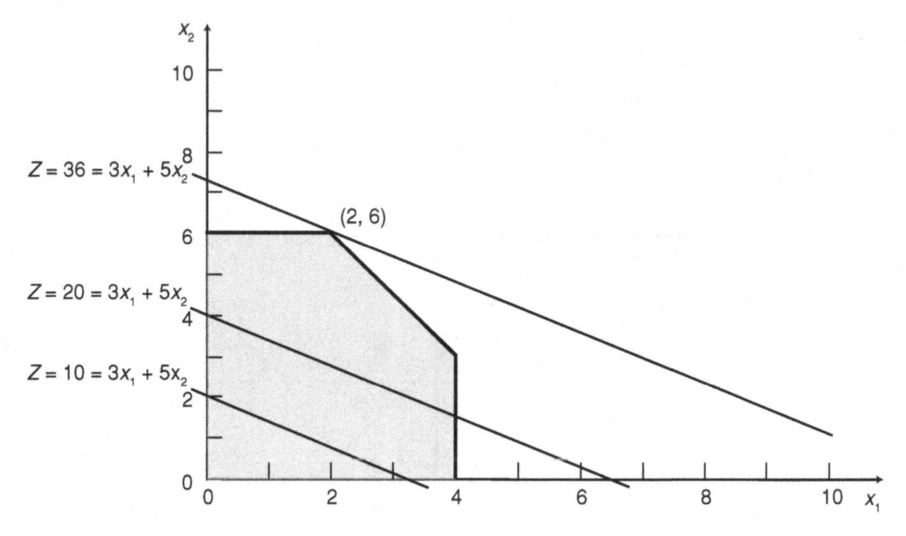

Figure 6.3 Value of (x_1, x_2) that maximizes $3x_1 + 5x_2$.

The SIMON D&W example is on the CD-ROM, in the directory LINPRO, subdirectory Examples, Example 1.

6.1.2 Algebraic Representations of Linear Optimization Models

In many disaster management problems, the aim is to maximize or minimize some objective, but there are certain constraints on what can be done to this end. The term LP refers to a way of modeling many of these problems so that they have a special structure, and to the way of solving problems with such a structure. It is a technique that can be applied in many different problem domains. The process of formulating a model was discussed in Section 6.1.1 using SIMON D&W example. The first step is to decide which are the *decision variables*. These are the quantities that can be varied, and so affect the value of the objective. The second step in the formulation is to express the *objective* in terms of the decision variables. Lastly, we must write down the *constraints* that restrict the choices of decision variables. One common-sense constraint is that many variables cannot realistically be negative. We can summarize the mathematical representation of the LP model in the following way. Letting x_j be the level of activity j, for $j = 1, 2, \ldots, n$, we want to select a value for each x_j such that:

$$C_1x_1 + C_2x_2 + \cdots + C_nx_n$$

is maximized or minimized, depending on the context of the problem. The x_j are constrained by a number of relations, each of which is one of the following type:

$$a_1x_1 + a_2x_2 + \cdots + a_nx_n \leq a$$

$$b_1x_1 + b_2x_2 + \cdots + b_nx_n = b$$

$$c_1x_1 + c_2x_2 + \cdots + c_nx_n \geq c$$

The first relation includes the possible restriction $x_j \geq 0$. Such a constrained optimization problem may have:

- no feasible solution: that is, there may be no values of all the x_j, for $j = 1, 2, \ldots, n$, that satisfy every constraint;
- a unique optimal feasible solution;
- more than one optimal feasible solution;
- a feasible solution such that the objective function is unbounded—that is, the value of the function can be made as large as desired in a maximization problem, or as small in a minimization problem, by selecting an appropriate feasible solution.

Changing the Sense of the Optimization Any linear maximization model can be viewed as an equivalent linear minimization model, and vice versa, by accompanying the change in the optimization sense with a change in the signs of the objective function coefficients. Specifically,

$$\text{Max} \sum_{j=1}^{n} c_j x_j \text{ can be written as Min} \sum_{j=1}^{n} (-c_j) x_j \tag{6.2}$$

and vice versa.

Changing the Sense of An Inequality All inequalities in an LP model can be represented with the same directioned inequality since:

$$\sum_{j=1}^{n} a_j x_j \leq b_j \text{ can be written as} \sum_{j=1}^{n} (-a_j) x_j \geq -b_j \tag{6.3}$$

and vice versa.

Converting An Inequality to An Equality An inequality in a linear model can be represented as an equality by introducing a nonnegative variable as follows:

$$\begin{aligned} &\sum_{j=1}^{n} a_j x_j \leq b_j \text{ can be written as} \sum_{j=1}^{n} a_j x_j + 1s = b_j \\ &\text{where } s \geq 0 \\ &\sum_{j=1}^{n} a_j x_j \geq b_j \text{ can be written as} \sum_{j=1}^{n} a_j x_j - 1t = b_j \\ &\text{where } t \geq 0 \end{aligned} \tag{6.4}$$

It is common to refer to a variable such as s as a *slack variable* and t as a *surplus variable*.

Converting Equalities to Inequalities Any linear equality or set of linear equalities can be represented as a set of like directioned linear inequalities by imposing one additional constraint. The idea can be generalized as follows:

$$\sum_{j=1}^{n} a_{ij}x_j = b_i \text{ for } i = 1, 2, \ldots, m \text{ can be written as}$$

$$\sum_{j=1}^{n} a_{ij}x_j \leq b_i \text{ for } i = 1, 2, \ldots, m \text{ and } \sum_{j=1}^{n} a_j x_j \leq \beta \tag{6.5}$$

where

$$\alpha_j = -\sum_{i=1}^{m} a_{ij} \text{ and } \beta = -\sum_{i=1}^{m} b_i \tag{6.6}$$

Canonical Forms for Linear Optimization Models Sometimes it is convenient to be able to write *any* linear optimization model in a compact and unambiguous form. The various transformations presented above allow us to meet this objective, although it is now apparent that there is considerable freedom in the selection of a particular canonical form to employ. I illustrate one such representations here.

Any linear optimization model can be viewed as:

$$\begin{aligned}
&\text{Maximize } \sum_{j=1}^{n} c_j x_j \\
&\text{subject to} \\
&\sum_{j=1}^{n} a_{ij}x_j \leq b_i \text{ for } i = 1, 2, \ldots, m \\
&x_j \geq 0 \text{ for } j = 1, 2, \ldots, n
\end{aligned} \tag{6.7}$$

It is typical, although not required, that $n > m$.

Types of Linear Program Solutions Whenever an LP model is formulated and solved, the result will be one of four characteristic solution types. The graphical framework developed while solving the SIMON D&W problem is useful for visualizing these solution types.

Unique Optimal Solution The solution to the SIMON D&W problem was achieved by first graphing the feasible region in decision space, plotting the gradient of the objective function on the same graph, and then shifting the objective function gradient in the direction of improvement until it last intersected the feasible region. Figure 6.4a illustrates the similar situation. Note that point (6, 4) is the only point that satisfies all constraint equations simultaneously. Consequently, the optimal solution to the linear program is a unique one; the solution is said to have a *unique optima*

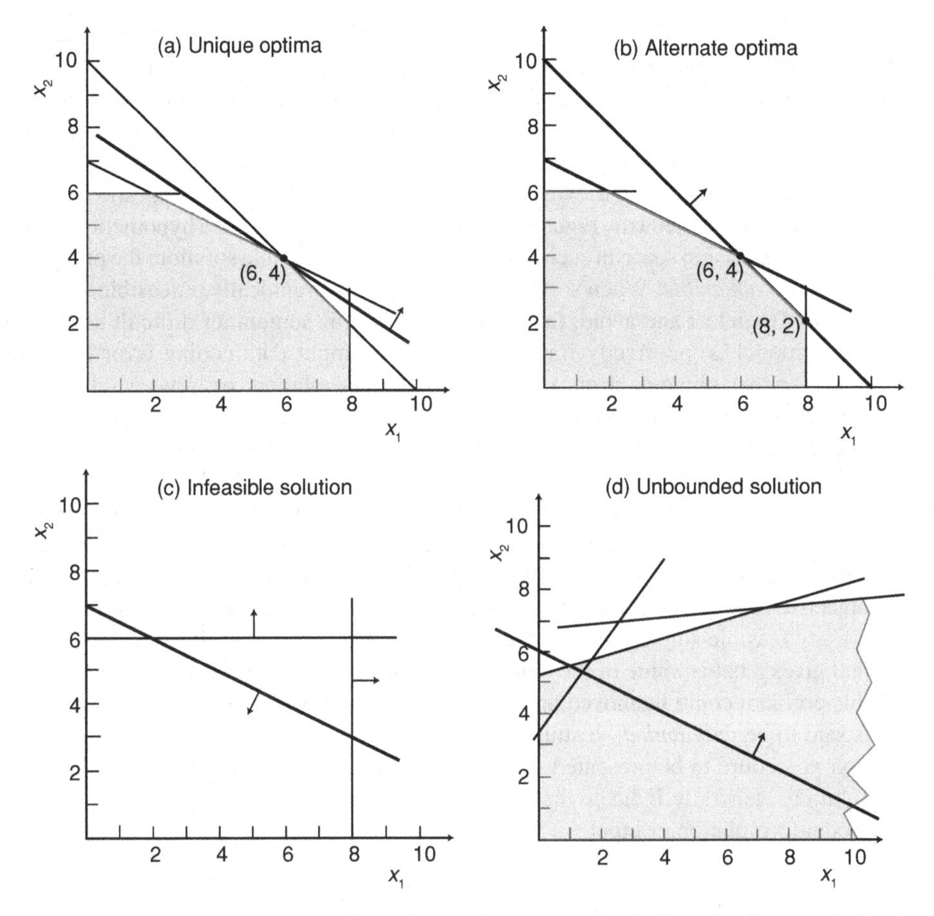

Figure 6.4 Linear program solutions: (a) unique optimal solution, (b) alternate optimal solution, (c) infeasible solution, and (d) unbounded optimal solution.

or *unique optimal solution.* It is possible, however, that more than one (perhaps an infinite number) solution would be optimal.

Alternate Optimal Solutions As demonstrated in the previous section, the orientation of the objective function in decision space is determined by the coefficients that multiply the decision variables. For example, if the coefficient on x_2 in the original objective function is decreased relative to the coefficient on x_1, the gradient becomes steeper (more negative). If the problem is as shown in Figure 6.4b, all points on the line segment connecting the points (6, 4) and (8, 2) yield the same value for the objective function and satisfy the constraint equation. This problem thus has an infinite number of optimal solutions, or is said to have alternate optima. Alternate optima are actually more common as the size of the LP problem (number of decision variables and constraints) increases, and it is important to be able to recognize their presence. An important consideration in our discussion of a simplex solution procedure for linear programs—the topic of Section 6.2—will be the identification of conditions that indicate the presence of alternate optimal solutions to LP problems.

Infeasible Solution It is possible that there are no feasible solutions for a given problem formulation, or due to errors in formulating logical constraints or mistakes in inputting a problem formulation to a model solver (computer program), the problem is over-constrained to the extent that there is no solution satisfying all constraint equations simultaneously. Figure 6.4c shows three constraints of a hypothetical model plotted in decision space in such a way that there is no feasible solution; the problem is said to be *infeasible*. When solving linear programs graphically, infeasible solutions are easy to detect and avoid. In larger problems, it is sometimes difficult to identify if the model is incorrectly formulated, or if an input data coding error has been made—either obvious errors such as an incorrect relation, or a more subtle error such as a typographical error in a variable name. We will discuss in Section 6.2 how infeasible solutions are detected by simplex solution procedures.

Unbounded Solution As infeasible problems result from problems that are over-constrained, we may also encounter problems that are underconstrained. For example, consider the feasible region of a hypothetical problem presented in Figure 6.4d. As before, the feasible region is shown shaded and constrained by three constraints. The objective function gradient and its direction of improvement are also shown. Note that for any feasible solution in this decision space, we can always find another solution that gives a better value of the objective function such that the objective function for this problem could be moved upward and to the right without limit. Such a problem is said to be *unbounded*—a situation that can also be identified by the simplex solution procedure to be presented in Section 6.2. Like infeasible solutions, unbounded solutions generally indicate that logical or typographical errors have been made in model formulation or input.

6.2 THE SIMPLEX METHOD FOR SOLVING LINEAR PROGRAMS

Probably you will never have to calculate manually the solution of an LP model in a real application, since a computer can do the work. Therefore you might ask, "Why do we need to know the underlying theory of linear optimization models?" In the light of considerable experience in applying LP to disaster management problems, I am convinced that a novice to the field must understand the principles explained here in order to make truly effective and sustained use of this optimization tool.

Many different algorithms have been proposed to solve LP problems, but the one below has proved to be the most effective in general. This is the general procedure:

Step 1: Select a set of m variables that yields a feasible starting trial solution. Eliminate the selected m variables from the objective function.

Step 2: Check the objective function to see whether there is a variable that is equal to 0 in the trial solution but would improve the objective function if made positive. If such a variable exists, go to step 3. Otherwise, stop.

Step 3: Determine how large the variable found in the previous step can be made until one of the m variables in the trial solution becomes 0. Eliminate the latter variable and let the next trial set contain the newly found variable instead.

Step 4: Solve for these m variables, and set the remaining variables equal to 0 in the next trial solution. Return to step 2.

The resulting algorithm does find an optimal solution to a general LP problem in a finite number of iterations. Often this method is termed *Dantzig's simplex algorithm*, in honor of the mathematician who devised the approach. Let us examine a "well-behaved" problem of SIMON D&W and explain the *simplex method* by means of this example.

Example 2
Consider the mathematical model developed in Example 1 for SIMON D&W:

$$\begin{aligned}&\text{Maximize } Z = 3x_1 + 5x_2\\&\text{subject to the restrictions}\\&x_1 \le 4\\&2x_2 \le 12\\&3x_1 + 2x_2 \le 18\\&x_1 \ge 0 \quad x_2 \ge 0\end{aligned} \tag{6.8}$$

Let x_0 be the value of the objective function, add slack variables, and convert the system (Eq. 6.8) into the system of linear equations:

$$\begin{array}{ll}\text{Row 0} & 1x_0 - 3x_1 - 5x_2 = 0\\ \text{Row 1} & x_1 + x_3 = 4\\ \text{Row 2} & 2x_2 + x_4 = 12\\ \text{Row 3} & 3x_1 + 2x_2 + x_5 = 18\end{array} \tag{6.9}$$

where all the variables must be nonnegative. Notice how the introduction of the variable x_0 in row 0 permits us to express the objective function in equation form. The example represented by the set of relationships (Eq. 6.9) is on the CD-ROM, in directory LINPRO, subdirectory Examples, Example 1.

The task of step 1 is to find a starting feasible solution to Equation (6.9). There are a large number of such solutions, but it is certainly most convenient to begin with $x_0 = 0, x_3 = 4, x_4 = 12$, and $x_5 = 18$, and all other variables equal to 0. In other words, we start with an all-slack solution. We term this an *initial feasible basic solution*, and x_0, x_3, x_4, and x_5 are known as the *basic variables*, sometimes shortened to the *basis*. The remaining variables we call *nonbasic*.

Interpretation of coefficients in row 0 is important. Each coefficient represents the increase (for negative coefficients) or decrease (for positive coefficients) in x_0 with a unit increase of the associated nonbasic variable. Now the iterative solution process can start. At each iteration, the simplex method moves from the current basic feasible solution to a better adjacent basic feasible solution. The movement involves converting one nonbasic variable into a basic variable (called the *entering basic variable*) and simultaneously converting a basic variable into a nonbasic variable (called *leaving basic variable*), and then identifying the new basic feasible solution.

Iteration 1: For step 2 the simplex method adopts the following easy-to-apply rule for deciding the variable to enter the next trial basis:

Simplex criterion I (maximization): If there are nonbasic variables with a negative coefficient in row 0, select the one with the most negative coefficient, that is, the best per unit potential gain (say x_j). If all nonbasic variables have positive or zero coefficients in row 0, an optimal solution has been obtained.

To decide which variable should leave the basis, we apply the following rule, or step 3:

Simplex criterion II: (a) Take the ratios of the current right-hand side to the coefficients of the entering variable x_j (ignore ratios with 0 or negative numbers in the denominator). (b) Select the minimum ratio—that ratio will equal the value of x_j in the next trial solution. The minimum ratio occurs for a variable x_k in the present solution; set $x_k = 0$ in the solution.

The process of applying criterion II is known as a *change-of-basis* calculation, or a *pivot operation*. A detailed calculation is presented in Table 6.2.

Iteration 2: At this point the first iteration of the simplex method has been completed. On returning to step 2, you are ready to determine whether an optimal solution has been obtained or another simplex iteration is required. Criterion I, which examines the nonbasic variables, indicates that a still better solution seems to exist. You might profitably enter into the basis x_1. Criterion I

TABLE 6.2 Simplex Tableau for the SIMON D&W Example 2 Problem

Iteration	Basis	Current	x_1	x_2	x_3	x_4	x_5	Row
1	x_0	0	−3	−5	0	0	0	0
	x_3	4	1	0	1	0	0	1
	x_4	12	0	2	0	1	0	2
	x_5	18	3	2	0	0	1	3
2	x_0	30	−3	0	0	5/2	0	0
	x_3	4	1	0	1	0	0	1
	x_2	6	0	1	0	1/2	0	2
	x_5	6	3	0	0	−1	1	3
3	x_0	36	0	0	0	3/2	1	0
	x_3	2	0	0	1	1/3	−1/3	1
	x_2	6	0	1	0	1/2	0	2
	x_1	2	1	0	0	−1/3	1/3	3

selects x_1 since it promises the greatest gain per unit increase. Next perform the step 3 calculations, using criterion II. From Table 6.2, notice that x_1 will replace x_5 in the next trial solution. At this iteration, you have just seen another aspect to the computational rule in criterion II. To sum up, criterion II ensures that each new basic solution results in only 0 or positive values for the trial values of the basis. Consequently, the solution remains feasible at every iteration.

Iteration 3: Having completed the second simplex iteration, once more examine the coefficients in row 0 to ascertain whether you have discovered an optimal solution. All the coefficients in row 0 are nonnegative, and consequently criterion I asserts that we have found an optimal solution. Thus, the calculations are terminated in step 2.

Summary In brief, the simplex method consists of four steps:

1. Selection of an initial basis.
2. Application of simplex criterion I. If the solution is not optimal go to step 3; otherwise, stop.
3. Application of simplex criterion II.
4. Change of basis, and return to second step.

The progress of the simplex method can easily be interpreted through the geometry of the solution space. Each basis corresponds to a cortex of the convex polyhedral set of feasible solutions. Going from one basis to the next represents going from one extreme point to an adjacent one. Thus, the simplex method can be said to seek an optimal solution by *climbing along the edges, from one vertex of the convex polyhedral solution set to a neighboring one*. Once we master the straightforward logic of the simplex iterations, considerable writing effort can be saved by organizing the computations in a convenient tabular form called a *simplex tableau* (Table 6.2).

Computer Implementation—LINPRO Computer Program Computer codes for the simplex method are widely available. They do not closely follow the algebraic or the tabular form of the simplex method presented here. The available software packages are used routinely to solve large LP problems (several thousands of constraints and decision variables). One difficulty in dealing with large LP problems is the tremendous amount of data involved. For example, a problem with just 1000 constraints and decision variables would have 1 million constraint coefficients to be specified. Therefore, the extensive use of computers is suggested for data processing both before and after applying the simplex method.

The CD-ROM accompanying this book includes the LINPRO software and all the LP examples developed in the text. Code of LINPRO package is developed with educational purpose in mind and completely replicates the tabular simplex procedure presented here. However, this does not diminish the power of LINPRO software that can be used for real applications too. The folder LINPRO contains two subfolders,

Linpro and Examples. The Readme file in the LINPRO folder contains instructions for the installation of the LINPRO software and a detailed tutorial for its use as a part of its Help menu.

6.2.1 Completeness of the Simplex Algorithm

In the application of criterion I, when two or more variables appear equally promising, as indicated by the values of their coefficients in row 0, an arbitrary rule may be adopted for selecting one of these. For example, use the lowest numbered variable, or the one suspected to be in the final basis.

In the application of criterion II, when two or more variables in the current basis are to fall simultaneously to the level 0 upon introducing the new variable, only one of these is to be removed from the basis. The others remain in the basis at 0 level. The resultant basis is termed *degenerate*. Unless some care is given to the method of deciding which variable to remove from the basis, there is no *proof* that the method always converges. However, long experience with simplex computations leads to the conclusion that for all *practical* purposes, the selection can be arbitrary and the associated danger of nonconvergence is negligible.

If at some iteration in applying criterion II there is no positive coefficient in any row for the entering variable, then there exists an *unbounded* optimal solution. In this event, the entering variable can be made arbitrarily large, the value of x_0 thereby increases without bound, and the current basis variables remain nonnegative. Thus, we may drop the earlier assumption that the optimal value of the objective function is finite. The simplex algorithm provides an indication of when an unbounded optimal solution occurs. Criterion II is easily reworded to cover this case.

I mentioned in Section 6.1.2 (under the definition of optimal solution) that a problem can have more than one optimal solution. This fact can be illustrated by changing the objective function in the SIMON D&W problem to $Z = 3x_1 + 2x_2$, so that every point on the line segment between (2, 6) and (4, 3) is optimal. It can be noted that every such problem has at least two optimal corner-point feasible solutions (basic feasible solutions). By taking *weighted averages*, these solutions can be used to identify every other optimal solution.

The simplex method automatically stops after finding *one* optimal solution. However, for many applications of LP, there are intangible factors not incorporated into the model that can be used to make meaningful choices between solutions that are alternative optimal solutions according to the model. In such cases, these other optimal solutions should be identified as well. After the simplex method finds one optimal basic feasible solution, how do we recognize when there are others, and how do we find them? The answer is summarized as follows:

> Whenever a problem has more than one optimal basic feasible solution, at least one of the nonbasic variables has a coefficient of *zero* in the final row 0, so increasing any such variable would not change the value of x_0. Therefore, these other optimal basic feasible solutions can be identified (if desired) by performing additional iterations of the simplex method, each time choosing a nonbasic variable with a zero coefficient as the entering basic variable.

Example 3

To illustrate, consider again the slightly modified case of the SIMON D&W problem with changed objective function to $Z = 3x_1 + 2x_2$. The simplex method obtains the first three tableaux shown in Table 6.3 and stops with an optimal basic feasible solution. However, because a nonbasic variable (x_3) then has a zero efficient in row 0, we perform one more iteration in Table 6.3 to identify the other optimal basic feasible solution. Thus, the two optimal basic feasible solutions are (4, 3, 0, 6, 0) and (2, 6, 2, 0, 0), each yielding $Z = 18$. Notice that the last tableau so has *a nonbasic* variable (x_4) with a zero coefficient in row 0. This situation is inevitable because the extra iteration(s) does not change row 0, so each leaving basic variable retains its zero coefficient. Making x_4 an entering basic variable now would only lead back to the third tableau. (I suggest that you check this.) Therefore, these two are the only basic feasible solutions that are optimal, and all *other* optimal solutions are weighted average of these two (see Figure 6.5).

TABLE 6.3 Simplex Tableau for the Modified SIMON D&W Problem

Iteration	Basis	Current	x_1	x_2	x_3	x_4	x_5	Row
1	x_0	0	−3	−2	0	0	0	0
	x_3	4	1	0	1	0	0	1
	x_4	12	0	2	0	1	0	2
	x_5	18	3	2	0	0	1	3
2	x_0	12	0	−2	3	0	0	0
	x_1	4	1	0	1	0	0	1
	x_4	12	0	2	0	1	0	2
	x_5	6	0	2	−3	0	1	3
3	x_0	18	0	0	0	0	1	0
	x_1	4	1	0	1	0	0	1
	x_4	6	0	0	3	1	−1	2
	x_2	3	0	1	−3/2	0	1/2	3
Extra	x_0	18	0	0	0	0	1	0
	x_1	2	1	0	0	−1/3	1/3	1
	x_3	2	0	0	1	1/3	−1/3	2
	x_2	6	0	1	0	1/2	0	3

The modified SIMON D&W example is on the CD-ROM, in the directory LINPRO, subdirectory Examples, Example 2.

Starting Basis and Other Model Forms Let me review the selection of an initial basis to begin the algorithm. Because each constraint in the example of the preceding section is of the form:

$$\sum_{j=1}^{n} a_{ij}x_j \leq b_i \quad \text{where } b_i \geq 0 \tag{6.10}$$

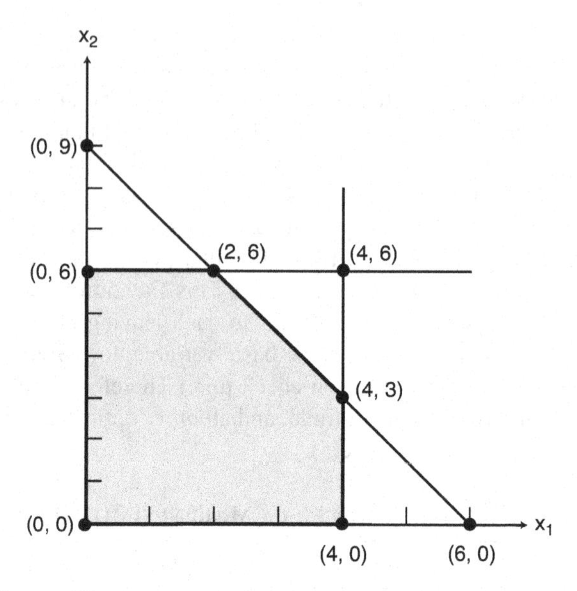

Figure 6.5 Corner point solutions for SIMON D&W problem.

adding a slack variable to each relation and starting with an all-slack basic solution provided a simple way of initiating the simplex algorithm. The only real problem that the other forms for functional constraints ($=$, $\geq$ or $b_i \leq 0$) introduce is in identifying an initial basic feasible solution. For example, let us consider the equality constraints in any LP model:

$$\sum_{j=1}^{n} a_{ij}x_j = b_i \quad \text{for } i = 1, 2, \ldots, m \text{ where } b_i \geq 0 \tag{6.11}$$

In this form, if a variable appears only in constraining relation i and has a coefficient of 1, as would be the case for a slack variable, it can be used as part of the initial basis. But relation i may not have such a variable. This can occur, for example, if the ith equation is linearly dependent on one or more of the other equations, such as being a sum of two equations. Then we can utilize the following approach.

Write the constraints as:

$$\sum_{j=1}^{n} a_{ij}x_j + 1y_i = b_i \quad \text{for } i = 1, 2, \ldots, m \text{ where } b_i \geq 0 \tag{6.12}$$

and where $y_i \geq 0$, use y_i as the basic variable for relation i. It is assumed here, for simplicity, that every constraint requires the addition of a y_i. The name *artificial variable* is given to y_i because it is added as an artifice in order to obtain an initial trial solution. Is this approach valid? The answer is yes, provided Condition A is satisfied.

Condition A. To ensure that the final solution is meaningful, every y_i must equal 0 at the terminal iteration of the simplex method.

If there is no feasible solution, it will be impossible to satisfy Condition A. At the final iteration of the simplex algorithm, at least one positive y_i will be in the solution indicating an *infeasible* optimal solution.

6.2.2 The Big *M* Method

There are a number of computational techniques for guaranteeing Condition A. One approach is to add to the maximizing objective function each y_i with a large penalty cost coefficient:

$$x_0 - \sum_{j=1}^{n} c_j x_j + \sum_{i=1}^{m} M\ y_i = 0 \tag{6.13}$$

where M is a relatively large number. Thus, each y_i variable is very costly compared with any of the x_j variables. To initiate the algorithm, first we eliminate each y_i from Equation (6.13) by using Equation (6.12). This gives:

$$x_0 - \sum_{j=1}^{n} c_j x_j + M \sum_{i=1}^{m} \sum_{j=1}^{n} a_{ij} x_j = -M \sum_{i=1}^{m} b_i \tag{6.14}$$

which simplifies to:

$$x_0 - \sum_{j=1}^{n} \left(c_j + M \sum_{i=1}^{m} a_{ij} \right) x_j = -M \sum_{i=1}^{m} b_i \tag{6.15}$$

Because the y_i variables are so expensive, the optimization technique drives the y_i variables to 0, *provided* there exists a feasible solution. Whenever a y_i drops from a basis at some iteration, we need never consider using it again, and can eliminate it from further computations. The following example will clarify the approach.

Example 4

Suppose that the SIMON D&W from Example 1 is modified to *require* that plant 3 be used at full capacity. The only resulting change in the LP model is that the third constraint, $3x_1 + 2x_2 \leq 18$, becomes an *equality* constraint, $3x_1 + 2x_2 = 18$.

Therefore, the feasible region for this problem (see Figure 6.2) now consists of *just* the line segment connecting $(2, 6)$ and $(4, 3)$. After introducing the slack variables still needed for the *inequality* constraints, the *augmented form* of the problem becomes:

$$\begin{array}{ll} \text{Row 0} & 1x_0 - 3x_1 - 5x_2 = 0 \\ \text{Row 1} & x_1 + x_3 = 4 \\ \text{Row 2} & 2x_2 + x_4 = 12 \\ \text{Row 3} & 3x_1 + 2x_2 = 18 \end{array} \tag{6.16}$$

Unfortunately, these equations do not have an obvious initial basic feasible solution because there is no longer a slack variable to use as the initial basic variable for equality constraint in row 3. The artificial variable technique circumvents this difficulty by introducing nonnegative artificial variable (y_1) into this equation. So the modified row 3 is:

$$\text{Row 3} \quad 3x_1 + 2x_2 + y_1 = 18$$

along with the nonnegativity constraint, $y_1 \geq 0$.

We now have an initial basic feasible solution (for the *revised problem*), $(x_1, x_2, x_3, x_4, y_1) = (0, 0, 4, 12, 18)$. The effect of introducing an artificial variable is to *enlarge* the feasible region. In this case, the feasible region expands from just the line segment connecting (2, 6) and (4, 3) to the entire shaded area shown in Figure 6.2. A feasible solution for the *revised* problem (with $3x_1 + 2x_2 + y_1 = 18$ and $y_1 \geq 0$) is also feasible for *original* problem (with $3x_1 + 2x_2 = 18$) if the artificial variable equals zero ($y_1 = 0$).

Now suppose that the simplex method is permitted to proceed and obtain *optimal* solution for the *revised* problem and that this solution happens to be *feasible* for the *original* problem. It can then be concluded that this solution must also be *optimal* for the *original* problem, so we are finished. (The reason is that this solution is the best one in the *entire* feasible region for the revised problem, which includes the feasible region for the original problem.)

Unfortunately, there is no guarantee that the optimal solution to the revised problem also will be feasible for the original problem; that is, there is no guarantee until another revision is made. Using the Big M method, this new revision amounts to assigning such an overwhelming penalty to being outside the feasible region for the original problem that the optimal solution to the revised problem *must* lie within this region. Recall that the revised problem coincides with the original problem when $y_1 = 0$. Therefore, if the original objective function, $Z = 3x_1 + 5x_2$, is changed to $Z = 3x_1 + 5x_2 - My_1$, where M denotes some large positive number, then the maximum value of Z must occur when $y_1 = 0$ (y_1 cannot be negative). After a few more transformations (discussed next), applying the simplex method to this revised problem automatically leads to the desired solution.

Using the revised objective function, row 0 becomes:

$$\text{Row 0} \quad 1x_0 - 3x_1 - 5x_2 + My_1 = 0$$

or in tabular form, the iteration 1 row 0 becomes [−3, −5, 0, 0, M, 0]. However, this row 0 cannot be used in the initial tableau for applying the simplex method because it is not in proper form from Gaussian elimination. The proper form requires that every basic variable (excluding x_0, which only pretends to be a basic variable) has been eliminated from row 0, whereas y_1 now is a basic variable with a coefficient of M. Restoring this proper form is essential for the application of simplex criteria I and II (optimality test and the procedure for determining the entering basic variable). To initiate the simplex algorithm, we have to eliminate every artificial variable from row 0 as shown in Equations (6.13) to (6.15). The elimination process is as follows:

$$\begin{array}{rlllllr} \text{Row } 0 & [-3 & -5 & 0 & 0 & M & 0] \\ -M & [3 & 2 & 0 & 0 & 1 & 18] \\ \text{New Row } 0 & [(-3M-3) & (-2M-5) & 0 & 0 & 0 & -18M] \end{array}$$

This completes the additional work required in the initialization step for problems of this type, and the rest of the simplex method proceeds just as before. The quantities involving M never appear anywhere except in row 0, so they need to be taken into account only in the optimality test (criterion I) and when determining an entering basic variable (criterion II). One way of dealing with these quantities is to assign some particular large numerical value to M and use the resulting numbers in row 0 in the usual way.

Using this approach on the SIMON D&W example yields the simplex tableaux shown in Table 6.4. The optimal solution ($x_1 = 2$, $x_2 = 6$) is the same as for the first version the SIMON D&W problem (see Table 6.2 for its tableaux).

TABLE 6.4 Simplex Tableau for the SIMON D&W Problem with Equality Constraint

Iteration	Basis	Current	x_1	x_2	x_3	x_4	y_1	Row
1	x_0	$-18M$	$-3M-3$	$-2M-5$	0	0	0	0
	x_3	4	1	0	1	0	0	1
	x_4	12	0	2	0	1	0	2
	y_1	18	3	2	0	0	1	3
2	x_0	$-6M+12$	0	$-2M-5$	$3M+3$	0	0	0
	x_1	4	1	0	1	0	0	1
	x_4	12	0	2	0	1	0	2
	y_1	6	0	2	−3	0	1	3
3	x_0	27	0	0	−9/2	0	$M+5/2$	0
	x_1	4	1	0	1	0	0	1
	x_4	6	0	0	3	1	−1	2
	x_2	3	0	1	−3/2	0	1/2	3
4	x_0	36	0	0	0	3/2	$M+1$	0
	x_1	2	1	0	0	−1/3	1/3	1
	x_3	2	0	0	1	1/3	−1/3	2
	x_2	6	0	1	0	1/2	0	3

This example involved only one equality constraint. If an LP model has more than one equality constraint, each should be handled in the same way. If the right-hand side is negative, we should multiply both sides by (-1) first.

The approach to other kinds of constraints requiring artificial variables is completely analogous. The modified SIMON D&W example with equality constraint is on the CD-ROM, in the directory LINPRO, subdirectory Examples, Example 3.

6.3 DUALITY IN LP

LP offers much more than the numerical values of an optimal solution. The mathematical structure of a linear model means, for example, that if a linear optimization model has a finite optimal solution, there exists an optimal *basic* solution. Many important postoptimality questions are easily answered, given the numerical information at the final simplex iteration. However, before we proceed with the discussion of sensitivity analysis let us introduce a unifying concept, known as *duality*, which establishes the interconnections for all of the sensitivity analysis techniques.

Consider the pair of LP models:

$$
\begin{aligned}
&\text{Maximize } \sum_{j=1}^{n} c_j x_j \\
&\text{subject to} \\
&\sum_{j=1}^{n} a_{ij} x_j \leq b_i \quad \text{for } i = 1, 2, \ldots, m \\
&x_j \geq 0 \quad \text{for } j = 1, 2, \ldots, n
\end{aligned}
\tag{6.17}
$$

and

$$
\begin{aligned}
&\text{Minimize } \sum_{i=1}^{m} b_i y_i \\
&\text{subject to} \\
&\sum_{i=1}^{m} a_{ij} y_i \geq c_j \quad \text{for } j = 1, 2, \ldots, n \\
&y_i \geq 0 \quad \text{for } i = 1, 2, \ldots, m
\end{aligned}
\tag{6.18}
$$

We arbitrarily call (Eq. 6.17) the *primal problem* and (Eq. 6.18) its *dual problem.*

The dual problem can be viewed as the primal model flipped on its side:

- The jth column of coefficients in the primal model is the same as the jth row of coefficients in the dual model.
- The row of coefficients of the primal objective function is the same as the column of constants on the right-hand side of the dual model.

- The column of constants on the right-hand side of the primal model is the same as the row of coefficients of the dual objective function.
- The direction of the inequalities and sense of optimization is reversed in the pair of problems.

Now we can define more closely the significant aspects of the primal–dual relationship:

Dual theorem: (a) In the event that both the primal and dual problems possess feasible solutions, then the primal problem has an optimal solution x_j^*, for $j = 1, 2, \ldots, n$, the dual problem has an optimal solution y_i^*, for $i = 1, 2, \ldots, m$, and

$$\sum_{j=1}^{n} c_j x_j^* = \sum_{i=1}^{m} b_i y_i^* \tag{6.19}$$

(b) If either the primal or dual problem possesses a feasible solution with a finite optimal objective function value, then the other problem possesses a feasible solution with the same optimal objective-function value.

The duality relationships are summarized in Table 6.5.

TABLE 6.5 Relationship between Primal and Dual Problems

Primal (Maximize)	Dual (Minimize)
Objective function	Right-hand side
Right-hand side	Objective function
*j*th column of coefficients	*j*th row of coefficients
*i*th row of coefficients	*i*th column of coefficients
*j*th variable nonnegative	*j*th relation an inequality
*j*th variable unrestricted in sign	*j*th relation an equality
*i*th relation an inequality	*i*th variable nonnegative
*i*th relation an equality	*i*th variable unrestricted in sign

In addition, we can observe the following relationship:

Optimal values of dual variables: (a) The coefficients of the slack variables in row 0 of the final simplex iteration of a maximizing problem are the optimal values of the dual variables. (b) The coefficient of variable x_j in row 0 of the final simplex iteration represents the difference between the left- and right-hand sides of the *j*th dual constraint for the associated optimal dual solution.

By reference to the notion of duality, we deepen our understanding of what is really happening in the simplex method. The coefficients of the slack variables in row 0 of the primal problem at each iteration can be interpreted as trial values of the dual variables. So the simplex method can be seen as an approach that seeks feasibility for the dual problem while maintaining feasibility in the primal problem. As soon as feasible solutions to both problems are obtained, the simplex iterations terminate.

Example 5

Let us consider the SIMON D&W problem from Example 2. Using LINPRO software (Example 1) we see that the optimal solution of the primal problem:

$$
\begin{aligned}
&\text{Maximize } Z = 3x_1 + 5x_2 \\
&\text{subject to the restrictions} \\
&x_1 \leq 4 \\
&2x_2 \leq 12 \\
&3x_1 + 2x_2 \leq 18 \\
&x_1 > 0 \quad x_2 > 0
\end{aligned}
\tag{6.8}
$$

and Table 6.2 is $Z^* = 36$, $x_1^* = 2$, and $x_2^* = 6$. Dual formulation of the problem is:

$$
\begin{aligned}
&\text{Minimize } Z = 4y_1 + 12y_2 + 18y_3 \\
&\text{subject to the restrictions} \\
&y_1 + 3y_3 \geq 3 \\
&2y_2 + 2y_3 \geq 5 \\
&y_1 \geq 0 \quad y_2 \geq 0 \quad y_3 \geq 0
\end{aligned}
\tag{6.20}
$$

Example 4 on the CD-ROM, in the directory LINPRO, subdirectory Examples, is a dual formulation of the SIMON D&W problem. Compare the LINPRO solutions of Example 1 and Example 4. Checking the coefficients of the three slack variables in row 0 of the final iteration of Example 1, we see that the optimal solutions of the dual problem of Example 4 are $Z^* = 36$, $y_1^* = 0$, $y_2^* = 1.5$, and $y_3^* = 1$. Check the solution of LINPRO problem in Example 4. Using the optimal solutions, we can verify that both constraints in Equation (6.20) are satisfied:

$$
\begin{aligned}
y_1 + 3y_3 &= 0 + 3 = 3 \geq 3 \\
2y_2 + 2y_3 &= 3 + 2 = 5 \geq 5
\end{aligned}
\tag{6.21}
$$

We can also see that the optimal value of the objective function in both cases is the same, $Z^* = 36$. So the values $y_2^* = 0$, $y_2^* = 1.5$, and $y_3^* = 1$ are optimal, since they satisfy all the dual constraints and yield an objective function value equal to the optimal primal value.

6.3.1 Sensitivity Analysis

Sensitivity analysis is the study of how the optimal solution and the value of the optimal solution to a linear program change, given changes in the various coefficients of the problem. That is, we are interested in answering questions such as the following:

What effect will a change in the coefficients in the objective function (c_j) have?

What effect will a change in the right-hand side values (b_i) have?

What effect will a change in the coefficients in the constraining equations (a_{ij}) have?

Since sensitivity analysis is concerned with how these changes affect the optimal solution, the analysis begins only after the optimal solution to the original LP problem has been obtained. Hence, sensitivity analysis can be referred to as *postoptimality analysis*. The mechanics for all postoptimality analyses are straightforward extensions of the simplex arithmetic, but duality is the key idea that ensures the mechanics are correct. Every LP model has a dual formulation. By solving one of these we automatically solve the other.

There are several reasons why sensitivity analysis is considered important from a disaster management point of view. First, consider the fact that disaster management occurs in a dynamic environment. Basic physical variables (number of people, flow of flood water, supply of aids, flow of money) change over time. If an LP model has been used in a decision-making situation and later we find changes in some of the coefficients associated with the initial LP formulation, we would like to determine how these changes affect the optimal solution to our original LP problem. Sensitivity analysis provides us with this information without requiring us to completely solve a new linear program.

Thus, through sensitivity analysis, we will be able to provide valuable information for the disaster managers. We begin our study of sensitivity analysis with the coefficients of the objective function.

Sensitivity Analysis: The Coefficients of the Objective Function Recall in Example 5 that the dual constraint corresponding to the primal variable x_1 is:

$$y_1 + 3y_3 \geq 3 \tag{6.22}$$

If the objective function coefficient of x_1 becomes $(3 + p_1)$, then $(3 + p_1)$ appears on the right-hand side of Equation (6.22). Substituting the optimal values of the dual variables into Equation (6.22), where $(3 + p_1)$ is used, yields:

$$0 \times 1 + 3 \times 1 \geq 3 + p_1 \tag{6.23}$$

or

$$0 \geq p_1 \tag{6.24}$$

Thus, the current dual solution remains feasible provided p_1 does not exceed 0. If p_1 is made larger than 0, the dual solution is no longer feasible, and consequently the primal solution is no longer optimal.

In summary, consider the following management interpretation of sensitivity analysis for the objective function coefficients. Let us use SIMON D&W example. Think of the basic variables as corresponding to a current product line and the nonbasic variables as representing other products a company might produce. Within bounds, changes in the profit associated with one of the products in the current product line would not cause the company to change its product mix or the amounts produced, but the changes would have an effect on its total profit. Of course, if the profit associated with one of the products changed drastically, it would change the product line (i.e., move to a different basic solution). For products that are not currently being produced (nonbasic variables), it is obvious that a decrease in per unit profit would not make the company want to produce them. However, if the per unit profit for one of these products became large enough, it would want to consider adding that product to the product line. The same meaning of sensitivity analysis for the objective function coefficients can be given to different decision problems represented using LP formulation.

Sensitivity Analysis: The Right-Hand Sides From the basic interpretation of the simplex algorithm, we know that the coefficient of a slack variable in row 0 of an optimal solution represents the incremental value of another unit of the resource associated with that variable. In the preceding section, we stated that the optimal value of a dual variable is the very same coefficient. Putting the two statements together, we have the following interpretation of the dual variables:

> The optimal value of a dual variable indicates how much the objective function changes with a unit change in the associated right-hand-side constant, provided the current optimal basis remains feasible.

This interpretation is in agreement with the fundamental equality relation in the dual theorem presented in Section 6.1.5, which states:

$$\text{Optimal value of } x_0 = \sum \text{(right-hand-side constants)} \times \text{(optimal dual variables)} \tag{6.25}$$

Since this is such an important property, let me state it again. Associated with every constraint of the primal LP problem is a dual variable. The value of the dual variable indicates how much the objective function of the primal problem will increase as the value of the right-hand side of the associated primal constraint is increased by one unit.

In other words, the value of the dual variable indicates the value of one additional unit of a particular resource. Hence, this value can be interpreted as the maximum value or price we would be willing to pay to obtain one additional unit of the resource.

Because of this interpretation, the value of one additional unit of a resource is often called the *shadow price* of the resource. Thus, the optimal values of the dual variables are shadow prices. When the right-hand-side constants represent quantities of scarce resources, the shadow price indicates the unit worth of each resource as predicated on an optimal solution to the primal problem.

For the LP problem from Example 5, the value of 0 is the shadow price for the first constraint, and similarly 1.5 holds for the second constraint, and 1 for the third constraint. Therefore, an additional unit of resource represented by the second constraint increases the objective function value by 1.5, and an additional unit of resources represented by the third constraint increases the objective function value by 1, but an additional unit of resource represented by the first constraint does not improve the objective function value. Why? Because this resource is already in excess supply, as evidenced by the slack variable x_3 being in the optimal basis (see Table 6.3).

In general, a resource in excess supply is indicated by the slack variable for that resource appearing in the final basis at a positive level. The corresponding shadow price is 0, because additional excess supply is of no value. Since the slack is in the basis, its final row 0 coefficient is 0.

Interpreting the values of the dual variables as shadow prices leads to an insightful view into the meaning of the dual problem. In the context of Example 5, think of each dual variable as representing the true marginal value of its associated resource, assuming that the decision is made optimally. Then Equation (6.25) indicates that the total value of the objective function is the same as evaluating the total worth of all the resources in scarce supply. Interpret each coefficient a_{ij} as the consumption of the ith resource by the jth activity. The summation:

$$\sum_{i=1}^{m} a_{ij} y_i$$

represents the underlying cost of using the jth activity, evaluated according to the shadow prices. The constraints of the dual problem ensure that an optimal solution never exceeds its true worth.

Sensitivity Analysis: The Matrix of Left-Hand Side Coefficients To begin consideration of sensitivity tests on a_{ij}, suppose an entirely new activity is added to the model. Is it advantageous to enter it into the basis? The easiest test is to check whether the associated dual constraint is satisfied by the current values of the dual variables. If not, the new activity should be introduced.

Consider SIMON D&W problem from Example 5, as given by Equation (6.16):

$$\begin{array}{ll} \text{Row 0} & 1x_0 - 3x_1 - 5x_2 = 0 \\ \text{Row 1} & x_1 + x_3 = 4 \\ \text{Row 2} & 2x_2 + x_4 = 12 \\ \text{Row 3} & 3x_1 + 2x_2 = 18 \end{array} \tag{6.16}$$

Suppose we add another variable:

$$\begin{array}{l} \text{Row 1 } +2/7x_5 \\ \text{Row 2 } +2x_5 \\ \text{Row 3 } +17x_5 \end{array} \tag{6.26}$$

Let the objective function coefficient of x_5 be c_5. At what value of c_5 is it attractive to enter x_5? The associated dual relation is:

$$2/7y_1 + 2y_2 + 17y_3 \geq c_5 \tag{6.27}$$

Substitute the current optimal values of the dual variables in Equation (6.27) to obtain:

$$2/7 \times 0 + 2 \times 1.5 + 17 \times 1 \geq c_5 \tag{6.28}$$

or

$$20 \geq c_5 \tag{6.29}$$

Therefore, if c_5 exceeds 20, we should enter x_5 into the basis.

If x_j is a nonbasic variable, we can examine the effect of changing its coefficients a_{ij} in exactly the same fashion. If x_j is a basic variable, analyzing the effect of changing a matrix coefficient is more complex. The analysis involves a simultaneous consideration of both the primal and dual problems. The derivation goes beyond the scope of this text, but can be found in several advanced books on LP like Hillier and Lieberman (1990).

6.4 SPECIAL TYPES OF LP PROBLEMS—TRANSPORTATION PROBLEM

The transportation problem is just a special type of LP problem, probably the most important one. It can be solved by applying the simplex method presented in Section 6.2. However, very significant computational shortcuts can be achieved by exploiting the special structure of the transportation problem. The special procedure used to solve the transportation LP problem is known as *the transportation simplex method.*

6.4.1 Formulation of the Transportation Problem

The general transportation problem is concerned with distributing any commodity from any group of supply centers *sources*, to any group of receiving centers, called *destinations*, in such a way as to minimize the total distribution cost. Thus, in general, source i ($i = 1, 2, \ldots, m$) has a supply of s_i units to distribute to the destinations, and destination j ($j = 1, 2, \ldots, n$) has a demand for d_j units to be received from the sources. A basic assumption is that the cost of distributing units from source i to destination j is directly proportional to the number distributed, where c_j denotes the cost per unit distributed. These input data can be summarized conveniently in the cost and requirements table shown in Table 6.6.

TABLE 6.6 Cost and Requirements Table for the Transportation Problem

		Cost per Unit Distributed				
		Destination				
		1	2	...	n	Supply
Source	1	c_{11}	c_{12}	...	c_{1n}	s_1
	2	c_{21}	c_{22}	...	c_{2n}	s_2
	.	.	.	...	.	.
	.	.	.	...	.	.
	.	.	.	...	.	.
	m	c_{m1}	c_{m2}	...	c_{mn}	s_m
Demand		d_1	d_2	...	d_n	

Letting Z be total distribution cost and x_{ij} $(i = 1, 2, \ldots, m; j = 1, 2, \ldots, n)$ be the number of units to be distributed from source i to destination j, the LP formulation of this problem becomes:

$$\text{Minimize } Z = \sum_{i=1}^{m} \sum_{j=1}^{n} c_{ij} x_{ij}$$

subject to

$$\begin{aligned} &\sum_{j=1}^{n} x_{ij} = s_i \quad \text{for } i = 1, 2, \ldots, m \\ &\sum_{i=1}^{m} x_{ij} = d_j \quad \text{for } j = 1, 2, \ldots, n \\ &x_{ij} \geq 0 \quad \text{for all } i \text{ and } j \end{aligned} \tag{6.30}$$

Note that the resulting table of constraint coefficients has the special structure—all coefficients are equal to 1. Any LP problem that fits this special formulation is of the transportation problem type, regardless of its physical context. For many applications, the supply and demand quantities in the model (the s_i and d_i) have integer values, and implementation will require that the distribution quantities (the x_{ij}) also have integer values. Fortunately, because of their special structure, the transportation problems have the following property.

Integer solutions property: For transportation problems where every s_i and d_j has an integer value, all the basic variables (allocations) in every basic feasible solution (including an optimal one) also have integer values.

In order to have an optimal solution of any kind, a transportation problem must possess feasible solutions. The following property indicates when this will occur.

Feasible solutions property: A necessary and sufficient condition for a transportation problem to have any feasible solutions is that:

$$\sum_{j=1}^{n} d_j = \sum_{i=1}^{m} s_i$$

This property may be verified by observing that the constraints require that both:

$$\sum_{j=1}^{n} d_j \quad \text{and} \quad \sum_{i=1}^{m} s_i \quad \text{equal} \quad \sum_{i=1}^{m}\sum_{j=1}^{n} x_{ij}$$

This condition that the total supply must equal the total demand merely requires that the system be in balance. If the problem has physical significance and this condition is not met, it usually means that either s_i or d_j actually represents a bound rather than an exact requirement. If this is the case, a fictitious "source" or "destination" (called the *dummy source* or the *dummy destination*) can be introduced to take up the slack in order to convert the inequalities into equalities and satisfy the feasibility condition.

Example 6

Let us look at the problem of water distribution to the disaster-devastated community. In order to provide the basic needs of the community the water has to be imported. There are three available locations: A, B, and C. The water is to be distributed to four locations within the disaster-hit area that are still connected by the existing pipelines: I, II, III, and IV. It is possible to supply any of these locations with water brought in from any of the three sources, with the exception that no connection is available between source C and location IV. The pipeline is damaged and not functional. However, because of the geographic layouts of the pipelines and the demand locations in the region, the cost of supplying water depends upon both the source of the water and the location being supplied. The variable cost per cubic meter of water (in dollars) for each combination of source and demand point is given in Table 6.7.

The disaster management command center is now faced with the problem of how to allocate the available water during the upcoming rebuilding period. Using units of 1 million cubic meter, the amounts available from the three sources are given in the right-hand column of Table 6.7. The command is committed to providing a certain minimum amount to meet the essential needs of population at each location (with the exception of location III, which has an independent source of water), as shown in the *Minimum needed* row of the table. The *Requested* row indicates that location II desires no more than the minimum amount, but that location I would like as much as

TABLE 6.7 Input Data for the Emergency Water Distribution Problem

		Cost per Cubic Meter				
		Destination Point				
		I	II	III	IV	Supply
Source	A	16	13	22	17	50
	B	14	13	19	15	60
	C	19	20	23	/	50
Minimum needed		30	70	0	10	In million m^3
Requested		50	70	30	∞	

20 more, location III would need up to 30 more, and location IV will take as much as it can get.

Disaster management wishes to allocate *all* the available water from the three sources to the four locations in such a way as to at least meet the essential needs of each location while minimizing the total cost involved in water supply.

Problem formulation: Table 6.7 already is close to the proper form for a cost and requirements table, with the one basic difficulty that it is not clear what the demands at the destinations should be. The amount to be received at each destination (except location II) actually is a decision variable, with both a lower and an upper bound. This upper bound is the amount requested unless the request exceeds the total supply remaining after meeting the minimum needs of the other locations, in which case this *remaining supply* becomes the upper bound. Thus, the location IV has an upper bound of $(50 + 60 + 50) - (30 + 70 + 0) = 60$. Unfortunately, the demand quantities in the transportation problem must be constants, not bounded decision variables. To begin resolving this difficulty, temporarily suppose that it is not necessary to satisfy the minimum needs, so that the upper bounds are the only constraints on amounts to be allocated to the demand locations. In this circumstance, the requested allocations can be viewed as the demand quantities for a transportation problem formulation after one adjustment.

The adjustment needed here is to introduce *a dummy source* to "send" the *unused demand capacity.* The imaginary supply quantity for this dummy source would be the amount by which the sum of the demands exceeds the sum of the real supplies: $(50 + 70 + 30 + 60) - (50 + 60 + 50) = 50$. This formulation yields the cost and requirements table shown in Table 6.8, which uses units of million cubic meter and million dollars. The cost entries in the *Dummy* row (shaded row in the Table 6.8) are *zero* because there is no cost incurred by the fictional allocations from this dummy source. On the other hand, a huge unit cost of M is assigned to the source C—location IV spot. The reason is that source C water cannot be used to supply location IV and assigning a cost of M will prevent any such allocation.

Now let us see how we can take each location's minimum needs into account in this kind of formulation. Because location III has no minimum need, it is already all set.

TABLE 6.8 Input Data for the Emergency Water Distribution Problem without Minimum Needs

		Cost per Cubic Meter				
		Destination Point				
		I	II	III	IV	Supply
Source	A	16	13	22	17	50
	B	14	13	19	15	60
	C	19	20	23	*M*	50
	Dummy	0	0	0	0	50
Requested		50	70	30	60	

Similarly, the formulation for location IV does not require any adjustments because its demand (60) exceeds the dummy source's supply (50) by 10, so the amount supplied to location IV from the real sources will be at least 10 in any feasible solution. Consequently, its minimum need of 10 from the sources is guaranteed. Location II minimum need equals its requested allocation, so its entire demand of 70 must be filled from the real sources rather than the dummy source. This requirement calls for the Big *M* method. Assigning a huge unit cost of *M* to the allocation from the dummy source to location II ensures that this allocation will be zero in an optimal solution. Finally, consider location I. In contrast to the case of location IV, the dummy source has an adequate (fictional) supply to "provide" at least some of location I minimum need in addition to its extra requested amount. Therefore, since location I minimum need is 30, adjustments must be made to prevent the dummy source from contributing more than 20 to the total demand of 50 for location I. This adjustment is accomplished by splitting location I into two destinations, one having a demand of 30 with a unit cost of *M* for any allocation from the dummy source and the other having a demand of 20 with a unit cost of zero for the dummy source allocation. This formulation gives the final cost and requirements table shown in Table 6.9.

TABLE 6.9 Input Data for the Emergency Water Distribution Problem without Minimum Needs

		Cost per Cubic Meter					
		Destination Point					
		I (min)	I (extra)	II	III	IV	Supply
Source	A	16	16	13	22	17	50
	B	14	14	13	19	15	60
	C	19	19	20	23	*M*	50
	Dummy	*M*	0	0	0	0	50
Requested		30	20	70	30	60	

This problem is solved in the next section to illustrate the special solution procedure for solving transportation problems.

6.4.2 Solution of the Transportation Problem

Because of the specific formulation of the transportation LP problem, almost the entire simplex tableau can be eliminated. To demonstrate the modified solution method for the transportation problem, I will use a transportation simplex tableau shown in Figure 6.6. Full appreciation for the great difference in efficiency and convenience between the simplex and the transportation simplex methods can be obtained by applying both to the same small problem. However, the difference becomes much more significant for large problems that must be solved on a computer.

		Destination					
		1	2	...	n	Supply	u_i
Source	1	c_{11}	c_{12}	...	c_{1n}	s_1	
	2	c_{21}	c_{22}	...	c_{2n}	s_2	
	⋮	⋮	⋮	⋮	⋮	⋮	
	m	c_{m1}	c_{m2}	...	c_{mn}	s_m	
Demand		d_1	d_2	...	d_n	$Z =$	
	v_j						

Additional information to be added to each cell:

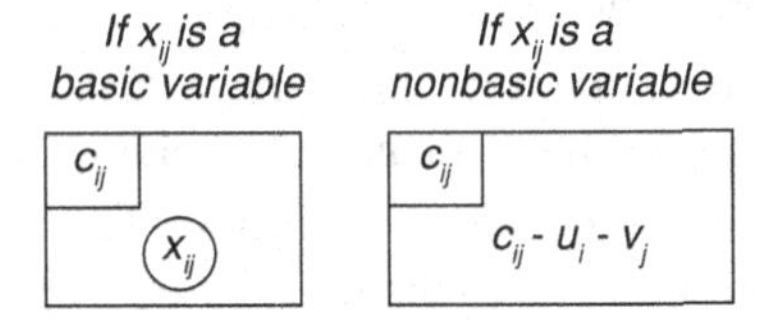

Figure 6.6 Format of transportation simplex tableau.

Implementation of the transportation simplex method proceeds as follows:

1. *Initialization step*: Construct an initial basic feasible solution. Go to the optimality test.
2. *Iterative step*:

   Part 1: Determine the entering basic variable. Select the nonbasic variable x_{ij} having the *largest* (in absolute terms) *negative* value of $(c_{ij} - u_i - v_j)$.

Part 2: Determine the leaving basic variable. Identify the chain reaction required to retain feasibility when the entering basic variable is increased. From among the donor cells, select the basic variable having the *smallest* value.

Part 3: Determine the new basic feasible solution. Add the value of the leaving basic variable to the allocation for each recipient cell. Subtract this value from the allocation for each donor cell.

3. *Optimality test*: Derive the u_i and v_j by selecting the row having the largest number of allocations and setting its $u_i = 0$ and then solving the set of equations $c_{ij} = u_i + v_j$ for each (i, j) such that x_{ij} is basic. If $(c_{ij} - u_i - v_j) \geq 0$ for every (i, j) such that x_{ij} is *nonbasic*, then the current solution is optimal, so stop. Otherwise, go back to step 2.

Initial Basic Feasible Solution The objective of the algorithm initialization step is to obtain an initial basic feasible solution. Because all functional constraints in the transportation problem are equality constraints, the simplex method would obtain the solution by introducing artificial variables as shown in Section 6.2.1. For transportation problem with m sources and n destinations, the number of functional constraints is $m + n$ but the number of basic variables is $m + n - 1$.

Therefore, any basic feasible solution appears in a transportation simplex tableau with exactly $(m + n - 1)$ circled *nonnegative locations*, where the sum of the allocations for each row or column equals its supply or demand. The procedure for constructing an initial basic feasible solution selects the $(m + n - 1)$ basic variables one at a time. A number of different criteria are available for selecting the basic variables. I will present and illustrate one of them after outlining the general procedure.

All source rows and destination columns of the transportation simplex tableau are initially under consideration for providing a basic variable (allocation).

Step 1: From among the rows and columns still under consideration, select the next basic variable (allocation) according to some criterion.

Step 2: Make that allocation large enough to exactly use up the remaining supply in its row or the remaining demand in its column (whichever is smaller).

Step 3: Eliminate that row or column (whichever had the smaller remaining supply or demand) from further consideration. (If the row and column have the same remaining supply and demand, then arbitrarily select the row as the one to be eliminated. The column will be used later to provide a degenerate basic variable, i.e., a circled allocation of zero.)

Step 4: If only one row or only one column remains under consideration, then the procedure is completed by selecting every remaining variable (i.e., those variables that were neither previously selected to be basic nor eliminated from consideration by eliminating their row or column) associated with that row or column to be basic with the only feasible allocation. Otherwise, return to step 1.

For the step 1, I will only introduce one alternative criterion known as the *Northwest corner rule*. For other criteria, the reader is advised to review Hillier and Lieberman (1990). Begin by selecting x_{11} (i.e., start in the northwest corner of the transportation simplex tableau). Thereafter, if x_{ij} was the last basic variable selected, then next select $x_{i,j+1}$ (i.e., move one column to the right) if source i has any supply remaining. Otherwise, next select $x_{i+1,j}$ (i.e., move one row down).

Example 7
Let us now demonstrate the application of the solution procedure by using the water distribution problem from Example 6 (shown in Table 6.9).

Because $m = 4$ and $n = 5$ in this case, the procedure would find an initial basic feasible solution having $m + n - 1 = 8$ basic variables. As shown in Figure 6.7, the first allocation is $x_{11} = 30$, which exactly uses up the demand in column 1 (and eliminates this column from further consideration). This first iteration leaves a supply of 20 remaining in row 1, so next select $x_{1,1+1} = x_{12}$ to be a basic variable. Because this supply is no larger than the demand of 20 in column 2, all of it is allocated, $x_{12} = 20$, and this row is eliminated from further consideration. Therefore, select $x_{1+1,2} = x_{22}$ next. Because the remaining demand of 0 in column 2 is less than the supply of 60 in row 2, allocate $x_{22} = 0$ and eliminate column 2.

Continuing in this manner, we eventually obtain the entire *initial basic feasible solution* shown in Figure 6.7, where the circled numbers are the values of the basic variables ($x_{11} = 30, \ldots, x_{45} = 50$) and all the other variables (x_{13}, etc.) are

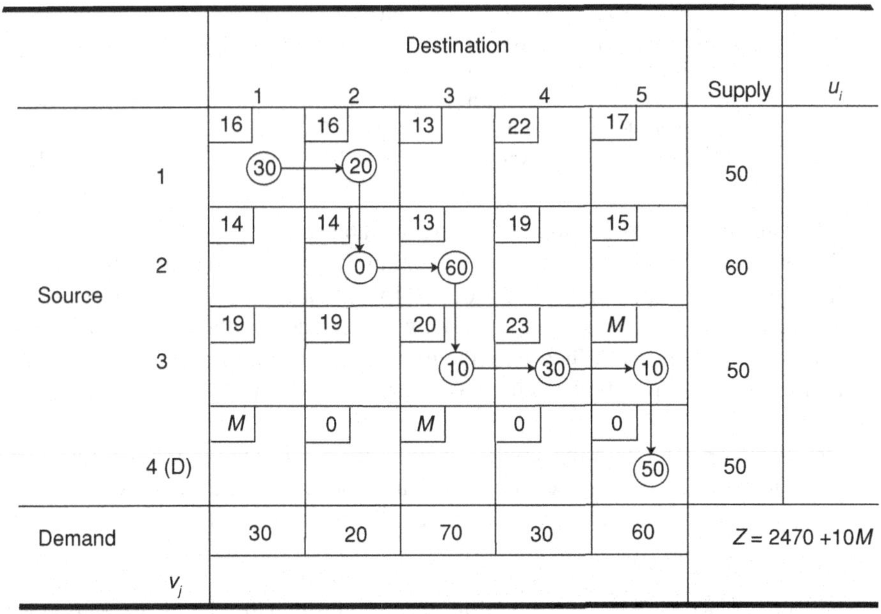

		Destination						
		1	2	3	4	5	Supply	u_i
Source	1	16 (30) →	16 (20) ↓	13	22	17	50	
	2	14	14 (0) →	13 (60) ↓	19	15	60	
	3	19	19	20 (10) →	23 (30) →	M (10) ↓	50	
	4 (D)	M	0	M	0	0 (50)	50	
Demand		30	20	70	30	60	$Z = 2470 + 10M$	
	v_j							

Figure 6.7 Initial basic feasible solution from northwest corner rule.

nonbasic variables equal to zero. Arrows have been added to show the order in which the basic variables (allocations) were selected. The value of Z for this solution is $Z = 16(30) + 16(20) + \ldots + 0(50) = 2470 + 10M$.

Optimality Test The standard optimality test for the simplex method (presented in Section 6.2) can be modified for the transportation problem in the following way:

> **Optimality test**: A basic feasible solution is optimal if and only if $(c_{ij} - u_i - v_j) \geq 0$ for every (i, j) such that x_{ij} is nonbasic.

Thus, the only check required by the optimality test is the derivation of the values of the u_i and v_j for the current basic feasible solution and then the calculation of these $(c_{ij} - u_i - v_j)$. Since $(c_{ij} - u_i - v_j)$ is required to be zero if x_{ij} is a basic variable, the u_i and v_j, satisfy the set of equations $c_{ij} = u_i + v_j$ for each (i, j) such that x_{ij} is basic.

There are $(m + n - 1)$ basic variables, and so there are $(m + n - 1)$ of these equations. Since the number of unknowns (the u_i and v_j) is $(m + n)$, one of these variables can be assigned a value arbitrarily without violating the equations. The rule I am suggesting is to select the u_i that has the largest number of allocations in its row and assign it the value of zero. It is then very simple to solve for the remaining variables algebraically.

To demonstrate, let us look at each equation that corresponds to a basic variable in our initial basic feasible solution.

x_{31}:	$19 = u_3 + v_1$	Set	$u_3 = 0$	so	$v_1 = 19$
x_{32}:	$19 = u_3 + v_2$	Set	$u_3 = 0$	so	$v_2 = 19$
x_{34}:	$23 = u_3 + v_4$	Set	$u_3 = 0$	so	$v_4 = 23$
x_{21}:	$14 = u_2 + v_1$	Known	$v_1 = 19$	so	$u_2 = -5$
x_{23}:	$13 = u_2 + v_3$	Known	$u_2 = -5$	so	$v_3 = 18$
x_{13}:	$13 = u_1 + v_3$	Known	$v_3 = 18$	so	$u_1 = -5$
x_{15}:	$17 = u_1 + v_5$	Known	$u_1 = -5$	so	$v_5 = 22$
x_{45}:	$0 = u_4 + v_5$	Known	$v_5 = 22$	so	$u_4 = -22$

The completed initial transportation simplex tableau is in Figure 6.8.

We are now in a position to apply the optimality test by checking the value of the $(c_{ij} - u_i - v_j)$ given in Figure 6.8. Because two of these values, $(c_{25} - u_2 - v_5) = -2$ and $(c_{44} - u_4 - v_4) = -1$, are negative, we conclude that the current basic feasible solution is not optimal. Therefore, the transportation simplex method must proceed to the iterative step to find a better basic feasible solution.

Iterative Step As with the simplex method from Section 6.2, the iterative step for the transportation simplex version must determine an entering basic variable (part 1), a leaving basic variable (part 2), and then identify the resulting new basic feasible solution (part 3).

Part 1: Since $(c_{ij} - u_i - v_j)$ represents the rate at which the objective function would change as the nonbasic variable x_{ij} is increased, the entering basic variable must

Iteration 0		Destination 1	2	3	4	5	Supply	u_i
Source	1	16 +2	16 +2	13 (40)	22 +4	17 (10)	50	−5
	2	14 (30)	14 0	13 (30)	19 +1	15 −2	60	−5
	3	19 (0)	19 (20)	20 +2	23 (30)	M $M-22$	50	0
	4 (D)	M $M+3$	0 +3	M $M+4$	0 −1	0 (50)	50	−22
Demand		30	20	70	30	60	$Z = 2570$	
	v_j	19	19	18	23	22		

Figure 6.8 Completed initial transportation simplex tableau.

have a negative $(c_{ij} - u_i - v_j)$ to decrease the total cost Z. Thus, the candidates in Figure 6.8 are x_{25} and x_{44}. To choose between the candidates, select the one having the largest (in absolute terms) negative value of $(c_{ij} - u_i - v_j)$ to be the entering basic variable, which is x_{25} in this case.

Part 2: Increasing the entering basic variable from zero sets off a chain reaction of compensating changes in other basic variables (allocations) in order to continue satisfying the supply and demand constraints. The first basic variable to be decreased to zero then becomes the leaving basic variable.

With x_{25} as the entering basic variable, the chain reaction in Figure 6.8 is the relatively simple one summarized in Figure 6.9. The entering basic variable should be always marked by placing a boxed + sign in its cell. Thus, increasing x_{25} requires decreasing x_{15} by the same amount to restore the demand of 60 in column 5, which in turn requires increasing x_{13} by this amount to restore the supply of 50 in row 1, which in turn requires decreasing x_{23} by this amount to restore the demand of 70 in column 3. This decrease in x_{23} successfully completes the chain reaction because it also restores the supply of 60 in row 2.

The net result is that cells (2, 5) and (1, 3) become recipient cells, each receiving its additional allocation from one of the donor cells, (1, 5) and (2, 3). These cells are indicated in Figure 6.9 by the + and − signs. Note that cell (1, 5) had to be the donor cell for column 5 rather than cell (4, 5), because cell (4, 5) would have no recipient cell in row 4 to continue the chain reaction.

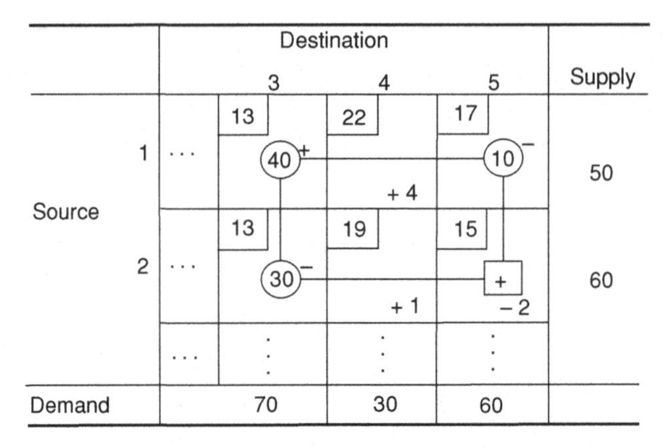

Figure 6.9 Chain reaction caused by increasing the entering basic variable x_{25}.

Each donor cell decreases its allocation by exactly the same amount that the entering basic variable (and other recipient cells) is increased. Therefore, the donor cell that starts with the smallest allocation—cell (1, 5) in this case (since $10 < 30$ in Figure 6.9)—must reach a zero allocation first as the entering basic variable x_{25} is increased. Thus, x_{15} becomes the leaving basic variable.

In general, there is always just one chain reaction (in either direction) that can be completed successfully to maintain feasibility when the entering basic variable is increased from zero. This chain reaction can be identified by selecting among the cells having a basic variable: first the donor cell in the column having the entering basic variable, then the recipient cell in the row having this donor cell, then the donor cell in the column having this recipient cell, and so on until the chain reaction yields a donor cell in the *row* having the entering basic variable.

Part 3: The new basic feasible solution is identified simply by adding the value of the leaving basic variable (before any change) to the allocation for each recipient cell and subtracting this same amount from the allocation for each donor cell. In Figure 6.9 the value of the leaving basic variable x_{15} is 10, so this portion of the transportation simplex tableau changes as shown in Figure 6.10 for the new solution. Since x_{15} is nonbasic in the new solution, its new allocation of zero is no longer shown in this new tableau.

We can now highlight a useful interpretation of the $(c_{ij} - u_i - v_j)$ quantities derived during the optimality test. Because of the shift of 10 allocation units from the donor cells to the recipient cells (shown in Figures 6.9 and 6.10), the total cost changes by:

$$\Delta Z = 10(15 - 17 + 13 - 13) = 10(-2) = 10(c_{25} - u_2 - v_5).$$

Thus, the effect of increasing the entering basic variable x_{25} from zero has been a cost change at the rate of -2 per unit increase in x_{25}. This is precisely what the value of $(c_{25} - u_2 - v_5) = -2$ in Figure 6.8 indicates would happen.

			Destination		
			3	4	5	Supply
Source	1	...	13 (50)	22	17	50
	2	...	13 (20)	19	15 (10)	60
		...	⋮	⋮	⋮	
Demand			70	30	60	

Figure 6.10 Change in the basic feasible solution.

Example 8

Let us now completely solve the postdisaster water distribution problem from Example 6. The complete set of transportation simplex tableaux is shown in Figure 6.11. Since all the $(c_{ij} - u_i - v_j)$ are nonnegative in the fourth tableau, the optimality test identifies the set of allocations in this tableau as being optimal, which concludes the algorithm.

6.5 SPECIAL TYPES OF LP PROBLEMS—NETWORK PROBLEMS

In this section, I am focusing on the network analysis as a special branch of linear optimization methods. Networks arise in numerous disaster management settings including transportation, power supply, and communications among others. Network representations are widely used in distribution, project planning, facilities location resource management, and financial planning. A network representation provides the most powerful visual and conceptual aid for describing the relationships between the components of systems.

This section provides introduction to three network algorithms: the shortest path, the minimum spanning tree, and the maximum flow. Each of these three network problem types has a very specific structure that arises frequently in disaster management applications. The following example will serve as a prototype example to introduce all three network algorithms.

Example 9

The City of Port-au-Prince in Haiti has been hit by a powerful earthquake. Many roads have been destroyed and population around the City left without shelter, water, and food. Most of the aid arrived by air. One of the main logistical problems was to establish functional links between the point of aid entry (airport) and people around the city.

Iteration 0

Source	Destination 1	2	3	4	5	Supply	u_i
1	16; +2	16; +2	13; (40)	22; +4	17; (10)	50	−5
2	14; (30)	14; 0	13; (30)	19; +1	15; [+]; −2	60	−5
3	19; (0)	19; (20)	20; +2	23; (30)	M; $M-22$	50	0
4(D)	M; $M+3$	0; +3	M; $M+4$	0; −1	0; (50)	50	−22
Demand	30	20	70	30	60	$Z = 2570$	
v_j	19	19	18	23	22		

Iteration 1

Source	Destination 1	2	3	4	5	Supply	u_i
1	16; +2	16; +2	13; (50)	22; +4	17; +2	50	−5
2	14; (30)−	14; 0	13; (20)	19; +1	15; (10)+	60	−5
3	19; (0)+	19; (20)	20; +2	23; (30)−	M; $M-20$	50	0
4(D)	M; $M+1$	0; +1	M; $M+2$	0; [+]; −3	0; (50)−	50	−20
Demand	30	20	70	30	60	$Z = 2550$	
v_j	19	19	18	23	22		

Iteration 2

Source	Destination 1	2	3	4	5	Supply	u_i
1	16; +5	16; +5	13; (50)	22; +7	17; +2	50	−8
2	14; +3	14; +3	13; (20)−	19; +4	15; (40)+	60	−8
3	19; (30)+	19; (20)	20; [+]; −1	23; (0)−	M; $M-23$	50	0
4(D)	M; $M+4$	0; +4	M; $M+2$	0; (30)+	0; (20)−	50	−23
Demand	30	20	70	30	60	$Z = 2460$	
v_j	19	19	21	23	23		

Iteration 3

Source	Destination 1	2	3	4	5	Supply	u_i
1	16; +4	16; +4	13; (50)	22; +7	17; +2	50	−7
2	14; +2	14; +2	13; (20)	19; +4	15; (40)	60	−7
3	19; (30)	19; (20)	20; (0)	23; +1	M; $M-22$	50	0
4(D)	M; $M+3$	0; +3	M; $M+2$	0; (30)	0; (20)	50	−22
Demand	30	20	70	30	60	$Z = 2460$	
v_j	19	19	20	22	22		

Figure 6.11 Complete set of transportation simplex tableaux for the postdisaster water distribution problem.

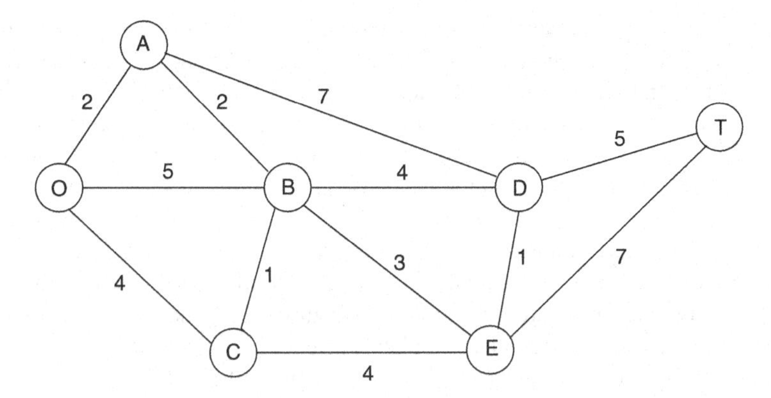

Figure 6.12 The road system for the prototype example.

The following are three problems created to illustrate potential application of optimization in managing disasters of that type. All data have been invented for illustration purposes.

In spite of the road damage, a limited movement is possible along a narrow, winding road system. This road system is shown (without the curves) in Figure 6.12, where location 0 is the entrance into the city (airport). Other letters designate the locations of limited facilities that are providing assistance to those involved in aid distribution. The numbers give the distances of these winding roads in kilometers. The point T is the main distribution point, where a majority of the affected people can access medical help, shelter, water, and food. A small number of vehicles can transport aid from the airport to the main distribution station T and back.

The disaster relief management currently faces three problems:

(i) One is to determine which route from the airport to station T has the *smallest total distance* for the operation of the vehicles. This is an example of the *shortest path problem* to be discussed in Section 6.5.1.

(ii) A second problem is that emergency electricity lines must be installed along the roads to support aid distribution and activate limited facilities at all the stations (emergency shelter, medical support, water, and food distribution). The electricity source is assumed to be at the airport. Because the installation is both expensive and temporary, lines will be installed along just enough roads to provide some connection between every pair of stations. The question is where the lines should be laid to accomplish this with *a minimum* total number of kilometers of line installed. This is an example of the *minimum spanning tree problem* to be discussed in Section 6.5.2.

(iii) The third problem is that more vehicles are available to take the ride from the airport to station T than can be accommodated by the road capacity. To avoid further destruction of the roads a strict limit has been placed on the number of vehicle trips that can be made on each of the roads per day. These limits differ for the different roads. Therefore, during the early stages of disaster recovery, various routes might be followed regardless of distance to increase the number of vehicle trips that can be made each day. The question is how to route the various trips to *maximize* the number of trips that can be made per day without violating the capacity limits on any individual road. This is an example of the *maximum flow problem* to be discussed in Section 6.5.3.

Before I start the introduction of various network algorithms let me review the terminology of networks. A relatively extensive terminology has been developed to describe the various kinds of networks and their components. Although I tried to avoid as much of this special vocabulary as I could, there is still a need to introduce a considerable number of terms for use throughout the following three sections of the chapter. I suggest that you read through this introduction once at the beginning, and then refresh your memory as the terms are used in subsequent sections. To assist you, each term is highlighted in **boldface** at the point where it is defined.

A network consists of a set of points and a set of lines connecting certain pairs of the points. The points are called **nodes** (or vertices). The lines are called **arcs** (or links or edges or branches). Arcs are labeled by naming the nodes at either end; *AB* is the arc between nodes *A* and *B*. The arcs of a network may have a flow of some type through them. If flow through an arc is allowed only in one direction, the arc is said to be a **directed arc.** If the flow through an arc is allowed in both directions, the arc is said to be an **undirected arc** (often referred to as a link). A network that has only directed arcs is called a **directed network**. Similarly, if all of its arcs are undirected, the network is said to be an **undirected network**. A network with a mixture of directed and undirected arcs can be converted into a directed network, if desired, by replacing each undirected arc by a pair of directed arcs in opposite directions.

When two nodes are not connected by an arc, they may be connected by a **path** that is defined as a sequence of distinct arcs connecting these nodes. When some or all of the arcs in the network are directed arcs, we then distinguish between directed paths and undirected paths. A **directed path** from node i to node j is a sequence of connecting arcs whose direction (if any) is toward node j, so that flow from node i to node j along this path is feasible. An **undirected path** from node i to node j is a sequence of connecting arcs whose direction can be either toward or away from node j.

To illustrate these definitions, Figure 6.13 shows a typical directed network. The sequence of arcs *AB-BC-CE* is a directed path from node *A* to node *E* since flow toward node *E* along this entire path is feasible. On the other hand, *BC-AC-AD* is not a directed path from node *B* to node *D*, because the direction of arc *AC* is away from node *D* (on this path).

A path that begins and ends at the same node is called a **cycle**. In a directed network, a cycle is either a directed or an undirected cycle, depending on whether the path involved is a directed or an undirected path. In Figure 6.13, for example, *DE-ED* is a directed cycle. By contrast, *AB-BC-AC* is not a directed cycle, because the direction of arc *AC* opposes the direction of arcs *AB* and *BC*. On the other hand, *AB-BC-AC* is an undirected cycle.

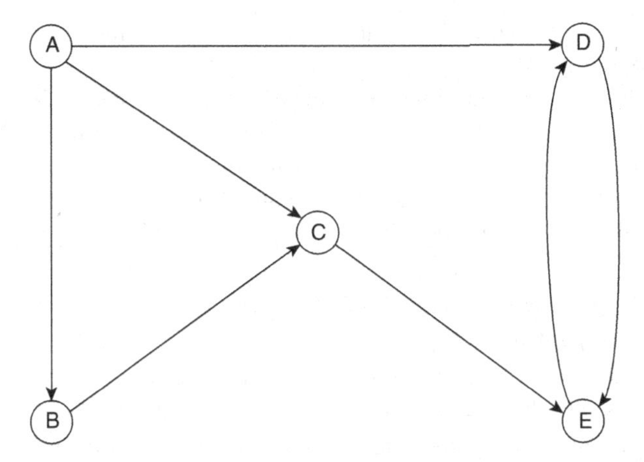

Figure 6.13 Example of a directed network.

Two nodes are said to be **connected** if the network contains at least one undirected path between them. A **connected network** is a network where every pair of nodes is connected. Thus, the networks in Figures 6.12 and 6.13 are both connected.

Consider a set of n nodes (the $n = 5$ nodes in Figure 6.13) without arcs. A "tree" can then be "grown" by adding one arc (or branch) at a time in a certain way. The first arc can go anywhere to connect some pair of nodes. Thereafter, each new arc should be between a node that already is connected to other nodes and a new node not previously connected to any other nodes. Adding an arc in this way avoids creating a cycle, and also ensures that the number of connected nodes is greater than the number of arcs. Each new arc creates a larger **tree**, which is a connected network (for some subset of the n nodes) that contains no undirected cycles. Once the $(n - 1)$ arc has been added, the process stops because the resulting tree spans (connects) all n nodes. This tree is called a **spanning tree**, that is, a connected network for all n nodes that contains no undirected cycles. Every spanning tree has exactly $(n - 1)$ arcs.

Finally, let us consider a little additional terminology about flows in networks. The maximum amount of flow that can be carried on a directed arc is referred to as the **arc capacity**. For nodes, a distinction is made among those that are net generators of flow, net absorbers of flow, or neither. A **supply node** (or source node or source) has the property that the flow out of the node exceeds the flow into the node. A **demand node** (or sink node or sink) has the flow into the node that exceeds the flow out of the node. A **transshipment node** (or intermediate node) satisfies conservation of flow, so flow in equals flow out.

6.5.1 The Shortest Path Problem

There are various versions of the shortest path problem (Hillier and Lieberman, 1990), and in this text, I will focus on the one simple version. Consider an undirected and connected network with the origin and the destination nodes. Associated with each of the links is a nonnegative distance. The objective is to find the shortest path from the origin to the destination.

A straightforward algorithm is available for this problem. The essence of the procedure is that it fans out from the origin, successively identifying the shortest path to each of the nodes of the network in the ascending order of their (shortest) distances from the origin, thereby solving the problem when the destination node is reached. The algorithm for shortest path problem can be summarized as follows:

1. *Objective of nth iteration*: Find nth nearest node to origin.
2. *Input for nth iteration*: $(n - 1)$ nearest nodes to origin (solved for at previous iterations), including their shortest path and distance from the origin. These nodes, plus the origin, are called solved nodes; the others are unsolved nodes.
3. *Candidates for nth nearest node*: Each solved node that is directly connected by a link to one or more unsolved nodes provides one candidate—the unsolved node with the *shortest* connecting link.
4. *Calculation of nth nearest node*: For each such solved node and its candidate, add the distance between them and the distance of the shortest path from the origin to this solved node. The candidate with the smallest such total distance is the nth nearest node, and its shortest path is the one generating this distance.

Example 10

Solve problem (i) of the Example 9. In solving this problem, we will use the shortest path algorithm presented above.

We need to find the shortest path from the airport (node 0) to the main distribution point (node T) through the road system shown in Figure 6.12. Applying the preceding algorithm to this problem yields the results shown in Table 6.10 (where the tie for

TABLE 6.10 Shortest Path Problem Solution for the Example 10

n	Solved Nodes	Closest Connected Unsolved Node	Total Distance	nth Nearest Node	Minimum Distance	Last Connection
1	O	A	2	A	2	OA
2, 3	O	C	4	C	4	OC
	A	B	$2 + 2 = 4$	B	4	AB
4	A	D	$2 + 7 = 9$			
	B	E	$4 + 3 = 7$	E	7	BE
	C	E	$4 + 4 = 8$			
5	A	D	$2 + 7 = 9$			
	B	D	$4 + 4 = 8$	D	8	BD
	E	D	$7 + 1 = 8$	D	8	ED
6	D	T	$8 + 5 = 13$	T	13	DT
	E	T	$7 + 7 = 14$			

the second nearest node allows skipping directly to seeking the fourth nearest node next).

The first column (n) indicates the iteration. The second column simply lists the solved nodes for beginning the current iteration. The third column then gives the candidates for the nth nearest node. The fourth column calculates the distance of the shortest path from the origin to each of these candidates. The candidate with the smallest such distance is the nth nearest node to the origin, as listed in the fifth column. The last two columns summarize the information for this new solved node that is needed to proceed to subsequent iterations.

The shortest path *from the airport to the main distribution point* can now be traced back through the last column of Table 6.10 as either T→D→E→B→A→O or T→D→B→A →O with a total distance of 13 km on either path.

6.5.2 The Minimum Spanning Tree Problem

The minimum spanning tree problem has some similarities to the main version of shortest path problem presented in Section 6.5.1. In both cases, an undirected network is being considered, where the given information includes the nodes and the distances between pairs of nodes. However, for the minimum spanning tree problem the links between the nodes are no longer specified. So the problem now involves choosing for the network the links that have the shortest length while providing a path between each pair of nodes. The links need to be chosen in such a way that the resulting network forms a tree that spans (connects) all the given nodes.

Figure 6.14 illustrates this concept of a spanning tree for a prototype problem from Example 9.

Thus, Figure 6.14a is not a spanning tree because the (O, A, B, C) nodes are not connected with the (D, E, T) nodes. The links in Figure 6.14b do span the network but do not form a tree because there are two cycles (O-A-B-C-O and D-T-E-D). Because the prototype problem has $n = 7$ nodes, the network must have exactly $(n - 1) =$ 6 links, with no cycles, to qualify as a spanning tree. This condition is achieved in Figure 6.14c, so this network is a feasible solution (with a value of 24 km for the total length of the links) for the minimum spanning tree problem.

This problem has a number of important practical disaster management applications. For example, it can sometimes be helpful in planning temporary transportation networks that will not be used much, where the primary consideration is to provide some path between all pairs of nodes in the most economical way. The nodes would be the locations that require access to the other locations, the branches would be transportation lanes (highways, railroad tracks, air lanes, etc.), and the "distances" (link values) would be the costs of providing the transportation lanes. In this context, the minimum spanning tree problem is to determine which transportation lanes would service all the locations with a minimum total cost. Other examples where a comparable decision arises include the planning of disaster communication networks and disaster aid distribution networks. Both represent important application areas.

The minimum spanning tree problem can be solved in a very straightforward way because it happens to be one of the few problems where being "greedy" at each stage

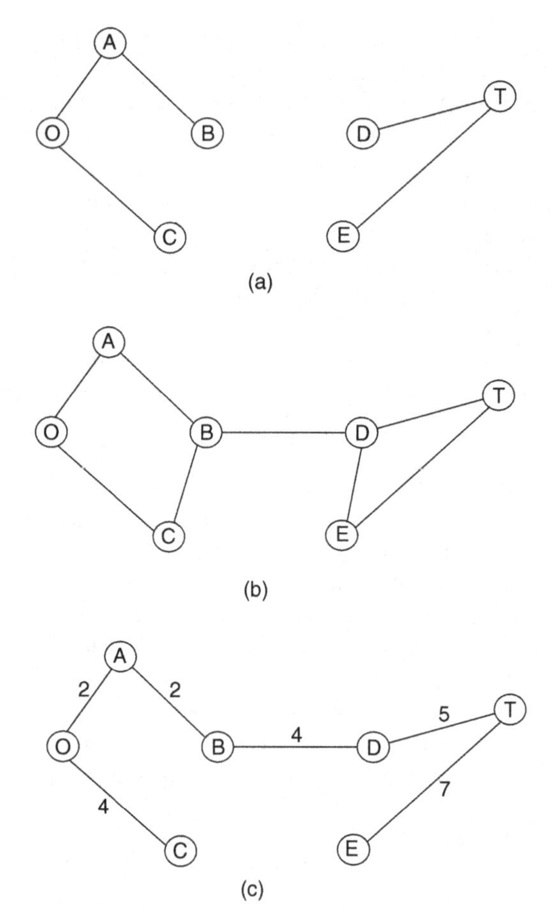

Figure 6.14 Illustration of the spanning tree concept for the prototype problem: (a) not a spanning tree, (b) not a spanning tree, and (c) a spanning tree.

of the solution procedure still leads to an overall optimal solution at the end. Thus, beginning with any node, the first stage involves choosing the shortest possible link to another node. The second stage involves identifying the unconnected node that is closest to either of these connected nodes and then adding the corresponding link to the network. This process would be repeated until all the nodes have been connected. The resulting network is guaranteed to be a minimum spanning tree. The algorithm for minimum spanning tree problem can be summarized as follows:

1. Select any node arbitrarily, and then connect it (add a link) to the nearest distinct node.
2. Identify the unconnected node that is closest to a connected node, and then connect these two nodes (i.e., add a link between them). Repeat this step until all nodes have been connected.

In the application of the minimum spanning tree algorithm, we may encounter a tie in both steps. Ties for the nearest distinct node (step 1) or the closest unconnected node (step 2) may be broken arbitrarily, and the algorithm must still yield an optimal solution. However, such ties are a signal that there may be (but need not be) multiple optimal solutions. All such optimal solutions can be identified by pursuing all ways of breaking ties to their conclusion. The fastest way of executing this algorithm manually is the graphical approach that will be used in our next example.

Example 11

Solve problem (ii) of Example 9. The disaster management team needs to determine along which roads electricity lines should be installed to connect all limited facility nodes with a minimum total length of line. Using the data given in Figure 6.12, let us outline the step-by-step solution of this problem. Nodes and distances for the problem are presented in Figure 6.12, where the lines connecting nodes now represent potential links.

Arbitrarily select node O to start. The unconnected node closest to node O is node A. Connect these two nodes as shown in Figure 6.15.

The unconnected node closest to either node O or A is node B (closest to node A). Connect node B to node A as shown in Figure 6.16.

The unconnected node closest to node O, A, or B is node C (closest to node B). Connect node C to node B as shown in Figure 6.17.

The unconnected node closest to node O, A, B, or C is node E (closest to node B). Connect node E to node B as shown in Figure 6.18.

The unconnected node closest to node O, A, B, C, or E is node D (closest to node E). Connect node D to node E as shown in Figure 6.19.

The only remaining unconnected node is node T. It is closest to D. Connect node D to node T as shown in Figure 6.20.

All nodes are now connected and the tree in Figure 6.20 is the optimal solution of the problem with total length of the links equal to 14 km. Note that the choice of the initial node is not affecting the optimal solution.

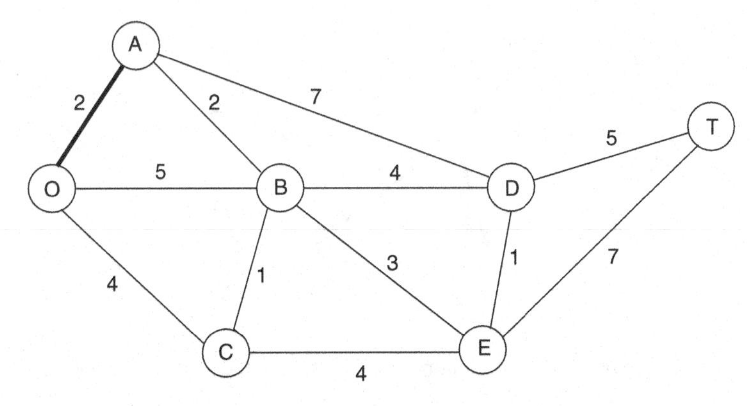

Figure 6.15 Minimum spanning tree for the prototype problem.

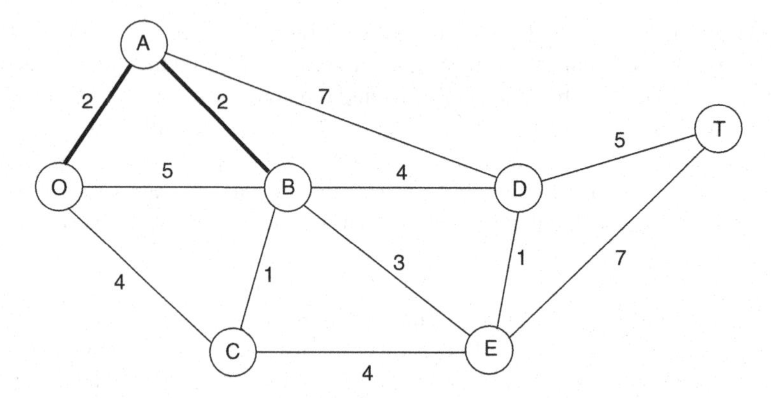

Figure 6.16 Minimum spanning tree for the prototype problem.

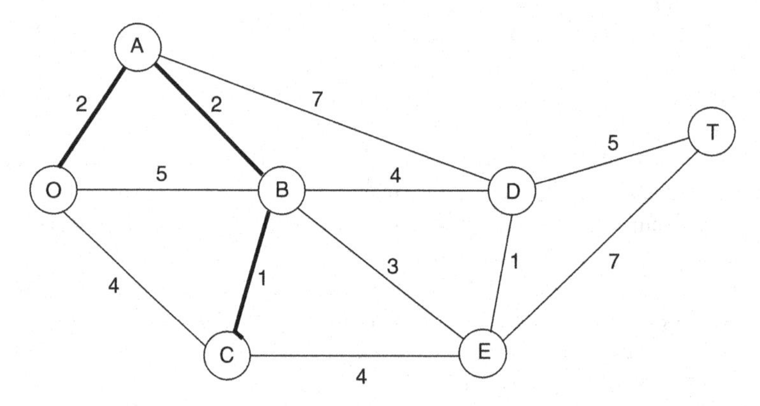

Figure 6.17 Minimum spanning tree for the prototype problem.

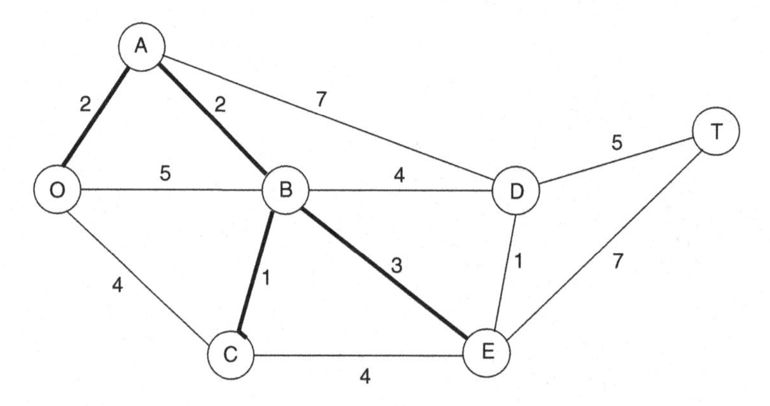

Figure 6.18 Minimum spanning tree for the prototype problem.

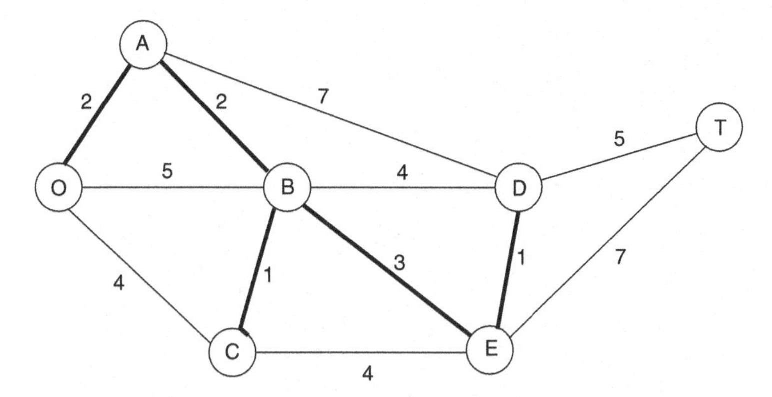

Figure 6.19 Minimum spanning tree for the prototype problem.

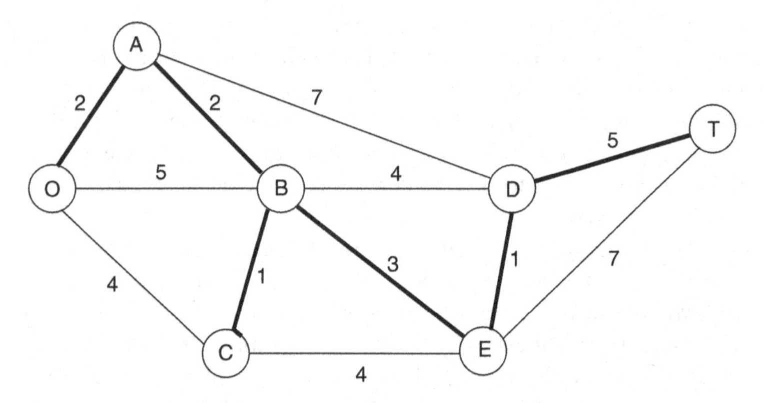

Figure 6.20 Minimum spanning tree for the prototype problem.

The minimum spanning tree problem addressed in this section falls into the category of network design problems. The main objective here is to design the most appropriate network for the given application rather than to analyze an already designed network.

6.5.3 The Maximum Flow Problem

Recall the problem (iii) of our prototype Example 9 shown in Figure 6.12. The third problem is how to increase the number of vehicle trips that can be made each day without violating the road capacity limits and without restriction on the distance traveled by aid vehicles. The question is how to route the various trips and maximize the number of trips that can be made per day starting from the airport (node O) and ending at the main distribution point (node T). Each vehicle will return by the same route it took on outgoing trip, so the analysis focuses on outgoing trips only. In order to accommodate the capacity of roads and not cause additional damage, strict

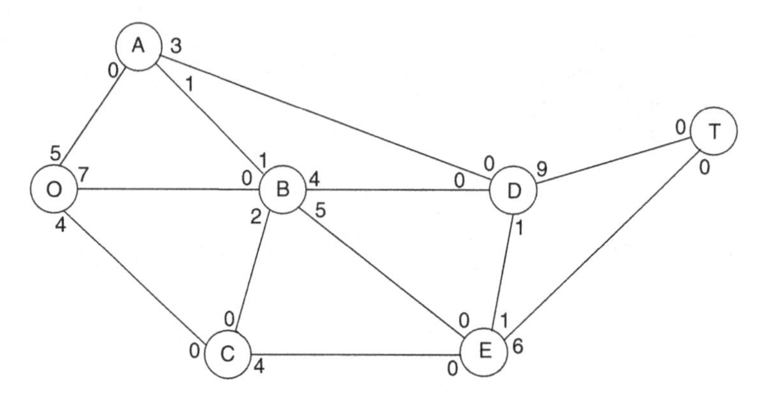

Figure 6.21 Limits on the number of trips per day for the prototype problem.

upper limits have been imposed on the number of outgoing trips allowed per day in each direction on each individual road. These limits are shown in Figure 6.21, where the numbers next to each node and road give the limit for that road in the direction leading away from that node. For example, only one loaded trip per day is allowed from station A to station B. Many combinations of routes (and the number of trips to assign to each one) need to be considered to find the one that maximizes the number of trips made per day.

Using the terminology introduced in the introductory part of this section, the maximum flow problem can be described as follows. Consider a directed and connected network where just one node is a supply node, one node is a demand node, and the rest are transshipment nodes. Given the arc capacities, the objective is to determine the feasible pattern of flows through the network that maximizes the total flow from the supply node to the demand node.

To adopt our prototype problem to the format with a directed network, each link in Figure 6.21a with 0 at one end would be replaced by a directed arc in the direction of feasible flow. For example, the link between nodes O and A would be replaced by a directed arc from node O to node A with an arc capacity of 5. The two links with a 1 at either end (AB and DE) would be replaced by a pair of directed arcs in opposite directions, each with an arc capacity of 1.

Because the maximum flow problem can be formulated as an LP problem, it can be solved by the simplex method. However, I will present here an even more efficient augmenting path algorithm that is available for solving this problem.

The algorithm requires introduction of the residual network and augmenting path concepts. The **residual network** is one in which each directed arc ($i \to j$) lacking a directed arc in the opposite direction ($j \to i$) has such an arc added with zero arc capacity. Subsequently, the arc capacities in the residual network (called residual capacities) are adjusted as follows: each time some amount of flow Δ is added to arc ($i \to j$) in the original network, the residual capacity of arc ($i \to j$) is decreased by Δ but the residual capacity of arc ($j \to i$) is increased by Δ. Thus, the residual capacity represents the unused arc capacity in the original network or the amount of flow in the opposite direction in this network that can be cancelled. Therefore, after assigning various flows to the original network, the residual network shows how much more can be done either by increasing flows further or by canceling previously assigned flows.

An **augmenting path** is a directed path from the supply node to the demand node in the residual network such that every arc on this path has strictly positive residual capacity. The minimum of these residual capacities is called the residual capacity of the augmenting path because it represents the amount of flow that can feasibly be added to the entire path. Therefore, each augmenting path provides an opportunity to further augment the flow through the original network.

The augmenting path algorithm repeatedly selects some augmenting path and adds a flow equal to its residual capacity to that path in the original network. This process continues until there are no more augmenting paths, so the flow from the supply node to the demand node cannot be increased further.

To summarize, the maximum flow problem algorithm consists of the following three steps:

1. Identify an augmenting path by finding some directed path from the supply node to the demand node in the residual network such that every arc on this path has strictly positive residual capacity. (If no augmenting path exists; the net flows already assigned constitute an optimal flow pattern.)
2. Identify the residual capacity c^* of this augmenting path by finding the minimum of the residual capacities of the arcs on this path. Increase the flow in this path by c^*.
3. Decrease by c^* the residual capacity of each arc on this augmenting path. Increase by c^* the residual capacity of each arc in the opposite direction on this augmenting path. Return to step 1.

When performing step 1, there often will be a number of alternative augmenting paths to choose from. There are various sophisticated algorithms for making this selection. In the following example, the selection will just be made arbitrarily.

Example 12

Solve problem (iii) of Example 9 with original network shown in Figure 6.12. For each iteration, we show residual network after completing all three steps, where a single line is used to represent the pair of directed arcs in opposite directions between each pair of nodes. The residual capacity of arc $(i \rightarrow j)$ is shown next to node i, whereas the residual capacity of arc $(j \rightarrow i)$ is shown next to node j. After each iteration, the **boldface** number (next to nodes O and T) is used to show the total amount of flow achieved thus far.

Iteration 1: Starting with Figure 6.21, one of several augmenting paths is O→B→E→T, which has a residual capacity of min{7, 5, 6} = 5. Assigning a flow of 5 to this path, the resulting residual network is in Figure 6.22.

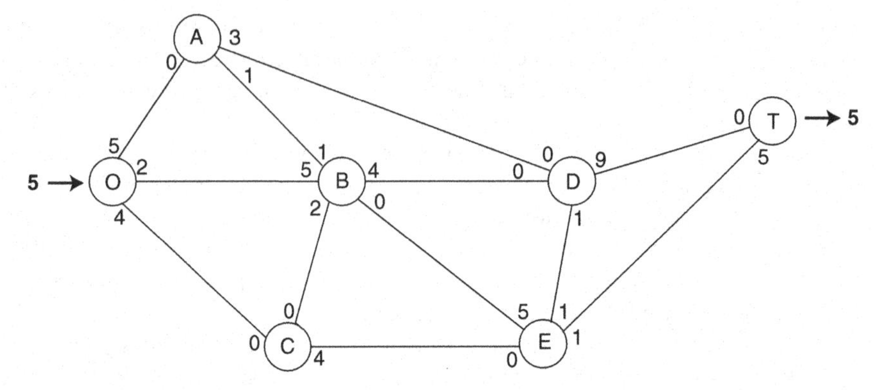

Figure 6.22 The residual network for the prototype problem after iteration 1.

Iteration 2: Assign a flow of 3 to augmenting path O→A→D→T. The resulting residual network is in Figure 6.23.

Iteration 3: Assign a flow of 1 to augmenting path O→A→B→D→T.

Iteration 4: Assign a flow of 2 to augmenting path O→B→D→T. The resulting residual network is in Figure 6.24.

Iteration 5: Assign a flow of 1 to augmenting path O→C→E→D→T.

Iteration 6: Assign a flow of 1 to augmenting path O→C→E→T. The resulting residual network is in Figure 6.25.

Iteration 7: Assign a flow of 1 to augmenting path O→C→E→B→T. The resulting residual network is in Figure 6.26.

There are no more augmenting paths and the solution shown in Figure 6.27 is optimal.

The most difficult part in the implementation of this algorithm to large networks is finding an augmenting path. This task may be simplified by the application of the

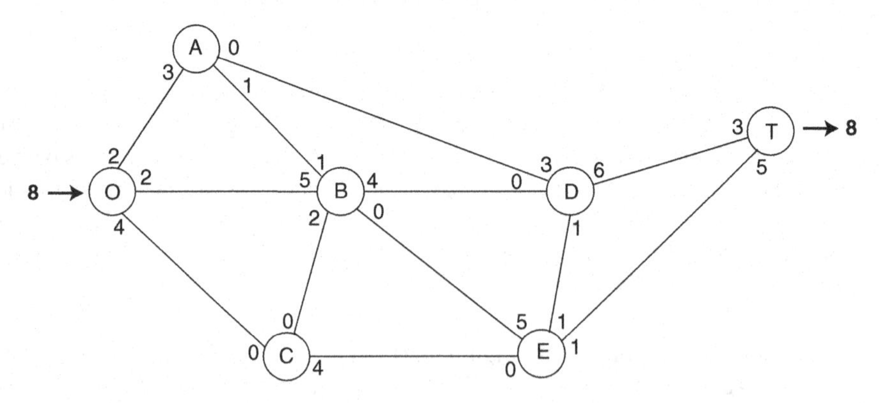

Figure 6.23 The residual network for the prototype problem after iteration 2.

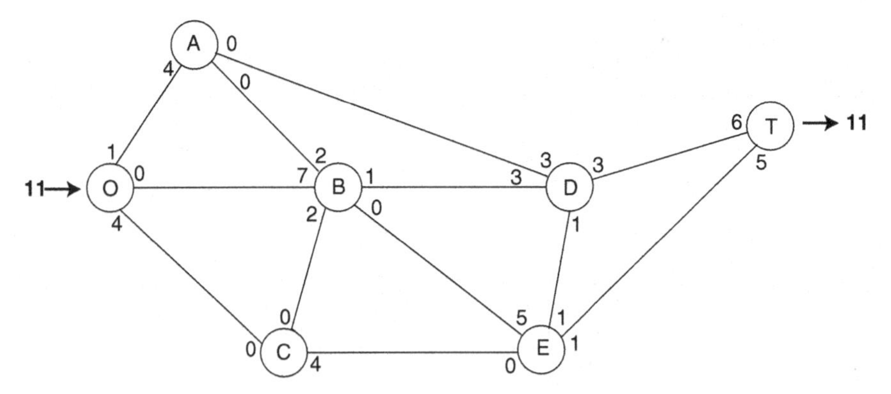

Figure 6.24 The residual network for the prototype problem after iteration 4.

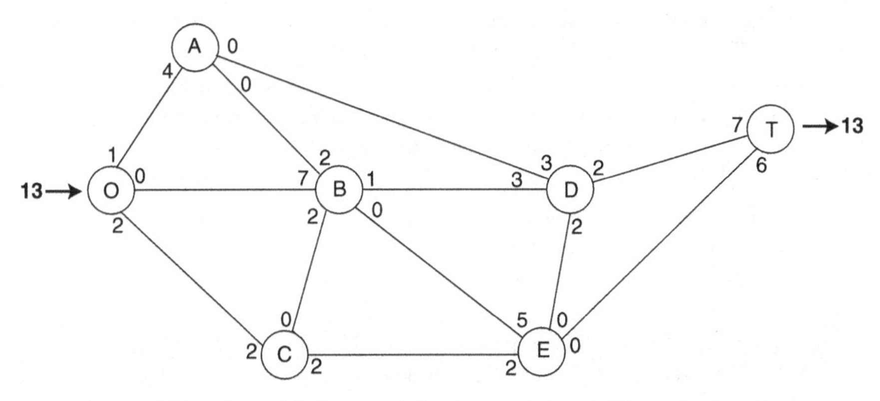

Figure 6.25 The residual network for the prototype problem after iteration 6.

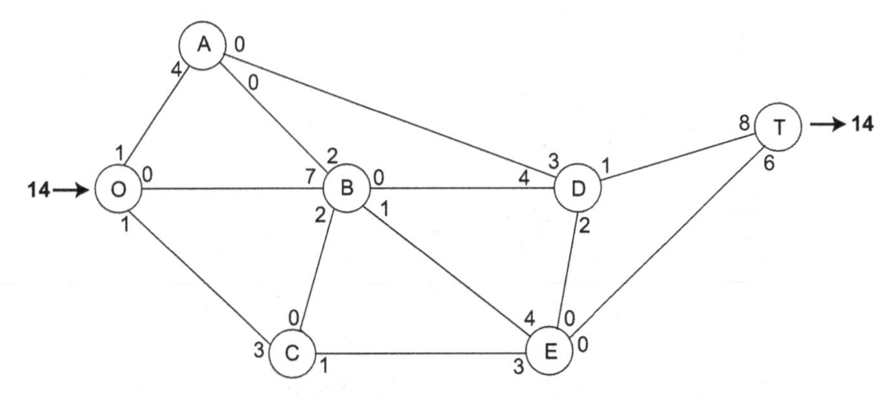

Figure 6.26 The residual network for the prototype problem after iteration 7.

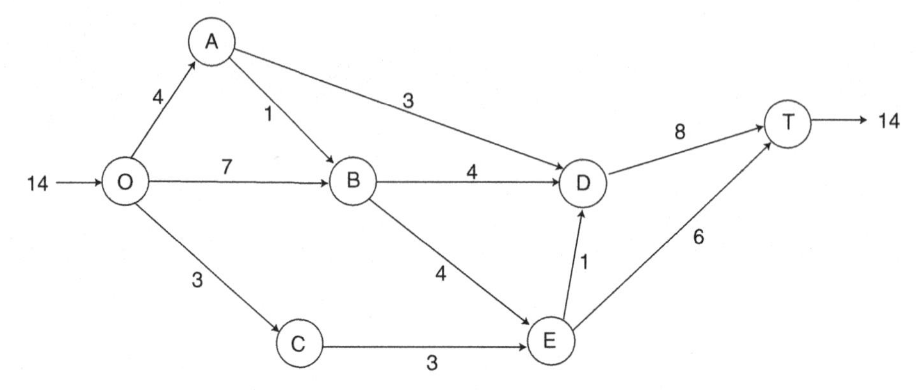

Figure 6.27 Optimal solution for the prototype problem.

following procedure that helps in recognizing when the optimal solution has been reached. The procedure is based on an important theorem of network theory known as the *max-flow min-cut theorem.* A **cut** is defined as any set of directed arcs containing at least one arc from every directed path from the supply node to the demand node. The cut value is the sum of the arc capacities of the arcs of the cut.

> The **max-flow min-cut theorem** states that, for any network with a single supply node and single demand node, the maximum feasible flow from the supply node to the demand node equals the minimum cut value for all of the cuts of the network.

Let us return to our prototype problem. Consider the cut in the network of Figure 6.21 that is shown in Figure 6.28. The value of the cut is $(3 + 4 + 1 + 6) = 14$, which was found to be the maximum value of flow, so this cut is a minimum cut. Also, in the residual network (Figure 6.26) resulting from iteration 7, where flow $F = 14$, the corresponding cut has a value of 0.

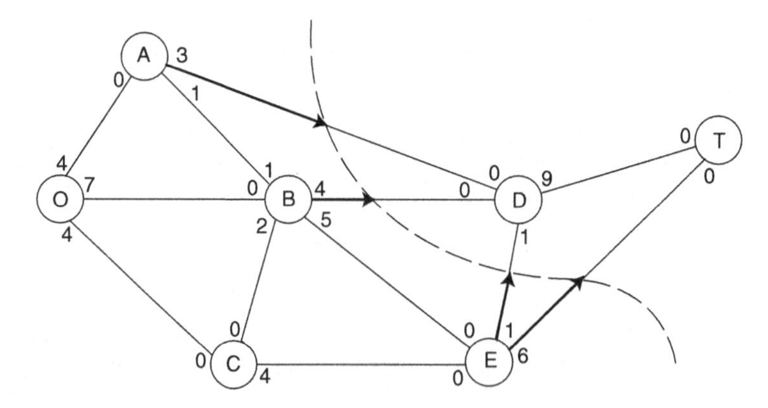

Figure 6.28 A minimum cut for the prototype problem.

6.6 AN EXAMPLE OF DISASTER MANAGEMENT OPTIMIZATION—THE OPTIMAL PLACEMENT OF CASUALTY EVACUATION ASSETS

This example is based on the work of Sundstrom et al. (1995). Through the use of LP techniques, the optimal number and positioning of disaster casualty evacuation assets within an area hit by a disaster may be determined to ensure the orderly transport of casualties from the affected area to the third-level medical treatment facilities. The optimal placement of casualty evacuation assets (OPTEVAC) model requires as input the dimensions of the disaster area, casualty location nodes, types of evacuation assets available, and preferred locations of medical treatment facilities. The OPTEVAC model then provides output as to the required numbers of ground and air ambulances as well as the optimal positioning of those evacuation assets and ambulance exchange points.

6.6.1 Introduction

The problem addressed here is one of casualties in an area that are treated at mobile medical facilities organized into a series of levels with the facilities at the higher levels having the greatest mobility but least surgical capability. The efficient evacuation of casualties from level to level is essential to ensure the wounded people reach a facility with the capability to render the required level of care. Distances between medical treatment facilities, as well as factors such as the type of terrain and mode of transportation, may all impact the evacuation process. Likewise, the number of treatment facilities deployed and where they are located greatly affects the casualty handling process and the ability to provide adequate casualty care.

Sundstrom et al. (1995) developed the model for military applications. In this section, model is presented for a application in disaster management. Location of treatment facilities has direct relationship to the number of casualties. If they are spaced further apart, some casualties requiring evacuation may end up out of help range or in locations not accessible by the available assets. Large-scale disasters may yield larger numbers of casualties with a greater potential for overwhelming the casualty evacuation system. Accurate assessments of the required numbers, types, and deployment locations of the evacuation assets ensure that wounded population is transported expeditiously from the point of injury through the various level of care to a highest level treatment facility, as dictated by the severity of the wound.

Because of the complexity of the problem, a probabilistic model has been considered for the description of the problem. However, as it will be shown later the probabilistic concept has been effectively handled in deterministic LP environment. The main objective of the optimization is the minimization of the total number of required evacuation assets within the area hit by a disaster, while at the same time incorporating randomness in vehicle availability. The literature offers many other models where the vehicles are sited on the basis of geographical coverage and not on the basis of availability. The assumption used in the model presented here is that the vehicles are continuously available for evacuation of casualties. In low casualty disasters, this assumption is not unreasonable; however, in high casualty disasters where frequent casualty calls keep evacuation assets on the roads, this assumption cannot be justified. Use of specific LP formulation will allow medical disaster personnel to input the α-*level* at which the optimization model is to run. This statistic represents the percentage of the maximum daily casualty load for which the proposed assets will meet the demand. The α-*level* is important because of the casualty "spikes" that occur during a disaster emergency operation. For instance, the planner may want to economize resources by deploying dedicated assets to ensure evacuation coverage for 80% of the maximum casualty load ($\alpha = .80$) and rely on "vehicles of opportunity" to transport the casualties that exceed the capabilities of these dedicated assets. Alternatively, if evacuation assets were programmed for the maximal projected daily casualty load, many assets would be used but a small percentage of the time, if at all.

Patient flows within the OPTEVAC model is determined from the user-specified input of casualty location nodes, with projections based on casualty distribution trends evidenced during previous disaster emergency operations.

The following section describes the LP model used for optimizing casualty evacuation assets and provides a detailed analysis of how the problem is mathematically represented. Essential to accurate assessment of the numbers of evacuation assets required are variables relating to the specific types of assets (transportation mode, range, mobility, etc.) as well as factors related to terrain and weather. The scope of this presentation is the linear program to be used in optimizing evacuation assets; the parameters relating to environmental conditions and ambulance characteristics are not addressed here.

6.6.2 The OPTEVAC Model

The OPTEVAC model provides medical emergency planners with the required number and optimal placement of evacuation assets to ensure sufficient casualty transport while minimizing oversupply of ground and air ambulances. The probabilistic module, incorporated into the casualty evacuation application, seeks the positions of Z evacuation assets in the disaster region of operations, which minimizes the number of vehicles (ground and air ambulances) required, so that there is a 95% certainty that the user-specified evacuation demands are met. Mathematical formulation of the casualty handling process is:

$$\begin{aligned}
&\text{Minimize } Z = \sum_{j \in J} x_j \\
&\text{subject to} \\
&\sum_{j \in N_i} x_j \geq b_i \quad \forall i, \\
&x_j \text{ are integers and } b_i \text{ is the smallest integer satisfying} \\
&1 - \left(\frac{F_i}{b_i}\right)^{b_i} \geq 0.95 \\
&F_i = \left(\frac{1}{24}\right) t \sum_{k \in M_i} f_k
\end{aligned} \tag{6.31}$$

where t is the average duration (hours) of a casualty call within the disaster-affected area (average distance from a demand node to a treatment facility/weighted average speed of available assets, using the ratios of available assets); f_k is frequency of casualty calls, or trips, at demand node k (casualty calls per day); and M_i is the set of demand nodes within S of demand node i. Model variables include Z—the total number of evacuation assets distributed over all of the facilities in the disaster area ($\sum b_i$); i, I—index and set of demand nodes; j, J—index and set of treatment facility sites; d_{ij}—the distance from facility site j to demand node i; S—the distance standard within which a treatment facility is desired to be found; N_i—$\{j \mid d_{ji} \leq S\}$ the set of facility sites within S of demand node i; x_j—the number of evacuation assets at facility site j; and b_i—the minimum number of evacuation assets required within M_i.

6.6.3 A Casualty Evacuation Example

Let us consider a 14-day rescue operation in which there are five second-level facilities, three rescue operation nodes (locations) with potential affected populations shown in Figure 6.29 (node 1: 12,000 people, node 2: 5000 people, and node 3:

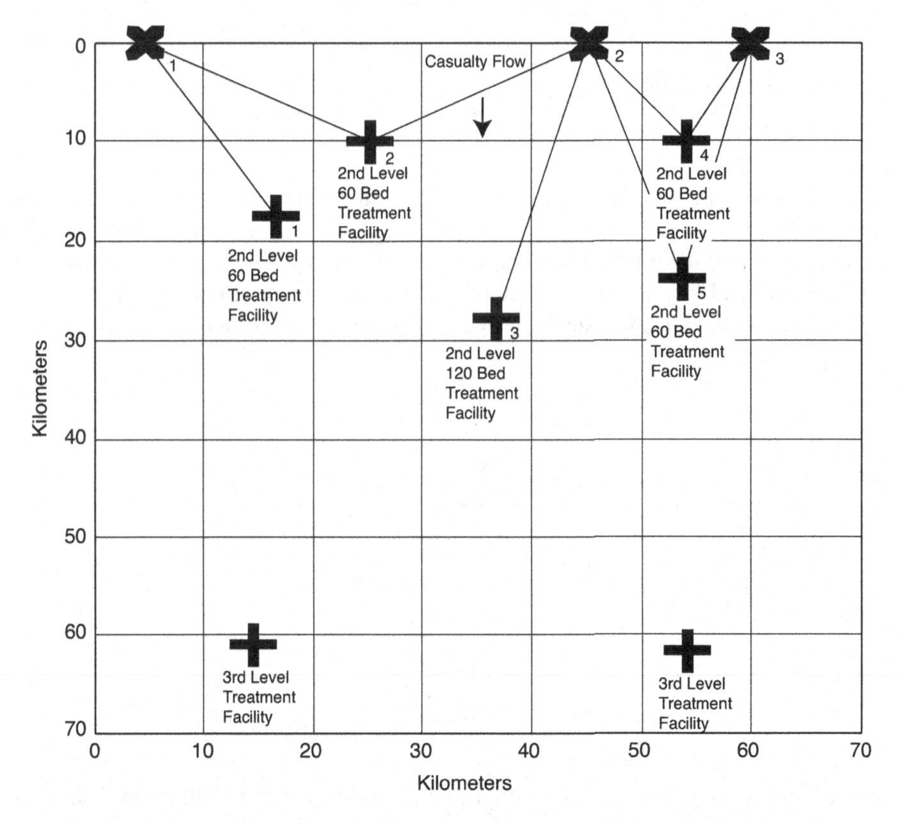

Figure 6.29 A simple input screen to the OPTIVEC model showing rescue operation nodes and medical treatment facilities.

10,000 casualties), and the maximum projected rescue range of seriously wounded casualties is 5.77 per 1000 people per day and the maximum projected rescue range of injured casualties is 4.22 per 1000 people per day. Also, we assume that 10 ground ambulances and 2 air ambulances are available for deployment. The first calculation performed is to determine the frequency of calls per day at each rescue operation node within the disaster area. The necessary algorithm must factor in the placement and allocation of evacuation assets for varying capacities (number of casualties per type of evacuation asset).

Therefore, the LP algorithm needs to be able to include separate frequencies ($f_{k,\text{ground}}$ and $f_{k,\text{air}}$) and separate demand areas' vehicles ($b_{i,\text{ground}}$ and $b_{i,\text{air}}$) computed for each type of vehicle. Below is an application of the algorithm that calculates the frequency of calls per day, when multiple vehicle assets are being used for evacuation (vehicle specifications are also listed). For instance, with only two types of vehicles (e.g., a single ground ambulance type and a single air ambulance type) this algorithm computes $f_{k,\text{ground}}$ and $f_{k,\text{air}}$ as presented in Table 6.11.

TABLE 6.11 Data for the Casualty Evacuation Example

	Capacity	Speed	Quantity Available
Ground ambulance	4	16 km/hr	10
Air ambulance	6	150 km/hr	2

Variables are defined as a/g ratio = air ambulance speed/ground ambulance speed = 150/16 = 9.375; g_{capac}—ground ambulance capacity; a_{capac}—air ambulance capacity; g_{avail}—number of ground ambulances available; a_{avail}—number of air ambulances available; and tot_{avail}—the total number of ground and air ambulances available.

So now we can calculate the remaining variables as follows:

$$g_{\text{factor}} = \frac{g_{\text{aval}}}{tot_{\text{aval}}} \times g_{\text{capac}} \tag{6.32}$$

$$a_{\text{factor}} = \frac{a_{\text{aval}}}{tot_{\text{aval}}} \times (g_{\text{capac}}) \times \left(\frac{a}{g} ratio\right) \tag{6.33}$$

$$Z_{\text{factor}} = \frac{maximum\ daily\ casualties}{g_{\text{aval}} + a_{\text{aval}}} \tag{6.34}$$

$$f_{k,\text{ground}} = \frac{Z_{\text{factor}} \times g_{\text{factor}}}{g_{\text{capac}}} \tag{6.35}$$

$$f_{k,\text{air}} = \frac{Z_{\text{factor}} \times g_{\text{factor}}}{a_{\text{capac}}} \tag{6.36}$$

The sum of the frequency of calls from all demand nodes within the disaster area ($f_{k,\text{ground}} + f_{k,\text{air}}$) will be used to calculate F_i, which will then lead to the calculations of the b_i's. By solving Equation (6.31), we can determine that 3 ground ambulances and

1 air ambulance are required within M_1 (demand area 1), 4 ground ambulances and 1 air ambulance are required within M_2 (demand area 2), and 4 ground ambulances and 1 air ambulance are required within M_3 (demand area 3) as well (symbolically, for ground ambulances: $b_{1,\text{ground}} = 3$, $b_{2,\text{ground}} = 4$, $b_{3,\text{ground}} = 4$; and for air ambulances: $b_{1,\text{air}} = b_{2,\text{air}} = b_{3,\text{air}} = 1$).

Using Figure 6.29, we can formulate an LP problem for the optimal number of ground ambulances in mathematical form as follows:

$$\begin{aligned}
&\text{Minimize } Z = x_{11} + x_{21} + x_{31} + x_{41} + x_{51} \\
&\text{subject to} \\
&x_{11} + x_{21} \geq 3 \\
&x_{21} + x_{31} + x_{41} + x_{51} \geq 4 \\
&x_{41} + x_{51} \geq 4
\end{aligned} \tag{6.37}$$

where x_j are positive integers.

When solving LP problem for the optimal number of air ambulances in the example scenario, the corresponding mathematical setup is the following:

$$\begin{aligned}
&\text{Minimize } Z = x_{12} + x_{22} + x_{32} + x_{42} + x_{52} \\
&\text{subject to} \\
&x_{12} + x_{22} + x_{32} + x_{42} + x_{52} \geq 1 \\
&x_{12} + x_{22} + x_{32} + x_{42} + x_{52} \geq 1 \\
&x_{12} + x_{22} + x_{32} + x_{42} + x_{52} \geq 1
\end{aligned} \tag{6.38}$$

where x_j are positive integers.

In Equations (6.37) and (6.38), the first subscript represents the second-level treatment facilities and the second represents the type of vehicle. The first constraint of Equation (6.37) may be interpreted as "casualty rescue node 1 is within range of second-level medical treatment facilities 1 and 2 by way of ground ambulance." Similarly, each constraint of Equation (6.38) indicates that none of the casualty rescue nodes are out of the air evacuation range for each of the five second-level treatment facilities.

We will use LINPRO program on the CD-ROM to solve these two LP problems. Example 5 on the CD-ROM, in the directory LINPRO, subdirectory Examples, is the formulation shown in Equation (6.37) and Example 6 is the formulation shown in Equation (6.38).

After solving the first LP problem in Equation (6.37), we see that the optimal solution is an alternate solution with placement of 3 ground ambulances at treatment facility 1, and 4 at treatment facility 5, for a total of 7 necessary ground ambulances ($x_{11} = 3$, $x_{41} = 4$, and $Z_{\text{ground}} = 7$).

After solving the second LP problem in Equation (6.38), we see that the optimal solution is an alternate solution with placement of only 1 air ambulance at any second-level facility, though placing it in the center facility would be the logical choice.

6.6.4 Summary

In this sample example, the LP optimization algorithms were able to minimize the number of evacuation assets required in the rescue operation. Ten ground ambulances and two air ambulances were available for the operation, but only seven land transports and one air asset were necessary to perform the evacuation demands of the defined scenario. OPTEVAC finds the optimum numbers of required dedicated assets for a given rescue operations, and in the example scenario, the planner saved on resources by 30% for ground ambulances and 50% for air ambulances. These results demonstrate that the OPTEVAC model allows the user to determine the minimum numbers and optimal positioning of patient transportation assets required to meet the casualty handling demands of the rescue operation.

REFERENCES

Angelis, V.D., M. Mecoli, C. Nikoi, and G. Storchi (2007), "Multiperiod integrated routing and scheduling of World Food Programme cargo planes in Angola," *Computational Operations Research*, 34(6):1601–1615.

Balcik, B. and B.M. Beamon (2008), "Facility location in humanitarian relief," *International Journal of Logistics: Research and Applications*, 11(2):101–121.

Bryson, K.-M., H. Millar, A. Joseph, and A. Mobolurin (2002), "Using formal MS/OR modeling to support disaster recovery planning," *European Journal of Operational Research*, 141:679–688.

Dantzig, G.B. (1963), *Linear Programming and Extension*, Princeton University Press, Princeton, NJ.

Haghani, A. and Oh, S.C. (1996), "Formulation and solution of a multi-commodity, multi-modal network flow model for disaster relief operations," *Transportation Research Part A-Policy and Practice*, 30(3):231–250.

Hillier, F.S. and G.J. Lieberman (1990), *Introduction to Mathematical Programming*, McGraw Hill, New York.

Jenkins, L. (2000), "Selecting scenarios for environmental disaster planning," *European Journal of Operational Research*, 121(2):275–286.

Knott, R. (1987), "The logistics of bulk relief supplies," *Disasters*, 11:113–115.

Ozdamar, L., E. Ekinci, and B. Kucukyazici (2004), "Emergency logistics planning in natural disasters," *Annals of Operations Research*, 129:217–245.

Sarker, B., S. Mahankali, L. Mann, and E. Triantaphyllou (1996), "Power restoration in emergency situations," *Computers and Industrial Engineering*, 31(1/2):367–370.

Sundstrom, S.C., C.G. Blood, and S.A. Matheny (1995), *The Optimal Placement of Casualty Evacuation Assets: A Linear Programming Model*, Naval Health Research Center, Report No. 95–39, San Diego.

Wagner, H.M. (1975), *Principles of Operations Research*, 2nd edn, Prentice-Hall, Englewood Cliffs, NJ.

EXERCISES

6.1 Consider the following problem:

$$\begin{aligned}
&\text{maximize } Z = 3x_1 + 2x_2\\
&\text{subject to :}\\
&2x_1 + x_2 \leq 6\\
&x_1 + x_2 \leq 6\\
&x_1, x_2 \geq 0
\end{aligned}$$

(a) Solve the problem graphically. Identify the corner-point feasible solutions.

(b) Identify all the sets of the two defining equations for this problem. For each one, solve for the corner-point solution, and classify it as a corner-point feasible or infeasible solution.

(c) Introduce slack variables in order to write the problem in the canonical form.

6.2 Consider the following problem:

$$\begin{aligned}
&\text{maximize } Z = 2x_1 + 4x_2 + 3x_3\\
&\text{subject to :}\\
&x_1 + 3x_2 + 2x_3 \leq 30\\
&x_1 + x_2 + x_3 \leq 24\\
&3x_1 + 5x_2 + 3x_3 \leq 60\\
&x_1, x_2, x_3 \geq 0
\end{aligned}$$

You are given the information that $x_1 > 0$, $x_2 = 0$, and $x_3 > 0$ in the optimal solution.

(a) How can you use this information to adapt the simplex method to solve the problem in the minimum possible number of iterations?

(b) Solve this problem using the simplex method.

6.3 Label each of the following statements as True or False, and then justify your answer.

(a) The Simplex Criteria I is used because it always leads to the best basic feasible solution.

(b) The Simplex Criteria II is used because making another choice normally would yield a basic solution that is not feasible.

(c) When an LP model has an equality constraint, an artificial variable is introduced into this constraint in order to start the simplex method with an obvious initial basic solution that is feasible for the original model.

6.4 Consider the following LP problem:

$$\begin{aligned}
&\text{maximize} = x_1 + x_2 \\
&\text{subject to}: \\
&-x_1 + x_2 \leq -1 \\
&x_1 - x_2 \leq -1 \\
&x_1 \geq 0 \text{ and } x_2 \geq 0.
\end{aligned}$$

(a) Find graphically the solution of the stated problem.

(b) Discuss the solution obtained in details.

(c) Modify the problem by changing the sign of both inequalities and adding one more constraint:

$$x_1 + x_2 = 6$$

and solve it using the LINPRO program on the CD-ROM.

6.5 After the disaster, a building block manufacturer is facing a high demand for two types of building blocks used in rebuilding the community: type A and type B. For each set of 100 type A blocks the manufacturer can make a profit of \$5, whereas for each set of 100 type B blocks he can make \$8. Because of the high demand, all blocks that are produced can be sold. It takes 1 hour to make 100 type A blocks and 3 hours to make 100 type B blocks. Each day, there are 12 hours available for block manufacturing. A set of 100 type A blocks requires 2 units of cement, 3 units of aggregate, and 4 units of water; a set of 100 type B blocks requires 1 unit of cement and 6 units of water. Each day, 18 units of cement and 24 units of water are available for block manufacturing. There is no restriction on the availability of aggregate. How many type A and type B blocks should be made during the day to maximize profits?

(a) Formulate the problem as an LP problem.

(b) Name the decision variables, objective function, and constraints.

(c) Solve the problem graphically.

(d) Use LINPRO program on the CD-ROM to solve the problem.

(e) What is the value of the dual variable, or shadow price, associated with the 24 units of available water?

6.6 (Modified after Wagner, 1975) The Emergency Flush Company has two spills to take care of pollutants located along a stream. Spill 1 is generating 20 units of pollutants daily and spill 2 is generating 14 units. Before the wastes are discharged into the river, part of these pollutants has to be removed by an emergency waste treatment facility that will be implemented at each spill location. The costs associated with removing a unit of pollutant are \$1000 and

\$800 for spills 1 and 2. The rates of flow in the streams are $Q_1 = 5$ m^3/sec and $Q_2 = 2$ m^3/sec, and the flows contain no pollutants until they pass the location of spills.

Emergency Flush Company has to respect the stream water quality standards. They require that the number of units of pollutants per cubic meter of flow should not exceed 2. Twenty percent of the pollutants entering the stream at spill 1 will be removed by natural processes before they reach location of spill 2. The company wants to determine the most economical operation of its emergency treatment facilities that will allow it to satisfy the stream standards.

(a) Formulate the problem as an LP problem.

(b) Name decision variables, objective function, and constraints.

(c) Solve the problem graphically.

(d) Use LINPRO program on the CD-ROM to solve the problem.

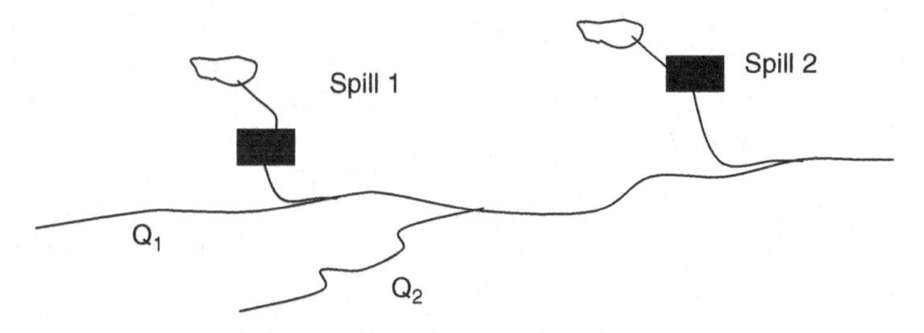

6.7 (Modified after Wagner, 1975) Disaster rescue operation has the following minimal daily requirements for rescue personnel:

Time of Day (24-hour clock)	Period	Minimal Number of Rescue Personnel During Period
2–6	1	22
6–10	2	55
10–14	3	88
14–18	4	110
18–22	5	44
22–1	6	33

Consider period 1 as following immediately after period 6. Each person works eight consecutive hours. Let x_t denote the number of persons starting the rescue operation in period t every day during the rescue operation. Your duty is to obtain a daily schedule that employs the least number of rescue personnel provided that each of the above requirements is met.

(a) Formulate an LP model to find an optimal schedule.

(b) Use LINPRO to solve the problem utilizing data provided above.

6.8 Solve the water distribution transportation problem from Example 6 using LINPRO program. Compare your results to the solution obtained in Section 6.4.2.

6.9 Derive the values of the u_i and v_j given in the second, third, and fourth transportation tableaux of postdisaster water distribution problem from Example 6. Try doing this by working directly on the tableaux in Example 8. Also, check out the chain reactions in the second and third tableaux.

6.10 (Modified after Hillier and Lieberman, 1990) Consider the transportation problem having the following cost and requirements table:

		Destination					
		1	2	3	4	5	Supply
Source	1	8	6	3	7	5	20
	2	5	M	8	4	7	30
	3	6	3	9	6	8	30
	4 (D)	0	0	0	0	0	20
Demand	25	25	20	10	20		

After several iterations of the transportation simplex method, the following transportation simplex tableau is obtained:

Iteration		Destination						
		1	2	3	4	5	Supply	u_i
Source	1	8	6	3 (20)	7	5	20	
	2	5 (25)	M	8	4 (5)	7	30	
	3	6	3 (25)	9	6 (5)	8	30	
	4(D)	0	0 (0)	0 (0)	0	0 (20)	20	
Demand		25	25	20	10	20		
	v_j							

Continue the transportation simplex method for two more iterations. After two iterations, state whether the solution is optimal and, if so, why.

6.11 (Modified after Hillier and Lieberman, 1990) Consider the directed network below for practice of some key network definitions from Section 6.5.

(a) What is the directed path from node A to node F? Find three other undirected paths from node A to node F.

(b) Identify three directed cycles and an undirected cycle that includes every node.

(c) Identify a set of arcs that forms a spanning tree.

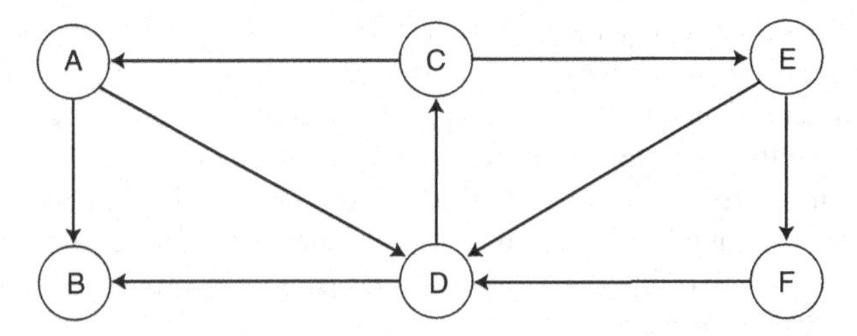

6.12 (Modified after Hillier and Lieberman, 1990) Algorithm for the shortest path through network is presented in Section 6.5.1. Using this algorithm find the shortest path through networks (a) and (b). Numbers along the arcs represent actual distance between the corresponding nodes.

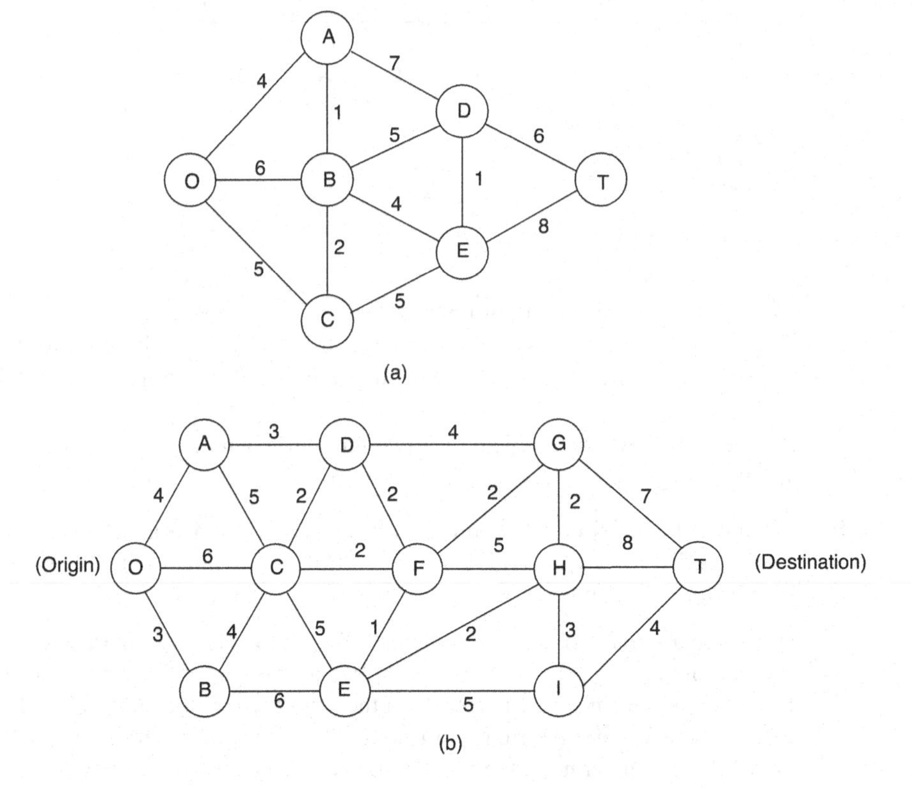

6.13 For each network from the previous exercise formulate the shortest path problem as an LP problem and use LINPRO program from the CD-ROM to solve it.

6.14 Verify the optimal solution in Example 11 by starting the spanning tree algorithm from node B.

6.15 (Modified after Hillier and Lieberman, 1990) A community recovering from the disaster is hooking up special telephone communication network between the disaster management center (DMC) and five makeshift hospitals (Hi) that are receiving disaster victims. The phone line from a hospital need not be connected directly to the disaster management center. It can be connected indirectly by being connected to another hospital that is connected (directly or indirectly) to the disaster management center. The only requirement is that every hospital be connected by some route to the disaster management center.

The cost for the special phone lines is directly proportional to the distance involved (provided below):

	Distance					
	DMC	H1	H2	H3	H4	H5
DMC	/	190	70	115	270	160
H1	190	/	100	240	215	50
H2	70	100	/	140	120	220
H3	115	240	140	/	175	80
H4	270	215	120	175	/	310
H5	160	50	220	80	310	/

(a) Determine which pairs of nodes (DMC and Hi) should be directly connected by special phone lines in order to connect every hospital to the disaster management center (directly or indirectly) at a minimum total cost.

(b) Describe how this problem fits the minimum spanning tree problem.

(c) Use the algorithm presented in Section 6.5.2 to solve the problem.

6.16 Formulate the maximum flow problem from Section 6.5.3 as an LP problem. Use LINPRO computer program and confirm the optimal solution of the problem from Example 12.

6.17 (Modified after Hillier and Lieberman, 1990) For the networks in (a) and (b) use the augmenting path algorithm described in Section 6.5.3 to find the flow pattern giving the maximum flow from the supply node (the left most node) to the demand node (the right most node). The arc capacity from node i to node j is the number nearest node I along the link between the nodes.

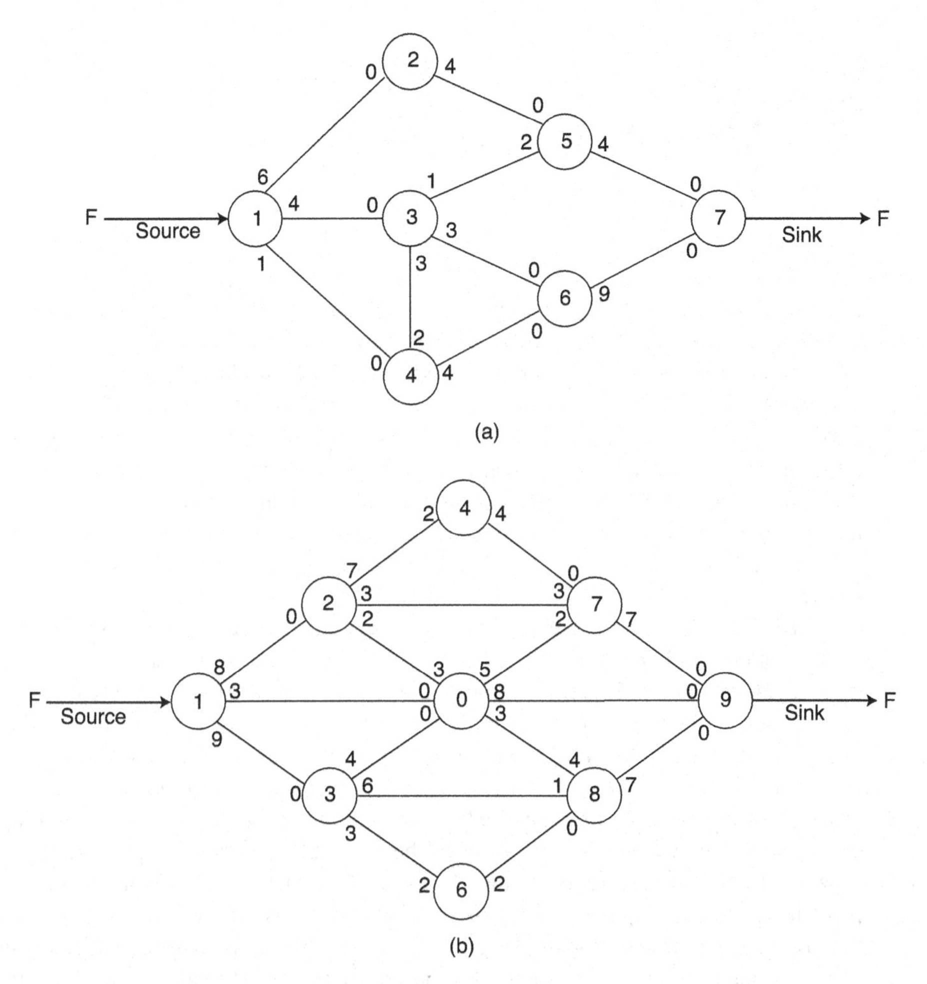

(a)

(b)

6.18 For each network in Problem 6.17, formulate the maximum flow problem as an LP problem and use LINPRO program from the CD-ROM to solve it.

7 Multiobjective Analysis

Disaster management is by nature multiorganizational, but organizations are only loosely connected, leading to managerial confusions and ambiguity of authority as discussed in Section 4.5. The multifunctional nature and political hierarchy in emergency management organizations are well suited for hierarchical planning and multiattribute multiobjective approaches, as various groups have different priorities before, during, and after a disaster hits. Chapter 6 focused on the presentation of the management tools that can be used in disaster management when the assessment of performance can be successfully achieved using a single-objective function—minimization of cost, minimization of the response time, maximization of resource allocation, and similar.

Since 1970s, however, there has been an increased awareness of the need to identify and consider simultaneously several objectives in the analysis and solution of some disaster management problems, in particular those derived from the study of large-scale disasters. For example, in a disaster recovery study we may not be satisfied with learning what actions lead to minimizing the recovery costs. Instead, the study may be conducted so that it identifies multiple objectives as these relate to recovery costs, short-term and long-term capital needs, population satisfaction and well-being, and future community resiliency. How all these objectives relate to one another, and how much of each one can be obtained subject to a common set of constraints, can provide disaster management with far greater insight into recovery operations than that provided by the adoption of a single objective. In disaster management, the design of projects and programs has focused traditionally on the estimation of benefits and costs. A more realistic analysis would include social, environmental, and regional objectives, as well.

Societal values and norms are shifting, however, from a position of unchecked economic growth and less attention on social and environmental issues to one of concern for people and the environment themselves, and how they relate to the quality of life. This shifting of societal values and norms has prompted the enactment of laws regulating management of natural and human-induced disasters with explicit references to multiple objectives (Mileti, 1999; IAP, 2009). In turn, the inclusion of multiple objectives in the study of resource-allocation problems has motivated the development of multiobjective analysis, which is the subject treated in this chapter.

7.1 INTRODUCTION

Although there is a growing commitment on the part of government agencies, research centers, and general disaster practitioners to conduct multiobjective analysis in a variety of disaster problem settings, as discussed in Section 4.5, it is not immediately clear how to proceed.

7.1.1 Toward Operational Framework for Multiobjective Analysis

A sequence of steps is suggested (Goicoechea et al., 1982) to provide an operational framework for decision making, which is general enough to accommodate many of the multiobjective techniques available in the literature and can be applied in disaster management:

1. Prepare a general statement of needs as perceived in the study of the problem at hand.
2. Formulate broad goals and specific objectives; these should reflect the needs stated above and societal values.
3. Identify pertinent decision variables.
4. Select a mathematical framework for multiobjective analysis (the nature of the disaster management problem will often suggest the choice of framework).
5. Formulate a set of objective functions—each function must address one or more of the goals and objectives stated in step 2 above and must be expressed in terms of the decision variables in step 3.
6. Formulate a set of physical constraints—these constraints must be functions of the decision variables and represent limitations on resources available.
7. Generate an alternative solution (plan)—the makeup of this alternative solution may be a value attained for each one of the objective functions.
8. Evaluate actual consequences, direct and associated—once a solution is generated, its consequences can be outlined in terms of actual resources utilized, and how well the goals stated in step 2 are met.
9. Determine whether alternative solution is acceptable to the decision maker—the decision maker (an individual or group of individuals) responsible for the project is asked to subjectively assess the value of the current solution to him or her—proceed to step 14 if affirmative.
10. Determine whether the decision maker is willing to relax some of his or her expectations—here the decision maker must ascertain whether he or she could accept less in some of the objective functions in the hope of receiving more in others and still consider the aggregated value acceptable—if affirmative, proceed to step 11, otherwise proceed to step 12.
11. Input is elicited from the decision maker to attempt to establish the relative worth to him or her of objective function units—schemes are available to structure these responses into *weights* that now can be incorporated in the mathematical framework to generate another alternative solution.

12. Determine whether additional resources/technology can be committed—this step asks whether additional resources can be committed to the project, that is, capital, time, manpower, and hardware. If affirmative, return to step 6, otherwise proceed to step 13.
13. No feasible plan is available.
14. Implement alternative solution.

Selection of Objective Functions The choice of the number and makeup of the multiple objectives for analysis can determine to a significant degree the success, or lack of it, of a planning effort. The nature of a project will very likely suggest at the beginning candidates for objectives. If the number of objectives is small, initially, a closer look at the magnitude of the problem may dictate additional objectives to consider. Very often, limited time and financial resources with which to conduct a study may tend to preclude relevant objectives from being considered. A balance should be struck so that the number of selected objectives helps ensure acceptance of an alternative plan. Furthermore, the selection must be realistic to the extent that all these objectives can be effectively addressed with the resources available. In many examples, we can see that the number and makeup of these objectives is, in itself, a major decision made in the planning effort.

Social Values and Preferences Inherent in the design of alternative plans is the statement and adoption of social values and preferences as articulated by both disaster managers and decision makers. Mechanisms for articulating these preferences may range from an enumeration of general concerns and question–answer formats to *weights* to multiply the various objectives by.

Choice of Mathematical Framework A wide variety of mathematical frameworks are available to develop alternative feasible plans. They include mathematical programming, utility assessment, input–output analysis, simulation techniques, classical regression, and many more. The remaining of this chapter will focus on mathematical programming as a very useful and flexible framework for multiobjective analysis when the objective and physical constraints of a problem can be expressed as functions of decision variables. A key characteristic of this framework is that it permits the selection of a range of reasonably good management decisions based on realistic assumptions and contribution of the various decision variables to the objectives themselves (Zeleny, 1973; Wagner, 1975; Hillier and Lieberman, 1990).

7.1.2 An Illustrative Example

Let me use one example of a disaster management problem situation where there is a need to formulate goals, develop alternative plans, and design criteria to be used in the selection of a plan. The example is in floodplane (FP) management from Novoa and Halff (1977).

The city of Dallas, Texas, uses a variety of methods to manage damaging floods caused by the region's hydrology and its rapid urbanization. Several FP management options are available to the city of Dallas in its efforts to curve down personal and property flood losses. These options range from limiting urban development and purchasing flood insurance to the clearing and relocation of every structure in the flood-prone areas.

The Current Problem and Existing Floodplane Figure 7.1 shows the Peaks Branch watershed, which lies within Dallas' city limits. Runoff from Upper Peaks Branch flows south, within storm sewers, from Mockingbird Lane to Jamaica Street. Then, on Jamaica Street an open channel conveys the storm water through the lower portion of the watershed to its discharge point into the White Rock Creek.

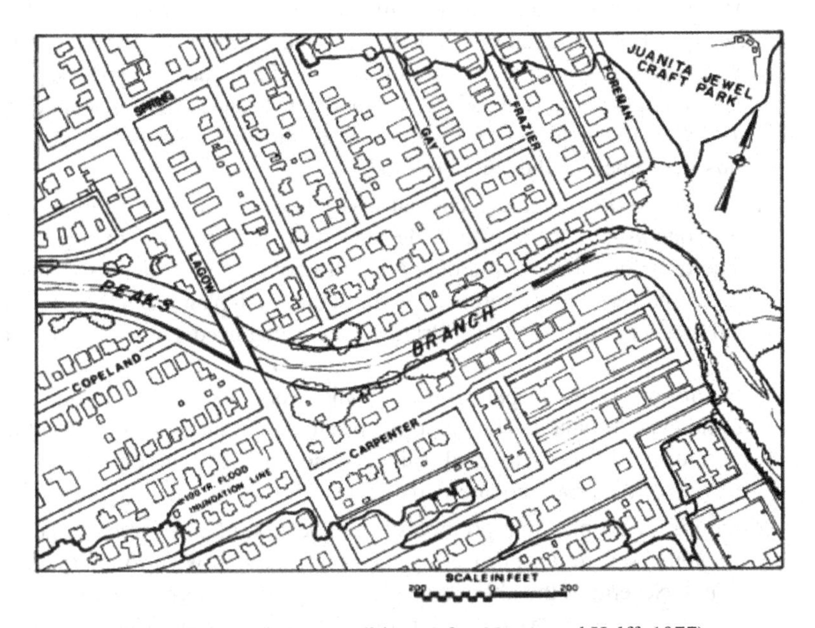

Figure 7.1 Existing conditions (after Novoa and Halff, 1977).

The present channel is designed to carry a 5-year-frequency flood only. To discourage construction in this area, city zoning ordinances limit the value of improvements on structures within the FP area to $300. This area may be used for (a) agriculture; (b) parks, community centers, public golf courses, and playgrounds; (c) sanitary fills for garbage disposal; and (d) local utilities. After many hearings and complaints from dissatisfied citizens who were not convinced that their homes could be flooded, the Board of Adjustment granted a 5-year blanket variance to allow improvements up to 50% of the market value of a structure. Shortly thereafter, the City Council authorized the consulting firm to conduct a study and identify the areas subject to flooding and develop a number of alternative measures for FP management.

Multiple Criteria Used in the Study Formulation of alternative flood management measures called for the enumeration of criteria that would address a variety of technical details, community preferences, and levels of protection to life and property:

Flood protection: The fundamental goal of this improvement project was flood protection. Each alternative plan was to be designed to sustain a 100-year flood with a minimum loss of property. To this end, open channel floodways provide for a greater margin of safety than closed conduits do. But whereas paved channels are more reliable than grass-covered channels, the latter are more aesthetically pleasing.

Neighborhood improvement: The prevention of flooding is beneficial to a neighborhood in that property values will generally increase over time, and the area can accept structures suitable for business and commerce as the danger of flooding is reduced. Multiple uses of floodway land provide open spaces, parks, and recreational facilities, in addition to flood conveyance. These aesthetic qualities contribute to the quality of life in the area, while increasing its property values, thus strengthening the neighborhood.

Neighborhood acceptance: Technical feasibility of an alternative plan represents only one aspect of the overall design effort. A floodplane management plan should include an ongoing dialogue with community residents to insure that social and aesthetic preferences are built into such plan. Both the city of Dallas and residents of the Peaks Branch area recognized the importance of this cooperation, and consequently, the southeast Dallas Neighborhood Club invited the city of Dallas' Department of Public Works to present preliminary plans to the general public. This dialogue provided the mechanism needed to elicit preferences from the affected residents and the use of these preferences in the ranking of the proposed alternative plans.

Relocations: Any plan that promotes the relocation of structures and people will encounter severe disapproval by the residents who wish to continue in their homes. Several of the alternative plans proposed require the removal of structures from the 100-year FP. On the one hand, the transfer of people from substandard housing to better housing can be desirable, provided financial assistance or credit programs are available so that these people can now afford the better housing. On the other hand, many families in the Peaks Branch watershed believed the flood hazard to be a remote possibility and would not welcome a plan that called for relocation.

Project and maintenance costs: Costs are always an important consideration in any type of project. Financing of a flood management plan would come from the sale of bonds and other monies that may become available from State and Federal programs. Certainly the amount of capital available will determine the extent of structural changes/additions, the number of family relocations that can be carried out, and ultimately, the level of flood protection that can be afforded.

Legal aspects: With the exception of a *no-action* alternative plan, the other plans call for a variety of structural changes (i.e., construction of water conveyance channels, flood management lakes, parks, and bike trails) and nonstructural changes (i.e., restrictions on the use of flood prone areas, and the planting of vegetal cover to reduce runoff by increasing interception and infiltration). Area residents feel very strongly about the desirability, or lack of it, of the proposed changes, and are willing to pursue legal action. Resulting litigation can be time consuming and expensive to all parties involved. Even when legal suits are not brought up, the sequence of steps to follow in requesting a change in a zoning ordinance can be time consuming. Removal of the FP prefix, for instance, requires a public hearing, the approval of the directors of the City's Public Works, Urban Planning, Parks and Recreation Departments, as well as the City Council.

In the final selection of an alternative plan, the parties involved ought to consider the number of legal obstacles associated with that plan.

Floodplane Management Alternatives The following eight alternative plans have been formulated:

1. *No action*: This alternative plan says "leave the FP area as it is" with no corrective measure. Potential losses, computed as a function of the depth of flooding and the value of housing, have been subtracted from the Residence Flood Damage tables of the U.S. Department of Agriculture. The 5-year flood losses would be small, but the 100-year flood would cause direct property damage of $1,300,000. Limited improvements would be possible through the purchase of flood insurance subsidized by the Federal Flood Insurance Program. In addition, residents would be eligible for home-improvement loans from federally insured lending institutions. Capital costs for alternative No. 1 are then zero, but the area would continue to be subject to flooding and general deterioration.
2. *Purchase of the FP*: Under this alternative, every structure that would be inundated by the 100-year flood would be purchased and cleared. This plan requires the acquisition of 367 single-family houses, 34 duplexes, 39 apartment units, and 60 commercial buildings; it would displace 800 families. The cost of purchasing the land involved—about 100 acre, privately owned—is 12 million dollars, based on 1976 prices, and includes relocation allowances averaging $2000 per family.
3. *A Park greenway—north*: Two lakes, several hike-and-bike trails, and parkways would be built; the Peaks Branch channel would be widened to its confluence with White Rock Creek; the west channel would be extended to Second Avenue and the east channel to Penelope Street (Figure 7.2). In the process, lands and structures would need to be purchased at an estimated cost of $9,300,000.
4. *A Park greenway—south*: Four lakes, hike-and-bike trails, and parkways would be built (Figure 7.3). This plan is similar to Park greenway—north, but calls for the removal of a larger number of apartments while saving more single-family

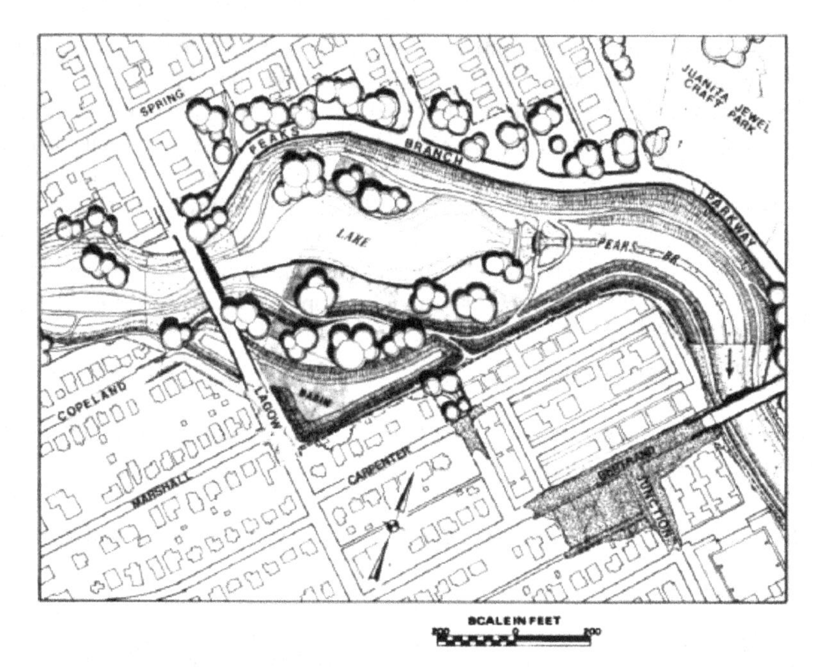

Figure 7.2 Park Greenway—North (after Novoa and Halff, 1977).

Figure 7.3 Park Greenway—South (after Novoa and Halff, 1977).

housing. The need for additional right-of-way would entail the purchase of 164 structures, displacing a total of 419 families, at a cost of $10,900,000.

5. *A basic greenway*: Calls for widening of the Peaks Branch channel from Frank Street to its confluence with White Rock Creek. Additional right-of-way for the expansion and the purchase of structures within the 100-year FP would entail the relocation of 164 families at an estimated cost of $7,700,000.
6. *A concrete channel*: Most of the affected houses occupy a low area, new Spring Avenue, and Lagow Street, which would be used for interior-drainage facilities and a sump for the storage of water. A gravity sluice would drain the sump when water in the Peaks Branch channel is low, and a pump station with a capacity of 10,000 gal/min would drain the sump when the high water in the channel prevents gravity drainage. The channel itself would be widened from Jamaica Street to its confluence with White Rock Creek. A total of 69 families and associated structures would be displaced at an estimated cost of $5,600,000.
7. *A bypass conduit*: All of the runoff from the upper Peaks Branch storm sewers would be intercepted and routed along the proposed State Highway 352 right-of-way, in a box culvert 14-ft wide by 14-ft high, and divided into five cells. The capacity of this culvert, 11,000 cfs, is the equivalent of the 100-year design discharge. The cost is $12,300,000, including right-of-way land purchases and relocation.
8. *Purchase and redevelopment of the FP*: Combines design aspects of alternatives 2 and 3; a new subdivision, including a 10-acre shopping center, 336 residential lots, and 40 acre of parks, would be built on the land not used for the greenbelt at a cost of $16,300,000.

Evaluation of Alternatives In this part of the study, the planners used the multiobjective analysis to analyze the problem. A preliminary payoff matrix summarizing the criteria/objectives is shown in Table 7.1. As it can be seen, some criteria are given quantitative (numeric) values and some are provided in qualitative (nonnumeric verbal) form. Absent are the criterion weights. Further discussion and solution of the problem presented here will follow in Section 7.5.

7.2 MULTIOBJECTIVE ANALYSIS METHODOLOGY

Multiple objective, in contrast to single objective, decisions concerning disaster management do not have an optimal solution. As a result, there have been great efforts to develop a methodology for assessing trade-offs between alternatives based on using more than one objective. In the last three decades of multiobjective research, efforts have been made in:

- objective quantification;
- the generation of alternatives; and
- selection of the preferred alternative.

TABLE 7.1 Evaluation of Alternative Plans

	Factors Used in Evaluation					
	Relative Flood Protection	Relative Neighborhood Improvement	Number of Family Relocations	Project Cost (Millions)	Maintenance Cost (Thousands)	Relative Number of Management Problems
Alternative 1	Bad	Bad	0	0	12	Many
Alternative 2	Excellent	Fair	800	12.0	0	Many
Alternative 3	Good	Very good	237	9.3	41	Some
Alternative 4	Good	Very good	419	10.9	41	Some
Alternative 5	Fair	Good	164	7.7	41	Few
Alternative 6	Excellent	Fair	69	5.6	16	Few
Alternative 7	Fair	Fair	30	12.3	0	Very few
Alternative 8	Excellent	Excellent	604	16.3	41	Very many

7.2.1 Change of Concept

Chapter 6 showed that a single-objective programming problem consists of optimizing one objective subject to a constraint set. On the other hand, a multiobjective programming problem is characterized by an r-dimensional vector of objective functions:

$$
\begin{aligned}
&Z(x) = Z_1(x), Z_2(x), Z_3(x), \ldots, Z_r(x) \\
&\text{subject to} \\
&\boldsymbol{x} \in \boldsymbol{X} \qquad (7.1) \\
&\text{where } \boldsymbol{X} \text{ is a feasible region:} \\
&\boldsymbol{X} = \{\boldsymbol{x} : \boldsymbol{x} \in \boldsymbol{R}^n, g_i(\boldsymbol{x}) \leq 0, x_j \geq 0 \; \forall \; i, j\}
\end{aligned}
$$

where $\boldsymbol{R}$ is the set of real numbers, $g_i(\boldsymbol{x})$ is the set of constraints, and $\boldsymbol{x}$ is the set of decision variables.

The word *optimization* has been deliberately kept out of the definition of a multiobjective programming problem since we cannot, in general, optimize a priori a vector of objective functions. The first step of the multiobjective analysis consists of identifying the set of nondominated solutions within the feasible region $\boldsymbol{X}$. So instead of seeking a single optimal solution, a set of noninferior solutions is sought. The essential difficulty with multiobjective analysis is that the meaning of the optimum is *not defined* as long as we deal with multiple objectives that are truly different. An illustrative example is provided in Section 4.5.

I pointed out in Section 4.5 that to obtain a single global optimum over all objectives requires that we either establish or impose some means of specifying the value of each of the different objectives. If all objectives can indeed be valued on a common basis, the optimization can be stated in terms of that single value. The multiobjective problem has then disappeared, and the optimization proceeds relatively smoothly in terms of a single objective.

In practice, it is frequently awkward if not impossible to give every objective a relative value. The relative worth of damages, lives lost, the environment, and other objectives are unlikely to be established easily by anyone, or to be accepted by all concerned. We cannot hope, then, to be able to determine an acceptable optimum analytically.

The focus of multiobjective analysis in practice is to sort out the mass of clearly dominated solutions, rather than determine the single best design. The result is the identification of a small subset of feasible solutions that are worthy of further consideration. Formally, this result is known as the set of *nondominated solutions*.

7.2.2 Nondominated Solutions

To understand the concept of nondominated solutions, it is necessary to look closely at the multiobjective problem (Eq. 7.1). (Note that nondominated solutions are sometimes referred to by other names: noninferior, Pareto-optimal, efficient, etc. Throughout this text different names are used with the same meaning.) The essential feature of the multiobjective problem is that the feasible region of production of the solutions is much more complex than for a single objective. In single optimization, any set of inputs, x, produces a set of results, Z, that can be represented by a straight line going from worst (typically 0 output) to best. In a multiobjective problem, any set of inputs, x, defines a multidimensional space of feasible solutions, as Figure 7.4 indicates. Then there is no exact equivalent of a single optimal solution.

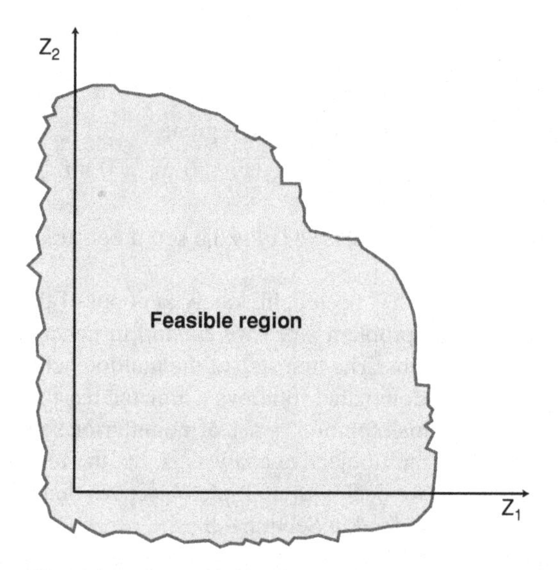

Figure 7.4 Feasible region of a multiobjective problem presented in the objective space.

The nondominated solutions are the conceptual equivalents, in multiobjective problems, of a single optimal solution in a single-objective problem. The main characteristic of the nondominated set of solutions is that for *each solution outside the set, there is a nondominated solution for which all objective functions are unchanged or improved and at least one is strictly improved*.

The preferred solution for any problem should be one of the nondominated solutions. So long as all objectives worth taking into account have been considered, no solution that is not among the nondominated solutions is worthwhile: it is dominated by some solutions that are preferable on all accounts. This is the reason why multiobjective analysis focuses on the determination of nondominated solutions.

It is often useful to group the nondominated solutions into major categories. The purpose of this exercise is to facilitate discussions about which solution to select. Indeed, to the extent that it is not possible to specify acceptable relative values for the objectives, and thus impossible to define the best solution analytically, it is necessary for the choice of the solution to rest on judgment. As individuals find it difficult to consider a large number of possibilities, it is helpful to focus attention on major categories.

If we introduce levels of acceptability for each of the objectives, the nondominated solutions are best divided into two categories: the major alternatives and the compromises. A *major alternative* group of nondominated solutions represents the best

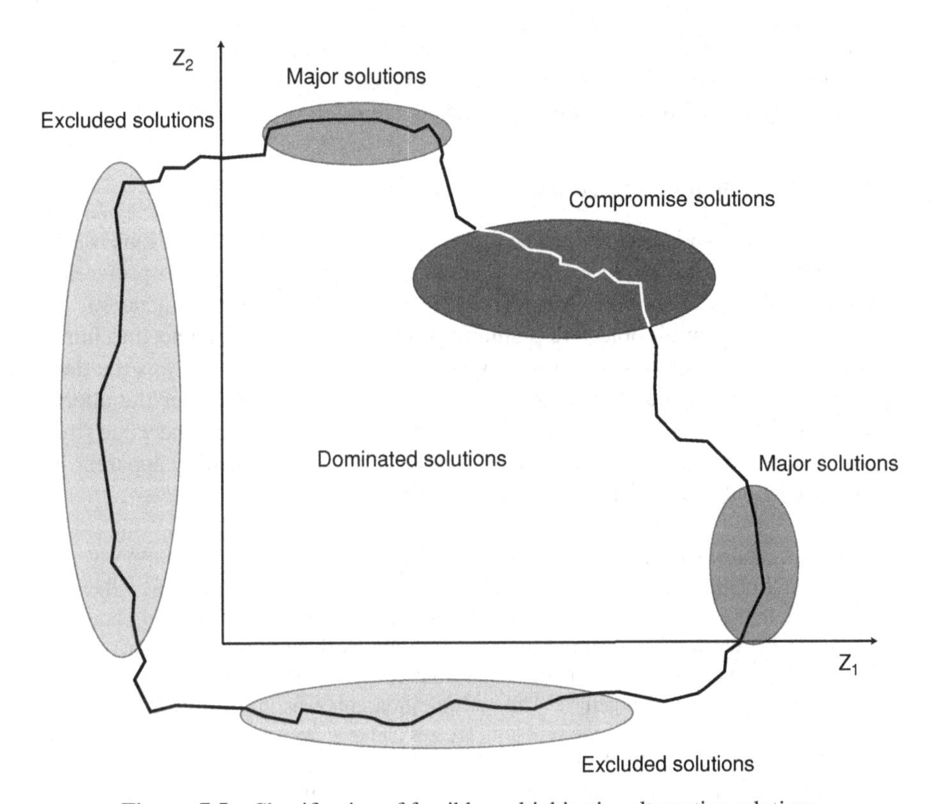

Figure 7.5 Classification of feasible multiobjective alternative solutions.

performance on some major objective. As Figure 7.5 indicates, the major alternatives represent polar extremes. A *compromise group* lies somewhere in between the major alternatives.

The remainder of the feasible region of solutions is likewise usefully categorized into dominated and excluded solutions. *Dominated* solutions are those that are inferior in all essential aspects to the other solutions. They can thus be set aside from further consideration. *Excluded* solutions are those that perform so badly on one or more objectives that they lie beneath the threshold of acceptability. Thus, they may be dropped from further consideration.

The concepts of nondominated solutions and of major categories are often highly useful in a practical sense. They organize the feasible solutions into a small number of manageable options, and draw attention to the choices that must be made. These options can be explored evenly when the feasible region is not defined analytically.

Given a set of feasible solutions $\boldsymbol{X}$, the set of nondominated solutions is denoted as $\boldsymbol{S}$ and defined as follows:

$$\begin{aligned}\boldsymbol{S} = \{\boldsymbol{x} : \boldsymbol{x} \in X, &\text{ there exists no other } \boldsymbol{x}' \in \boldsymbol{X} \quad \text{such} \\ &\text{that } Z_q(\boldsymbol{x}') > Z_q(\boldsymbol{x}) \text{ for some } q\{1, 2, \ldots, r\} \quad \text{and} \\ &Z_k(\boldsymbol{x}') \leq Z_k(\boldsymbol{x}) \text{ for all } k \neq q\}\end{aligned} \tag{7.2}$$

It is obvious from the definition of S that as we move from one nondominated solution to another nondominated solution and one objective function improves, then one or more of the other objective functions must decrease in value.

7.2.3 Participation of Decision Makers

From Figure 7.5 it is possible to see that the set of nondominated solutions is a subset of the initial set of feasible solutions, and that to determine this set the preferences of the decision maker are now required. Such partial orderings are characteristic of, but not restricted to, multiobjective planning problems, and imply the need to introduce value judgments into the solution process. At this point in the analysis the decision maker can be asked to articulate his or her value structure to order the alternative solutions in the nondominated set (Keeney and Raiffa, 1993). How the value structure of the decision maker is to be brought into the analysis is not readily apparent. In the theory of the displaced ideal, Zeleny (1974) points out:

> If one obtains an accurate measurement of the net attractiveness (or utility) of each available alternative, one can predict with reasonable accuracy that a person will choose the alternative which is 'most attractive'. So, the problem of prediction of choice becomes the technical problem of measurement and mechanical search. Furthermore, if the alternatives are complex and multi-attributed, then the measurement of utility could be too difficult to be practical. The real question concerns the process by which the decision-maker structures the problem, creates and evaluates the alternatives, identifies relevant criteria, adjusts their priorities and processes information ... It is important to realize that whenever we face a single attribute, an objective function, a utility function, or any other single aggregate measure, there is no decision making involved. The decision is implicit in the measurement and it is made by the search ... It is only when facing multiple attributes, objectives, criteria, functions, etc., that we can talk about decision making and its theory. (p. 482)

To define decision making is not simple. It is a process rather than an act. Although it involves choice from the set of feasible alternatives, it is also concerned with the generation of alternatives. Decision making is a dynamic process with all its components changing and evolving during its course: alternative solutions are added and removed, the criteria or objectives for their evaluation as well as the relative importance of the criteria are in a dynamic flux, the interpretation of outcomes varies, human values and preferences are reassessed.

Conflict An element of conflict is inherent in multiobjective problems because resources are limited. Conflict provides the decision-motivating tension, a period of frustration, and dissatisfaction with the status quo of a current situation. Conflict is a property of a situation in which the simultaneous attainment of all the objective functions, at desired levels, is not possible.

Conflict can be resolved in two ways: innovation and adaptation. *Innovation* refers to the development of previously unknown alternatives so that the original goals can be attained. Information plays an important role here, because it can suggest new avenues to search. *Adaptation* refers to changes in the current value structure of the individual so that he or she becomes content with one of the available alternatives. In reality, conflict is often resolved by both methods simultaneously. The decision maker often chooses to attain that which he or she can attain while still striving to broaden the range of the attainable. This problem of conflict resolution is also dealt with by Zeleny (1974), as he recognizes a predecision situation where the component values of the ideal alternative become clearly perceived, and an effort for conflict resolution is replaced by an attempt at conflict reduction. Also, a postdecision situation may exist where the attractiveness of chosen alternatives is enhanced, while that of the rejected alternatives is reduced.

Goals A goal can be defined as the "objective, condition, or activity toward which the motive is directed; in short, that which will satisfy or reduce the striving." A decision-making goal should be something desired, and it should be operational. In the single-objective optimization problem, the search is directed toward an optimal policy vector that possesses well-defined mathematical properties. However, in multiobjective analysis we are dealing with a collection of objective functions and an undefined preference function that, somehow, must be articulated to arrive at a "solution." If the concept of an "optimal solution" is no longer applicable, the concept of a "satisfactory solution," termed a *satisfactum*, must be examined. Simon (1976) states:

> Most human-decision making, whether individual or organizational, is concerned with the discovery and selection of satisfactory alternatives; only in exceptional cases is it concerned with the discovery and selection of optimal alternatives. (p. 272)

Acceptability is a value judgment derived from the individual's preference function. A satisfactum, then, is any value within an interval of acceptability on the range of an objective function. A multiple goal satisfactum implies acceptable values of all objective functions. Since it is assumed that more of each objective function is desirable, a satisfactum is always a member of the set of nondominated solutions.

7.2.4 Classification of Multiobjective Techniques

Multiobjective techniques are classified into three groups:

- methods for generating the nondominated set;
- methods with prior articulation of preferences; and
- methods with progressive articulation of preferences (Goicoechea et al., 1982).

Methods for Generating the Nondominated Set A generating method does just one thing: it considers a vector of objective functions and uses this vector to identify and generate the subset of nondominated solutions in the initial feasible region. In doing so, these methods deal strictly with the physical realities of the problem (i.e., the set of constraints), and make no attempt to consider the preferences of a decision maker. The desired outcome, then, is the identification of the set of nondominated solutions to help the decision maker gain insight into the physical realities of the problem at hand.

There are several methods available to generate the set of nondominated solutions, and four of these methods are widely known. These methods are:

- weighting method;
- ε-constraint method;
- Phillip's linear multiobjective method; and
- Zeleny's linear multiobjective method.

The first two methods transform the multiobjective problem into a single-objective programming format and then, by parametric variation of the parameters used to effect the transformation, the set of nondominated solutions can be generated. The weighting and constraint methods can be used to obtain nondominated solutions when the objective functions and/or constraints are nonlinear.

The last two methods generate the nondominated set for linear models only. However, these two approaches do not require the transformation of the problem into a single-objective programming format. These methods operate directly on the vector of objectives to obtain the nondominated solutions.

Methods with Prior Articulation of Preferences Methods in this class are further divided into continuous and discrete methods.

Continuous Once the set of nondominated solutions for a multiobjective problem has been identified using any of the methods mentioned in the first group, the decision maker is able to select one of those nondominated solutions as her or his final choice. This solution will be one that meets the physical constraints of the problem and happens to satisfy the value structure of the decision maker as well. However, a more likely situation is that the decision maker is unwilling or unable to accept one of those solutions made available. In that case, a good alternative is to solicit the decision maker's preferences regarding the various objective functions in the search for a solution. Various methods are available, where decision makers are asked to articulate their worth or preference structure, and these preferences are then built into the formulation of the mathematical model for the multiobjective problem.

To assist decision makers in articulating their preferences, a series of questions may be put to them, where they are asked to consider specific trade-offs among several objectives and so indicate a preference for a particular allocation for each objective. In the process, use is made of basic elements of utility theory (Keeney and Raiffa, 1993) and probability.

The net effect of articulating the preferences of the decision maker prior to solving the multiobjective problem is to reduce the set of nondominated solutions to a much smaller set of solutions, facilitating the task of selecting a final choice. Depending on the method used, this smaller set may contain several solutions, one solution or none at all. Preferences, then, provide an ordering of solutions stronger than that provided by the concept of nondominated solutions.

The best-known methods in the group of continuous methods with prior articulation of preferences are:

- goal programming,
- utility function assessment, and
- the surrogate worth trade-off method.

Discrete There are many decision situations in which the decision maker must choose among a finite number of alternatives that are evaluated on a common set of noncommensurable multiple objectives or criteria. Problems of this sort occur in many practical situations; for example, which of five candidate pipe sizes should be selected, which of ten water supply options should be selected, or which of eight communication systems should be chosen and implemented?

In problems of this type, the solution process can be described as follows. First, a statement of the general goals relating to the situation is made. Second, the alternatives must be identified or developed. Third, the common set of relevant criteria for evaluation purposes must be specified. Fourth, the levels of the criteria for each alternative must be determined. Finally a choice is made based on a formal or informal evaluation procedure.

The structure of the discrete problem can be represented in a payoff matrix as shown in Table 7.2. The rating of the ith objective/criteria on the jth alternative solution ($i = 1, 2, \ldots, m$ and $j = 1, 2, \ldots, n$) is represented by v_{ij}.

TABLE 7.2 Payoff Matrix

Alternatives		1	2	...	n
Criteria/objective	1	v_{11}	v_{12}	...	v_{1n}
	2	v_{21}	v_{22}	...	v_{2n}
	...	...	...	...	
	m	v_{m1}	v_{m2}	...	v_{mn}

Clearly the choice in a problem such as that represented in Table 7.2 is sufficiently complex to require some type of formal assistance. Determining the worth of alternative solutions that vary on many dimensions presents formidable cognitive difficulties. People faced with such complex decisions react by reducing the task complexity by using various heuristics. Unfortunately, it has been observed that decision makers who rely on heuristic decision rules systematically violate the expected utility principle. Moreover, decision makers tend to ignore many relevant variables in order to simplify their problem to a scale consistent with the limitations of the human intellect. While such simplification facilitates the actual decision making, it can clearly result in a suboptimal decision.

In effect, as the decision-making task increases, researchers have observed systematic discrepancies between rational theory and actual behavior. Evidence exists that even experts have great difficulty in intuitively combining information in appropriate ways. In fact, these studies and many others indicate that global judgments (i.e., combinations of attributes) are not nearly as accurate as analytical combinations. Because of the severe limitations of the intuitive decision-making process, it is evident that analytical methods are needed to help determine the worth of multiattributed alternatives.

Ideally, the alternative that maximizes the utility of the decision maker should be chosen. Therefore, the obvious first step in the application of any discrete multiattributed method is the elimination of all dominated alternatives. Occasionally, for discrete problems, this dominance analysis will yield only one nondominated alternative, in which case the problem is solved; no further analysis is needed.

The methods available in this group range from the very simple to the very complex. Some of the methods are:

- exclusionary screening,
- conjunctive ranking,
- weighted average,
- ELECTRE I and II,
- indifference trade-off method, and
- direct-rating method.

Methods of Progressive Articulation of Preferences The characteristic of the methods in this group is the following general algorithmic approach. First, a nondominated solution is identified. Second, the decision maker is solicited for trade-off information concerning this solution, and the problem is modified accordingly. These two steps are repeated until the decision maker indicates the acceptability of a current achievement level, provided one exists.

The methods typically require greater decision-maker involvement in the solution process. This has the advantage of allowing the decision maker to gain a greater understanding and feel for the structure of the problem. On the other hand, the required interaction has the disadvantage of being time consuming. The decision maker may not feel that the investment of the time required provides any better decision making than ad hoc approaches. That is, the decision maker may perceive the costs to be greater than any benefits. Some literature points out that decision makers have less confidence in the interactive algorithms and find them more difficult to use and understand than trial and error methods. Certainly, we cannot ignore these behavioral difficulties. What they seem to indicate is that more research is needed on how analysts can successfully interact with decision makers to implement improved but complex decision aids.

At any rate, knowledge of some of the advantages and disadvantages should help the managers or analysts to choose the appropriate decision aid. The use of any of the methods as aids to decision making will depend on the analyst's assessment of the personality and tastes of the decision maker. The methods of progressive articulation of preferences vary in the degree of sophistication and the degree of required interaction. Some of them are:

- step method (stem),
- Geoffrion's method,
- Compromise programming, and
- SEMOPS method.

I present two multiobjective analysis methods in this book. Section 7.3 presents the weighting method as one of the tools that can help in identifying nondominated solutions. Section 7.4 presents the compromise programming as a very useful method for practical application in disaster management domain for the solution of discrete problems. The CD-ROM accompanying this book includes the COMPRO software package for the implementation of compromise programming method.

7.2.5 Disaster Management Applications

Review of the most recent literature in disaster management (Altay and Green, 2006) reveals that only about 10% of research effort included in their review deals with multiobjective analysis. However, practical applications of multiobjective analysis in disaster management date back to early 1970s and show strong presence in everyday practice. Here is a brief review of the range of disaster management problems for which documented application of multiobjective analysis can be located in generally available literature.

Mitigation *FP Management* Novoa and Halff (1977) report the use of multiobjective in FP analysis of the lower reach of Peaks Branch, a stream in east Dallas, Texas. This is the example presented in more details in Section 7.1.2. The same example will be used later in this chapter to illustrate the application of the compromise programming multiobjective technique. The original work considered eight alternative flood management options, ranging from no action to stream channelization to complete redevelopment. The alternates are evaluated in terms of their relative safety, effects on neighborhoods, required relocations of families and businesses, initial costs, and maintenance costs. Weighted average method from the group of methods with prior articulation of preferences has been applied in finding the most preferred alternative. Creation of a stream-side greenway, offering lakes and parks, is recommended. This plan best balances costs and required relocations with community benefits, including flood protection.

Fire Station Location Disaster managers are often faced with the task of identifying sites that are suitable for additional facilities as the demand for services increases. Many examples of multiobjective analysis exist in the area of public facilities location (health clinics, fire stations, energy plants, schools, etc.). A 15-month study conducted by the Johns Hopkins University for the City of Baltimore has applied a multiobjective facility location (MOFLO) model to investigate the location of new fire stations and the relocation of existing fire companies. Using a 0–1 integer programming formulation, the city was represented by a 600-point network, each of which was located in the centroid of an area roughly 9 square mile in size. Objectives were to maximize property value covered by the new fire stations, population coverage, expected fires covered, and property hazard covered. Generating methods (first group of multiobjective methods) were utilized in the study. For a review of early facility location models, the reader is referred to a text by Cohon (1978). An interactive computer system for multiobjective facility location problems has been reported by Hultz et al. (1981) to conduct a heuristic search for nondominated solutions. A recent detailed review of multiobjective tools for location problems is available in Captivo and Climaco (2008).

Assigning Search and Rescue Aircraft to Bases Armstrong and Cook (1979) developed several variations of a goal programming multiobjective model (second group of multiobjective methods) for optimally allocating a fleet of search and rescue (SAR) aircraft to a fixed set of available and potentially available bases. Two types of models are presented—SAR resource allocation models, and fleet planning models. The strengths and weaknesses of these models as planning tools are discussed and the problems of data availability, data quality, impact of resources external to the system, are investigated. Although the models would apply in any SAR setting they use a Canadian problem as a particular case for the investigation. Virtually all the literature on search and rescue has been concerned with the phase II or micro problem of allocating search time for a given aircraft within a given search region. Work addressed in this research is concerned with the phase I or macro model of search, namely, the allocation of aircraft to bases and search areas. The basis of this model is goal programming.

Preparedness *Canadian Force Officer Recruitment and Promotion* This example is indirectly related to disaster management and also illustrates one broad area of multiobjective analysis application. Price and Piskor (1972) have constructed a goal programming model (second group of multiobjective methods) of the manpower system for officers within the Canadian Forces. The model forms part of a control system for fixing promotion quotas and strengths for the various rank levels in many occupational classifications over a planning horizon of 3 years. The form of the constraints is set out, and the method used by the authors to determine the objective function weights is outlined. The control system provides a fast response time and measurements indicate that it is successful in attaining the aims of the planners. An example of the use of the system to evaluate proposed manning policies is given. The authors conclude that goal programming is a most useful tool for manpower planning.

Manpower Scheduling in the Postal Service This work is also not directly related to disaster management, but it illustrates another broad application area of multiobjective analysis to manpower scheduling. In addition, an example of distribution of evacuation orders illustrates an indirect role of postal service in disaster management. Ritzman and Krajewski (1973) are weighting and aggregating individual objective functions for purpose of analyzing short-term manpower scheduling in a post office. In the process, a desirable balance between mail transit times and resource expenditures is achieved.

Nebraska State Patrol Manpower Allocation This work shows another application of goal programming multiobjective method to the problem that can be very easily transferred into disaster management domain. The primary function of a state's traffic law enforcement group is to patrol the state's road system. Lee et al. (1978) utilizes goal programming (second group of multiobjective methods) to assign division patrolmen to specific road segments in Nebraska. A force of 264 uniformed people must be allocated observing constraints on number of cars assigned to each road, maximum hours per work shift, overlapping of staggered shifts, and minimum level of protection. A total of 42 decision variables and 72 constraints are considered. Every disaster event in recent history did include a task of patrolling the roads/streets in order to provide stability and safety after the disaster event. Almost the same problem formulation as presented in this work applies straight to the disaster patrol problem.

Israeli Police Department The Israel police force is composed of several investigative department that are responsible for handling various investigations in a number of regions. In the past, a trial-and-error process has been used to allocate additional manpower. Passy and Levanon (1980) use a multiattribute utility function (second group of multiobjective methods) in a model to recommend the distribution of added personnel among five investigative groups, as the department force increases from 190 to 310 police officers.

Entertainment Building Evacuation Chow and Lui (2002) are addressing two key questions concerning (i) the effect of increasing the corridor width from 1.05 to 1.2 m and (ii) elimination of "dead ends" that are raised by the entertainment industry in order to meet the fire safety requirements. To answer these two questions, evacuation pattern in a typical karaoke establishment was studied numerically with the software building EXODUS. Six sets of scenarios were designed to study different layouts and occupancy levels. Evacuation times were studied for each scenario by changing the corridor width, exit access configurations such as the number of exits and their locations. Four corridor widths of 0.5, 1.0, 1.5, and 2.0 m were set and three simulations were carried out for each case, giving a total number of $6 \times 4 \times 3$ or 72 simulations. The simulations were of the normal emergency evacuation without specifying a fire. It is found that wider corridor might not necessarily give shorter evacuation time. The presence of dead ends might not affect the evacuation pattern, depending on their distances. However, to ensure safety, good fire safety management must be implemented. The software building EXODUS is developed to simulate the evacuation of a large number of individuals from large multifloor buildings. The individual trajectory of each occupant would be tracked while moving out of the building. In other words, the evacuation pattern can be traced. The evacuation time can then be determined from the personal elapsed time that the last person moved out. In this study, the simulation software is used in a multiobjective setup through the comparative analysis of six selected scenarios.

Evacuation Planning In an emergency situation, evacuation is conducted in order to displace people from a dangerous place to a safer place, and it usually needs to be done in a hurry. It is necessary to prepare evacuation plans in order to have a good response in an emergency situation. A central challenge in developing an evacuation plan is in determining the distribution of evacuees into the safe areas, that is, deciding where and from which road each evacuee should go. To achieve this aim, several objective functions should be brought into consideration and need to be satisfied simultaneously, though these objective functions may often conflict with each other. Saadatseresht et al. (2009) address the use of multiobjective evolutionary algorithms (MOEAs) and the geographical information system (GIS) for evacuation planning. The work proposes a three-step approach for evacuation planning. It explains that the last step, which corresponds to distribution of evacuees into the safe areas, is a spatial multiobjective optimization problem (MOP), because the objective functions and data required for solving the problem have a spatial component. To solve the MOP, two objective functions are defined, different algorithms for solving the problem are investigated, and the proper algorithm is selected. Finally, in the context of a case study project and based on the proposed approach and algorithm, evacuation planning is conducted in a GIS environment, and the results are tested.

Response *Emergency Response Planning* Narzisi et al. (2006) are presenting a new tool for multiobjective analysis based on agent-based models (ABMs)/multiagent systems (MASs) that are today one of the most widely used modeling–simulation–analysis approaches for understanding the dynamical behavior of complex systems. These models are often characterized by several parameters with nonlinear interactions that together determine the global system dynamics, usually measured by different conflicting objectives. The problem that emerges is that of tuning the controllable system parameters at the local level, in order to reach some desirable global behavior. In this research, the tuning of an ABM for emergency response planning is defined as a multiobjective analysis problem. The authors propose the use of MOEAs for exploration and optimization of the resultant search space. Then they employ two well-known MOEAs, the Nondominated Sorting Genetic Algorithm II and the Pareto Archived Evolution Strategy, and test their performance for different pairs of objectives for plan evaluation. In the experimental results, the approximate Pareto front of the nondominated solutions is effectively obtained. Additional robustness analysis is performed to help policymakers select a plan according to higher-level information or criteria not present in the original problem description.

Emergency Evacuation Georgiadou et al. (2007) developed a model for the temporal and spatial distribution of the population under evacuation around a major hazard facility. A discrete state stochastic Markov process simulates the movement of the evacuees. The area around the hazardous facility is divided into nodes connected among themselves with links representing the road system of the area. Transition from node-to-node is simulated as a random process where the probability of transition depends on the dynamically changed states of the destination and origin nodes and on the link between them. Solution of the Markov process provides the expected distribution of the evacuees in the nodes of the area as a function of time. A Monte Carlo solution of the model provides in addition a sample of actual trajectories of the evacuees. This information coupled with an accident analysis that provides the spatial and temporal distribution of the extreme phenomenon following an accident determines a sample of the actual doses received by the evacuees. Both the average dose and the actual distribution of doses are then used as measures in evaluating alternative emergency response strategies. It is shown that in some cases the estimation of the health consequences by the average dose might be either too conservative or too nonconservative relative to the one corresponding to the distribution of the received dose and hence not a suitable measure to evaluate alternative evacuation strategies.

Recovery *Postdisaster Temporary Housing Allocation* Natural disasters such as hurricanes, earthquakes, and tsunamis often cause large-scale destruction in residential areas. In the aftermath of these disasters, emergency management agencies need to urgently develop and implement a temporary housing plan that provides displaced families with satisfactory and safe accommodations. Work presented by El-Anwar et al. (2009a, 2009b) presents the computational implementation of a newly developed multiobjective optimization model to support decision makers in emergency management agencies in optimizing large-scale temporary housing arrangements. The model is capable of simultaneously minimizing (i) postdisaster social and economic disruptions suffered by displaced families, (ii) temporary housing vulnerabilities to postdisaster hazards, (iii) adverse environmental impacts on host communities, and (iv) public expenditures on temporary housing. The model is implemented in four main phases, and it incorporates four optimization modules to enable optimizing each of the aforementioned important objectives. A large-scale temporary housing application example is presented to demonstrate the unique capabilities of the model and illustrate the performed computations in each of the implementation phases. Disaster impact software systems, such as HAZUS-MH (FEMA) and MAEviz (MAE Center, 2006), enable emergency planners to estimate the expected displacement of families after natural disasters; however, they lack the capability of providing temporary housing solutions. An automated system developed in this work to support decision makers in optimizing postdisaster temporary housing arrangements has been integrated in MAEviz.

7.3 THE WEIGHTING METHOD

The weighting method belongs to the group of techniques for generating a nondominated set. It is based on the idea of assigning weights to the various objective functions, combining these into a single-objective function, and parametrically varying the weights to generate the nondominated set. I will use a presentation of the weighting method to further illustrate the concept of multiobjective analysis and provide a straightforward tool for its implementation. Mathematically, the weighting method can be stated as follows:

$$\begin{aligned} &\text{Max } \mathbf{Z}(\boldsymbol{x}) = w_1 Z_1(\boldsymbol{x}) + w_2 Z_2(\boldsymbol{x}) + \cdots + w_r Z_r(\boldsymbol{x}) \\ &\text{subject to} \\ &\boldsymbol{x} \in \boldsymbol{X} \end{aligned} \tag{7.3}$$

which can be thought of as an operational form of the formulation:

$$\begin{aligned}&\text{Max} - \text{dominate } \boldsymbol{Z}(\boldsymbol{x}) = [Z_1(\boldsymbol{x}), Z_2(\boldsymbol{x}), \ldots, Z_r(\boldsymbol{x})]\\&\text{subject to}\\&\boldsymbol{x} \in \boldsymbol{X}\end{aligned} \tag{7.4}$$

In other words, a multiobjective problem has been transformed through Equation (7.3) into a single-objective optimization problem for which solution methods exist. The coefficient w_i operating on the ith objective function, $Z_i(\boldsymbol{x})$, is called a weight and can be interpreted as *the relative weight or worth* of that objective compared with the other objectives. If the weights of the various objectives are interpreted as representing the relative preferences of some decision maker, then the solution to Equation (7.3) is equivalent to the best compromise solution; that is, the optimal solution relative to a particular preference structure. Additionally, the optimal solution to Equation (7.3) is a nondominated solution, provided all the weights are positive. The reasoning behind the nonnegativity requirement is as follows. Allowing negative weights would be equivalent to transforming the maximization problem into a minimization one for which a different set of nondominated solutions will exist. The trivial case where all the weights are 0 will simply identify every $\boldsymbol{x} \in \boldsymbol{X}$ as an optimal solution, and will not distinguish between the dominated and the nondominated solutions.

Conceptually, the generation of the nondominated set using the method of weights appears simple. However, in practice the generation procedure is quite demanding. Several weight sets can generate the same nondominated point (Hobbs, 1980). Furthermore, moving from one set of weights to another set of weights may result in skipping a nondominated extreme point. Subsequent linear combinations of the observed adjacent extreme points would, in many cases, yield a set of points that are only *close* to the nondominated border. In other words, in practice it is quite possible to miss the nondominated solution using weights that would lead to an extreme point. Therefore, the most that should be expected from the weighting method is an approximation of the nondominated set.

The sufficiency of the approximation obviously relates to the proportion of the total number of extreme points that are identified. For example, assume each weight is varied systematically between 0 and some upper limit using a predetermined step size. It seems reasonable to believe that the choice of a large increment will result in more skipped extreme points than the choice of a small increment. However, the smaller the increment, the greater the computational requirements. There is a trade-off between the accuracy of the specification of the nondominated set and the costs of the computation. Judgment must be exercised by the decision maker and the analyst to determine the desired balance.

Example 1

An emergency management team at the site of recent disaster is responsible for water distribution to population and hospitals. Water is taken from two functioning wells. If overused, the wells water quality will deteriorate to the level that it will not be safe for drinking. So the emergency water supply has to take into consideration:

- the water demand of affected population,
- proper functioning of the hospital units that are providing help to disaster victims, and
- the control of water quality in the wells.

Allocating the water to the population and hospitals, unfortunately, is in conflict with the third purpose. The emergency management team would like to minimize the negative effect on the water quality in the wells, and at the same time maximize the supply of water to population and hospitals.

Thus, there are two objectives: minimize the increase in water pollution and maximize water supply. Trade-offs between these two objectives are sought to assist the emergency management team in the decision-making process. The available data are listed in Table 7.3. Estimated water costs (energy for pumping, labor cost, etc.) and benefits (meeting supply needs, meeting hospital health standards, etc.) are obtained by the emergency team through the fast valuation of water supply impacts in "relative monetary units" in order to assist in the distribution of available resource.

TABLE 7.3 Available Data for Emergency Water Supply Example

	Population Water Supply	Hospitals Water Supply
Units of water delivered	x_1	x_2
Units of water required	1	5
Pump time (hour)	0.5	0.25
Labor time (person-hour)	0.2	0.2
Estimated water costs ($)	3	2
Estimated water benefits ($)	4	5

The following assumptions are made:

- Analysis is done for one time period $t = 0, 1$.
- The limiting pump capacity is 8 hours per period.
- The limiting labor capacity is 4 person-hours per period.
- The total amount of water in the wells available for the allocation during the time period is 72 units.
- The pollution in the wells increases by 3 units per 1 unit of water used for population water supply and 2 units per 1 unit of water used for hospital supply.

On the basis of the preceding information, we can formulate the objective functions and constraints of the problem.

Objective Functions The contribution margin of each allocation is provided as:

Population water supply = \$4.00 − \$3.00 = \$1.00 per unit of water delivered
Hospitals water supply = \$5.00 − \$2.00 = \$3.00 per unit of water delivered

The objective function for water supply becomes:

$$\text{Maximize } Z_1(\boldsymbol{x}) = x_1 + 3x_2$$

The objective function for well pollution is:

$$\text{Minimize } Z_2'(\boldsymbol{x}) = 3x_1 + 2x_2$$

or

$$\text{Maximize } Z_2(\boldsymbol{x}) = -3x_1 - 2x_2$$

The above transformation is performed so that the maximization criterion applies to both of the objective functions.

Finally, the technical constraints due to pump capacity, labor capacity, and water availability are respectively:

$$\begin{aligned} 0.5x_1 + 0.25x_2 &\leq 8 \\ 0.2x_1 + 0.2x_2 &\leq 4 \\ x_1 + 5x_2 &\leq 72 \end{aligned}$$

Now, using the operational form of the weighting method, the problem to solve is:

$$\begin{aligned} \text{Maximize } Z(\boldsymbol{x}) &= w_1 Z_1(\boldsymbol{x}) + w_2 Z_2(\boldsymbol{x}) \\ &= w_1(x_1 + 3x_2) + w_2(-3x_1 - 2x_2) \end{aligned}$$

$$\begin{aligned} &\text{subject to} \\ &0.5x_1 + 0.25x_2 - 8 \leq 0 \\ &0.2x_1 + 0.2x_2 - 4 \leq 0 \\ &x_1 + 5x_2 - 72 \leq 0 \\ &x_1 \geq 0, \quad x_2 \geq 0 \end{aligned} \tag{7.5}$$

Let us arbitrarily fix $w_1 = 1$ and increase w_2 at increments of 1 until all the non-dominated extreme points have been identified. For this example, the pairs of values selected for (w_1, w_2) are $(1, 0)$, $(1, 1)$, $(1, 2)$, $(1, 3)$, $(0, 1)$, as shown in Table 7.4. For example, for the pair of weights $(1, 0)$, the objective function to maximize is:

$$\begin{aligned} \text{Maximize } Z(\boldsymbol{x}) &= 1(x_1 + 3x_2) + 0(-3x_1 - 2x_2) \\ &= x_1 + 3x_2 \end{aligned}$$

subject to stated constraints. The objective functions and constraints of this problem are in linear form so that the methodology presented in Sections 6.1 and 6.2 could be implemented here. The solution can be obtained graphically by moving the line $Z(x) = x_1 + 3x_2$ out toward the boundary of the feasible region until it just touches the extreme point $x^* = (7, 13)$, yielding $Z(x^*) = 46$, $Z_1(x^*) = 46$ and $Z_2(x^*) = -47$ (Figure 7.6).

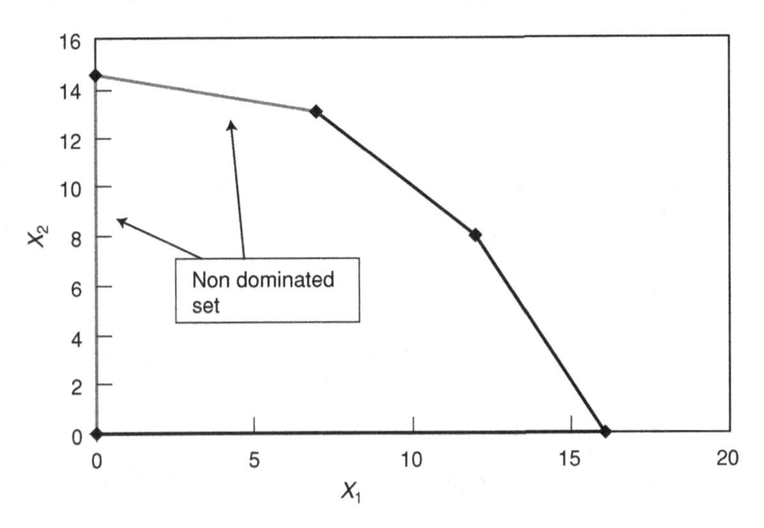

Figure 7.6 The feasible region and the nondominated set in the decision space.

Since all the objective function and constraints are in the linear form, we can use the LINPRO computer program on the CD-ROM to confirm our solution. This solution, however, is not unique to the pair of weights (1, 0) as can be observed from Table 7.4. After a graphical presentation of the solutions in the objective space (Figure 7.7), it is possible to identify nondominated points by visual inspection (gray lines in Figure 7.7). Since all of the nondominated extreme points are obviously identified, an exact representation of the nondominated set is achieved. However, for problems with a larger number of variables and constraints, representation of the nondominated set will be harder to visualize.

TABLE 7.4 Pairs of Weights and Associated Nondominated Solutions

Weight (w_1, w_2)	Nondominated Extreme Point $x^* = (x_1, x_2)$	$Z_1(x^*)$	$Z_2(x^*)$	$Z(x)$
(1, 0)	(7, 13)	46	−47	46
(1, 1)	(0, 72/5)	216/5	−144/5	72/5
(1, 2)	(0, 0)	0	0	0
(1, 3)	(0, 0)	0	0	0
(0, 1)	(0, 0)	0	0	0

Once the nondominated set is specified, the decision maker can use the information to select a preferred solution. The trade-offs are now readily apparent. For example, a decision to move from the solution (7, 13) to the solution (0, 14.4) results in a decrease

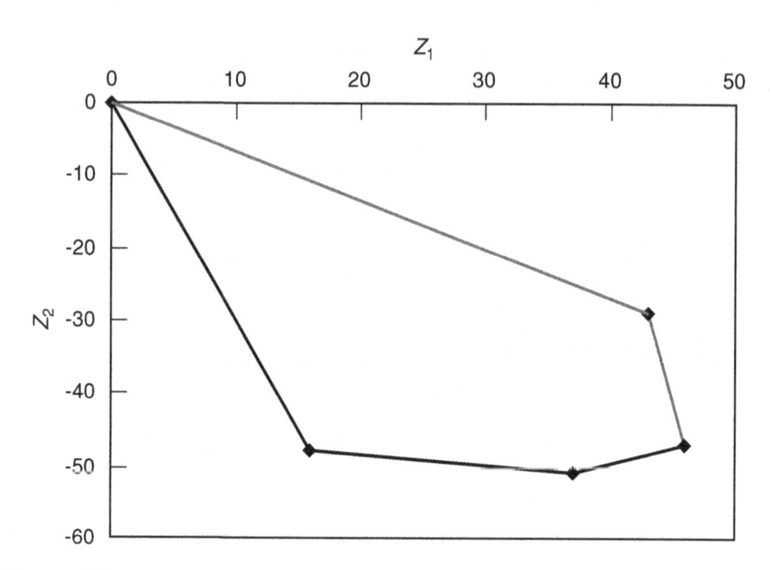

Figure 7.7 The feasible region and the nondominated set in the objective space.

of 2.8 units of water supply with the resulting benefit of a reduction of 18.2 units of pollution. This might be perceived as too great a sacrifice, and feasible production vectors in between these two extremes could be examined, with the corresponding trade-offs.

7.4 THE COMPROMISE PROGRAMMING METHOD

Compromise programming was originally developed as an interactive method with progressive articulation of decision-maker preferences. It is appropriately used in a multiple linear objective context (Zeleny, 1974). However, many variations of this method have also been used in the analysis of discrete objective problems with prior or progressive articulation of preferences. Compromise programming identifies solutions that are closest to the ideal solution, as determined by some measure of distance. Due to its simplicity, transparency, and easy adaptation to both continuous and discrete settings, compromise programming is recommended as the multiobjective analysis method of choice for application to disaster management.

7.4.1 Compromise Programming

Let us consider a two-objective problem, illustrated in Figure 7.8. The solution for which both objectives (Z_1, Z_2) are maximized is point I (Z_1^*, Z_2^*), where Z_i^* is the solution obtained by maximizing the objective i. It is clear that the solution I (named ideal point) belongs to the set of infeasible solutions. Let us consider a discrete case with four solutions available as a nondominated set $\{A, B, C, \text{ and } D\}$. The solutions

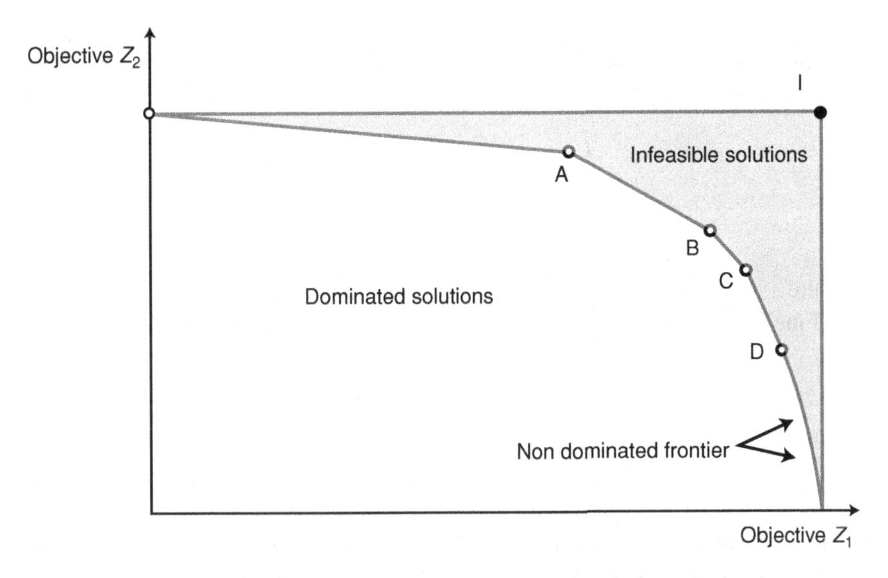

Figure 7.8 Illustration of compromise solutions.

identified as being closest to the ideal point (according to some measure of distance) are called *compromise solutions*, and constitute the *compromise set*. If we use a geometric distance, the set of compromise solutions may include a subset $\{A \text{ and } B\}$ of the nondominated set $\{A, B, C, \text{and } D\}$.

For a more explicit understanding of what is meant by a compromise solution, we must define what is meant by an ideal solution and specify the particular distance measure to be used. The ideal solution is defined as the vector $\boldsymbol{Z}^* = (Z_1^*, Z_2^*, \ldots, Z_r^*)$, where the Z_i^*, known as positive ideals, are the solutions of the following problems:

$$\begin{aligned} &\text{Max } Z_i(\boldsymbol{x}) \\ &\text{subject to} \\ &\boldsymbol{x} \in \boldsymbol{X} \qquad i = 1, 2, \ldots, r \end{aligned} \tag{7.6}$$

If there is a feasible solution vector $\boldsymbol{x}^*$ common to all r problems, then this solution would be the optimal one since the nondominated set (in the objective space) would consist of only one point, namely, $\boldsymbol{Z}^*(\boldsymbol{x}^*) = (Z_1^*(\boldsymbol{x}^*), Z_2^*(\boldsymbol{x}^*), \ldots, Z_r^*(\boldsymbol{x}^*))$. Obviously, this is most unlikely, and the ideal solution is generally not feasible. However, it can serve as a standard for evaluation of the possible nondominated solutions. Since all would prefer the ideal point if it was attainable (as long as the individual underlying utility functions are increasing), then it can be argued that finding solutions that are as close as possible to the ideal solution is a reasonable surrogate for utility-function maximization.

The procedure for evaluation of the set of nondominated solutions is to measure how close these points come to the ideal solution. One of the most frequently used measures of closeness, and the one we shall use, is a family of L_p metrics, defined in either of two operationally equivalent ways:

$$L_P = \left[\sum_{i=1}^{r} \alpha_i^P (Z_i^* - Z_i^*(x))^P\right]^{1/P}$$

or (7.7)

$$L_P = \left[\sum_{i=1}^{r} \alpha_i^P (Z_i^* - Z_i^*(x))\right]^P$$

where $1 \leq p \leq \infty$.

Finally, a compromise solution with respect to p is defined as such that:

$$\begin{aligned} &\text{Min } L_p(\boldsymbol{x}) = L_p(\boldsymbol{x}_p^*) \\ &\text{subject to} \\ &\boldsymbol{x} \in \boldsymbol{X} \end{aligned} \tag{7.8}$$

The compromise set is simply the set of all compromise solutions determined by solving Equation (7.8) for a given set of weights, $\{\alpha_1, \alpha_2, \ldots, \alpha_r\}$, and for all $1 \leq p \leq \infty$.

Operationally, three points of the compromise set are usually calculated, that is, those corresponding to $p = 1$, 2, and ∞. Varying the parameter p from 1 to ∞ allows us to move from minimizing the sum of individual regrets (i.e., having a perfect compensation among the objectives) to minimizing the maximum regret (i.e., having no compensation among the objectives) in the decision-making process. The choice of a particular value of this compensation parameter p depends on the type of problem and desired solution. In general, the greater the conflict between objectives, the smaller the possible compensation becomes. Clearly, the choice of p reflects the decision maker's concern with respect to the maximal deviation from the ideal solution. The larger the value of p, the greater the concern.

Introduction of α_i allows the expression of the decision maker's feelings concerning the relative importance of the various objectives. Thus, in compromise programming a double-weighting scheme exists: (i) the parameter p reflects the importance of the maximal deviation from the ideal point and (ii) the parameter α_i reflects the relative importance of the ith objective. After a few more mathematical transformations (Zeleny, 1973; Goicoechea et al., 1982; Simonović, 2009), we arrive at the operational definitions of a compromise solution for given value of p:

$$L_P(\boldsymbol{x}_p^*) = \text{Min}\left[L_p(\boldsymbol{x}) = \sum_{i=1}^{r} \alpha_i^P \left(\frac{Z_i^* - Z_i(\boldsymbol{x})}{Z_i^* - Z_i^{**}}\right)^P\right]$$

or

$$L_P(\boldsymbol{x}_p^*) = \text{Min}\left\{L_p(\boldsymbol{x}) = \left[\sum_{i=1}^{r} \alpha_i^P \left(\frac{Z_i^* - Z_i(\boldsymbol{x})}{Z_i^* - Z_i^{**}}\right)^P\right]^{\frac{1}{p}}\right\} \tag{7.9}$$

subject to

$\boldsymbol{x} \in \boldsymbol{X}$

Interestingly, solution of either formulation of Equation (7.9) always produces a nondominated point for $1 \leq p \leq \infty$. For $p = \infty$, there is at least one nondominated solution, $\boldsymbol{x}_p^*$.

The implementation of compromise programming results in a reduction of the nondominated set. If the compromise set is small enough to allow the decision maker to choose a satisfactory solution, then the algorithm stops. If not, the decision maker is asked to redefine the ideal point and the process is repeated. Accordingly, the interaction requirement of compromise programming is not very demanding.

Previous mathematical derivations are applicable to both continuous and discrete settings. There are many disaster management situations in which the decision maker must choose among a finite number of alternatives that are evaluated on a common set of objective functions. Problem of this sort occurs (as mentioned in Section 7.2.4) in many practical situations; for example, which one of five volunteering candidates should be selected, which of 10 suppliers of blankets should be selected, and which of 5 electricity supply sources should be used.

In a discrete setting the ideal solution is defined as the best value in a finite set of values of $Z_i(\boldsymbol{x})$ (see Figure 7.8). Essentially, the ideal solution in a discrete setting would be defined as the vector of best values selected from a payoff table like the one shown in Table 7.2. The vector of worst values, known as the *negative ideal*, defines the minimum objective function value, that is, the Z_i^{**}. With these values defined and α_i and p given, the compromise solution can be determined by calculating the distance of each alternative from the ideal solution, and selecting the alternative with the minimum distance as the compromise solution. In most cases, disaster management multiobjective problems are of a discrete nature. Therefore, our further discussion will be limited to discrete settings. Next example will convert the continuous problem considered in Example 1 into the discrete decision-making problem to demonstrate the application of compromise programming method.

Example 2

We will modify the well water allocation problem described in Example 1. Assume that the objectives from Example 1 are replaced with two new objectives:

$$\text{Maximize } Z_1(\boldsymbol{x}) = x_1 + 7x_2$$

and

$$\text{Maximize } Z_2(\boldsymbol{x}) = 10x_1 + 4x_2$$

representing the maximization of water allocation (Z_1) and maximization of water quality impacts (Z_2), respectively. Since the constraint set remains the same, we can identify four nondominated alternative solutions (A1 to A4). They are corner-point solutions of the feasible space defined by given constraints and shown in Figure 7.6. All four nondominated solutions are shown in Table 7.5.

Considering the discrete setting, our problem is now to identify compromise solutions from the set of four nondominated solutions presented in Table 7.5 (A1–A4).

As one would expect, the ideal point I corresponds to the maximum value of x_1 (point A4) and x_2 (point A1). To arrive at an approximation of the compromise set,

TABLE 7.5 Nondominated Solutions

Alternative	Nondominated Extreme Point $x^* = (x_1, x_2)$	$Z_1(x^*)$	$Z_2(x^*)$
A1	(0, 14.4)	100.8	57.6
A2	(7, 13)	98.0	122.0
A3	(12, 8)	68.0	152.0
A4	(16, 0)	16.0	160.0

we shall solve problem (Eq. 7.9) for $p = 1$, $p = 2$, and $p = \infty$, and the α_i weights must be specified.

Assuming the decision maker's views the objectives as equally important, we shall solve:

$$\text{Min} \left\{ L_p(x) = \left[\sum_{i=1}^{2} \alpha_i^p \left(\frac{Z_i^* - Z_i(x)}{Z_i^* - Z_i^{**}} \right)^p \right]^{\frac{1}{p}} \right\}$$

$$\text{subject to} \quad x \in X \tag{7.10}$$

The compromise solution for $p = \infty$ may or may not be a nondominated extreme point. To determine the compromise solution in practice, ∞ is replaced by a large number (e.g., 100). The calculations of the L_p matrix for all nondominated solutions are presented in Tables 7.6 and 7.7.

TABLE 7.6 Scaled Nondominated Solutions

Alternative	Nondominated Extreme Point $x^* = (x_1, x_2)$	$\frac{Z_1^* - Z_1(x)}{Z_1^* - Z_1^{**}}$	$\frac{Z_2^* - Z_2(x)}{Z_2^* - Z_2^{**}}$
A1	(0, 14.4)	0.00	1.00
A2	(7, 13)	0.03	0.37
A3	(12, 8)	0.39	0.08
A4	(16, 0)	1.00	0.00

Using the information from Table 7.6 and solving Equation (7.10) for $\alpha_1 = \alpha_2 = 0.5$ and $p = 1, 2$, and 100, the compromise set is identified. The final results are summarized in Table 7.7. The compromise set is a set of solutions closest to the ideal solution for different values of the parameter p. In our case for all three values of $p = 1, 2$, and 100, the alternative A2 is ranked first (closest to the ideal point). Therefore, our compromise set of solutions is reduced to alternative A2. In other words, this alternative is also the best compromise solution for the well water allocation problem.

You can use the COMPRO program provided on the CD-ROM (and described later in this chapter) to confirm the solution in Table 7.7. Example 2 data are in the folder COMPRO, subfolder Examples, file Example1.compro.

TABLE 7.7 Final Compromise Solutions

	$p = 1$		$p = 2$		$p = 100$	
Alternative	$L_p(\boldsymbol{x})$	Rank	$L_p(\boldsymbol{x})$	Rank	$L_p(\boldsymbol{x})$	Rank
A1	0.50	3	0.50	3	0.50	3
A2	0.20	1	0.19	1	0.18	1
A3	0.23	2	0.20	2	0.19	2
A4	0.50	4	0.50	4	0.50	4

7.4.2 Some Practical Recommendations

The original purpose of compromise programming is to reduce the nondominated set to the compromise set in direct interaction with decision makers. However, very often disaster management problems do need to identify one solution that should be a recommended solution. Based on extensive use of the compromise programming multiobjective method in practice, I suggest that the solution with the smallest Euclidean distance, corresponding to $p = 2$, be used as the first approximation of the "best compromise solution." In this case, an extensive sensitivity analysis of the final solution selection to the change in parameter p is advised.

Quite often in practice, the preferences of decision makers are not readily available. In some situations, they are not able to articulate them easily; in others, they may not be willing to openly express their values. In order to assist the decision-making process, I have developed a concept of *most robust compromise solution* as a replacement for the *best compromise solution* (Simonović, 2009; example in Section 10.6.1).

The best compromise solution is one closest to the ideal point for the fixed set of decision maker's preferences and one value of the distance parameter p. My recommendation is that $p = 2$ is used in identification of the best compromise solution. The most robust compromise solution is one that occupies a high rank (not always the highest), the most often for various sets of decision maker's preferences. So in this way we arrive at the solution that is not very sensitive to change in preferences, and therefore has a chance of a higher level of acceptance by the decision makers. The most robust solution is calculated through systematic sensitivity analysis, or repetitive solutions of Equation (7.9) for different values of α_i and one value of the distance parameter p (again $p = 2$ is suggested).

7.4.3 The COMPRO Computer Program

The CD-ROM accompanying this book includes the COMPRO software and all the examples developed in the text. The folder COMPRO contains three subfolders: Compro, ComproHelp, and Examples. The Readme file contains instructions for installation of the COMPRO software, and a detailed tutorial for its use is a part of the Help menu.

The COMPRO software package facilitates multiobjective analysis of discrete problems using compromise programming method. Compromise programming is the method for reducing the set of nondominated solutions according to their distance from the ideal solution. The distance from the ideal solution for each alternative is measured by a distance metric. An operational definition (Eq. 7.9) is used for the computation of the distance metric. This value, which is calculated for each alternative, is a function of the objective function/criteria values themselves, the relative importance of the various criteria to the decision makers (α_i), and the importance of the maximum deviation from the ideal solution (p).

The COMPRO package can be used in two ways: to narrow down the set of nondominated solutions (find a compromise set), or to identify the *best compromise alternative solution*. Narrowing down the compromise set is achieved by running the software for a set of parameter p values. Identification of the best compromise alternative is achieved by running the program for one preset value of the deviation parameter, $p = 2$.

7.5 AN EXAMPLE OF DISASTER MANAGEMENT MULTIOBJECTIVE ANALYSIS—SELECTION OF FLOOD MANAGEMENT ALTERNATIVE

Multiobjective models play an important role in disaster management. Some characteristic applications are introduced in Section 7.2.5.

Here, we shall look at one real problem in order to illustrate the value of compromise programming in solving disaster management problems in practice. We will focus on the selection of flood management alternative for the City of Dallas, Texas. This problem is taken from Novoa and Halff (1977). General problem formulation is provided in Section 7.1.2. Novoa and Halff (1977) solved this problem using weighted average method (second group of multiobjective methods). We will solve this problem using compromise programming. Three alternative solutions will be illustrated: (a) selection of the compromise set of solutions with progressive articulation of preferences (original form of the compromise programming method), (b) selection of the best compromise solution with prior articulation of preferences, and (c) selection of the most robust solution.

7.5.1 Preparation of Input Data

The City of Dallas authorized the consulting firm of Albert H. Halff and Associates to conduct a study, identify the areas subject to flooding, and provide a recommendation for floodplane management.

Eight alternative plans described in Section 7.1.2 were developed and initial evaluation took into consideration the following factors: relative flood protection, relative neighborhood improvement, number of family relocations, project cost, maintenance cost, and relative number of management problems (see Table 7.1).

Evaluation of alternative plans required a setup of the final payoff matrix. Starting with the data provided in Table 7.1 the first step was to expand and finalize the set of evaluation criteria/objectives. Factors considered in the initial evaluation were expanded into 10 final criteria: (i) maximum flood loss prevention, (ii) multiple use of floodway, (iii) enhancement of property values, (iv) maximum aesthetic value, (v) maximum strengthening of the neighborhood, (vi) minimum number of relocations (number of families), (vii) minimum project cost (millions of dollars), (viii) minimum operation and maintenance costs (thousands of dollars); (ix) ease of phased construction, and (x) minimum legal obstacles.

Criteria (vi), (vii), and (viii) are provided quantitative values, and the remaining seven criteria were evaluated using qualitative assessment. In the final payoff matrix (Table 7.8), qualitative assessments were shown using numeric scale from 1 to 8. Value of 1 shows a very low capacity of an alternative to meet an objective, and value of 8 shows a very high capacity of an alternative to meet an objective.

TABLE 7.8 Payoff Matrix for Floodplane Management Example

	Alternative								
Criteria	1	2	3	4	5	6	7	8	Weights
(i)	1	6	5	6	4	8	4	7	10
(ii)	3	7	5	6	4	2	3	8	7
(iii)	1	5	7	8	4	3	3	6	3
(iv)	1	4	7	8	5	2	3	6	5
(v)	1	2	7	8	5	4	4	6	6
(vi)	0	−800	−237	−419	−164	−69	−30	−604	8
(vii)	0	−12	−9.3	−10.9	−7.7	−5.6	−12.3	−16.3	9
(viii)	−12	0	−41	−41	−41	−16	0	−41	4
(ix)	8	7	4	4	3	2	1	6	2
(x)	2	2	7	6	5	4	8	1	1

The consultation process between the City of Dallas and residents was assisted by the consulting firm and through a number of public meetings; all 10 criteria were assigned relative importance weights on scale of 1 to 10. All the input data are shown in Table 7.8. The negative values are assigned to the objectives that require minimization. The last column of the table shows the relative weights of each criterion.

7.5.2 Solution of Flood Management Problem Using Compromise Programming

Three alternative solution procedures are provided to illustrate (a) selection of the compromise set of solutions with progressive articulation of preferences (original form of the compromise programming method), (b) selection of the best compromise solution with prior articulation of preferences, and (c) selection of the most robust solution. All the solutions are obtained using the COMPRO program provided on the CD-ROM. Operational form of distance metric shown by Equation (7.9) is used in ranking the discrete set of alternative flood management measures.

The Compromise Set of Solutions Original formulation of the Compromise method (Zeleny, 1973) calls for iterative procedure of progressive articulation of preferences by the decision makers. The solution process has been simulated in my graduate class, where the students in the class were acting as decision makers and were guiding the solution process. Three iterations were used.

The solution process started with the set of weights provided by the consulting firm as shown in Table 7.8. The COMPRO program has been set to solve the problem with three values of distance parameter $p = 1$, 2, and 100. The input data are on the CD-ROM, in the directory Compro, subdirectory Examples, file Example2. The computation results and the final ranks are summarized in Table 7.9.

TABLE 7.9 The Compromise Set of Solutions (Iteration 1)

Parameter p	1	2	100
	Distance Metric Value [Rank]		
ALT_1	5.79×10^{-1} [8]	2.612×10^{-1} [8]	1.820×10^{-1} [8]
ALT_2	5.796×10^{-1} [6]	2.258×10^{-1} [7]	1.450×10^{-1} [6]
ALT_3	4.106×10^{-1} [2]	1.642×10^{-1} [1]	9.360×10^{-2} [1]
ALT_4	3.787×10^{-1} [1]	1.675×10^{-1} [2]	1.097×10^{-1} [3]
ALT_5	5.194×10^{-1} [5]	1.887×10^{-1} [4]	1.040×10^{-1} [2]
ALT_6	4.451×10^{-1} [3]	1.812×10^{-1} [3]	1.270×10^{-1} [5]
ALT_7	5.416×10^{-1} [7]	2.198×10^{-1} [6]	1.238×10^{-1} [4]
ALT_8	4.736×10^{-1} [4]	2.173×10^{-1} [5]	1.640×10^{-1} [7]

After first iteration, the set of eight alternatives has been reduced to a compromise set including alternatives 4 and 3. These two alternatives were ranked first for the selected set of distance parameter p.

Students in the class presented with the results of first iteration came up with a new set of weights for the second iteration. The weights are again assigned on the scale 1 to 10 to all 10 criteria as follows: 8, 7, 3, 5, 6, 10, 9, 4, 2, 1 starting with criteria 1 to 10. The COMPRO data file for this iteration is on the CD-ROM, in the directory Compro, subdirectory Examples, file Example3. The computation results and the final ranks are summarized in Table 7.10.

TABLE 7.10 The Compromise Set of Solutions (Iteration 2)

Parameter p	1	2	100
	Distance Metric Value [Rank]		
ALT_1	5.426×10^{-1} [7]	2.369×10^{-1} [7]	1.450×10^{-1} [6]
ALT_2	5.549×10^{-1} [8]	2.492×10^{-1} [8]	1.820×10^{-1} [8]
ALT_3	4.057×10^{-1} [2]	1.606×10^{-1} [1]	9.360×10^{-2} [2]
ALT_4	3.875×10^{-1} [1]	1.744×10^{-1} [2]	1.097×10^{-1} [3]
ALT_5	5.059×10^{-1} [5]	1.794×10^{-1} [3]	8.480×10^{-2} [1]
ALT_6	4.482×10^{-1} [3]	1.815×10^{-1} [4]	1.27×10^{-1} [5]
ALT_7	5.218×10^{-1} [6]	2.106×10^{-1} [5]	1.238×10^{-1} [4]
ALT_8	4.963×10^{-1} [4]	2.321×10^{-1} [6]	1.640×10^{-1} [7]

After second iteration the set of eight alternatives has been reduced to a compromise set including alternatives 4, 3, and 5. These three alternatives were ranked first for the selected set of distance parameter p. Change of criteria preferences of the decision makers (students in the class) expanded the compromise set after the second iteration.

Students were given one more opportunity to adjust their preferences after reviewing the results in Table 7.10. The new set of weights (on the scale 1 to 10) is as follows: 10, 6, 3, 4, 5, 7, 9, 8, 2, 1 starting with criteria 1–10. The COMPRO data file for this iteration is on the CD-ROM, in the directory Compro, subdirectory Examples, file Example4. The computation results and the final ranks are summarized in Table 7.11.

TABLE 7.11 The Compromise Set of Solutions (Iteration 3)

Parameter p	1	2	100
		Distance Metric Value [Rank]	
ALT_1	5.497×10^{-1} [7]	2.450×10^{-1} [8]	1.820×10^{-1} [8]
ALT_2	4.818×10^{-1} [4]	2.059×10^{-1} [4]	1.270×10^{-1} [3]
ALT_3	4.631×10^{-1} [3]	2.025×10^{-1} [2]	1.450×10^{-1} [4]
ALT_4	4.352×10^{-1} [2]	2.048×10^{-1} [3]	1.450×10^{-1} [5]
ALT_5	5.603×10^{-1} [8]	2.191×10^{-1} [6]	1.450×10^{-1} [6]
ALT_6	4.279×10^{-1} [1]	1.661×10^{-1} [1]	1.090×10^{-1} [1]
ALT_7	5.028×10^{-1} [5]	2.066×10^{-1} [5]	1.238×10^{-1} [2]
ALT_8	5.217×10^{-1} [6]	2.441×10^{-1} [7]	1.640×10^{-1} [7]

After third iteration the set of eight alternatives has been reduced to a compromise set including alternative 6. This alternative was ranked first for the selected set of distance parameter p. The process has been terminated here. After very intensive class discussion, the students decided to accept as the compromise set alternatives 3, 4, 5, and 6 and suggested that the problem be revised to proceed with the evaluation of only these four alternatives in the second round.

The compromise programming method does not limit this process in any way. Iterations should be terminated when the decision makers are happy with the solution. Suggestion provided by the students, to continue with another round with smaller set of alternative solutions, is not uncommon in practice.

The Best Compromise Solution The compromise programming method can be used as a method with prior articulation of preferences. In this mode, the weights are preselected by the decision makers before the analysis and the ranking is obtained using the distance parameter $p = 2$ as suggested in Section 7.4.2.

Original set of weights provided by Novoa and Halff (1977) are used to demonstrate the selection of best compromise solution. COMPRO program is set to run with $p = 2$. The COMPRO data file for this run is on the CD-ROM, in the directory Compro, subdirectory Examples, file Example5. The computation results and the final rank are shown in Table 7.12.

TABLE 7.12 The Best Compromise Solution

Parameter p	2
	Distance Metric Value [Rank]
ALT_1	2.612×10^{-1} [8]
ALT_2	2.258×10^{-1} [7]
ALT_3	1.642×10^{-1} [1]
ALT_4	1.675×10^{-1} [2]
ALT_5	1.887×10^{-1} [4]
ALT_6	1.812×10^{-1} [3]
ALT_7	2.198×10^{-1} [6]
ALT_8	2.173×10^{-1} [5]

The original set of preferences provided by the consulting firm and used with the compromise programming multiobjective method as a method with prior articulation of preferences resulted in alternative 3 being the best compromise solution. It is interesting to look closely the distance metric values in Table 7.12. It is very clear that the distance metric values for alternatives 3 and 4 are not very different.

The Most Robust Solution One of the most difficult tasks in the application of multiobjective analysis, in my personal experience, is elicitation of decision-makers' preferences. The main difficulties are observed in decision situations with multiple decision makers (a) unable to articulate their preferences, or (b) unwilling to show their preferences to other decision makers. Therefore, the idea of the most robust solution is developed to assist the decision-making process resulting in the recommendations that are not the best solutions but the solutions that are least sensitive to the preferences of decision makers.

To demonstrate the process of identifying the most robust solution in our example, we have to modify the input data by adding a number of weighting sets that should capture the wide range of decision-maker's concerns (Table 7.13). For example, one set of weights may emphasize the economic factors (project cost, maintenance cost, legal obstacles), where the other may focus on the social factors (flood loss prevention, relocation of people, enhancement of property values). In the same way other emphases may be considered: environment, sustainability, security, etc.

TABLE 7.13 Alternative Weight Sets

	Criterion/Objective									
Weight Set	1	2	3	4	5	6	7	8	9	10
1	5.5	5.5	5.5	5.5	5.5	5.5	5.5	5.5	5.5	5.5
2	10	7	3	5	6	8	9	4	2	1
3	9	4	7	1	2	6	9	9	3	5
4	7	13	5	12	2	2	4	3	5	2
5	10	2	5	5	12	13	3	3	1	1

For the demonstration purposes, the flood management problem is solved with five sets of weights and the distance parameter $p = 2$. The weighting schemes used in this example are shown in Table 7.13. To simplify the comparison, total number of weight score points is selected to be 55. So each row of Table 7.13 should add to 55. In the implementation of COMPRO program, these weights are normalized to add to 1.

The COMPRO data files for these runs are on the CD-ROM, in the directory Compro, subdirectory Examples, file Example6 (weight set 1), Example7 (weight set 2), Example8 (weight set 3), Example9 (weight set 4), and Example10 (weight set 5). The computation results and the final ranks are integrated into one table and shown in Table 7.14.

TABLE 7.14 The Most Robust Solution

	Distance Metric Value [Rank]				
	Weight Set 1	Weight Set 2	Weight Set 3	Weight Set 4	Weight Set 5
ALT_1	2.348×10^{-1} [8]	2.612×10^{-1} [8]	2.382×10^{-1} [7]	3.363×10^{-1} [8]	3.140×10^{-1} [7]
ALT_2	1.911×10^{-1} [4]	2.258×10^{-1} [7]	1.974×10^{-1} [2]	1.617×10^{-1} [4]	3.154×10^{-1} [8]
ALT_3	1.502×10^{-1} [1]	1.642×10^{-1} [1]	2.109×10^{-1} [4]	1.602×10^{-1} [3]	1.293×10^{-1} [1]
ALT_4	1.524×10^{-1} [2]	1.675×10^{-1} [2]	2.159×10^{-1} [5]	1.268×10^{-1} [2]	1.505×10^{-1} [2]
ALT_5	1.850×10^{-1} [3]	1.887×10^{-1} [4]	2.306×10^{-1} [6]	2.246×10^{-1} [5]	1.750×10^{-1} [4]
ALT_6	1.977×10^{-1} [5]	1.812×10^{-1} [3]	1.628×10^{-1} [1]	3.207×10^{-1} [7]	1.694×10^{-1} [3]
ALT_7	1.985×10^{-1} [7]	2.198×10^{-1} [6]	1.991×10^{-1} [3]	2.901×10^{-1} [6]	1.944×10^{-1} [5]
ALT_8	1.979×10^{-1} [6]	2.173×10^{-1} [5]	2.666×10^{-1} [8]	1.267×10^{-1} [1]	2.099×10^{-1} [6]

According to the definition of the most robust solution from Section 7.4.2, alternative 4 is showing a very stable behavior—four times ranked second and ones fifth. Alternative 3 is again very close.

7.5.3 Summary

The analyses of the floodplane management problem in this section show the applicability of the compromise programming multiobjective method for a real disaster management decision making. Three different ways of implementing the compromise programming have been demonstrated.

All the analyses shown in this example are based on the basic data from the Novoa and Halff (1977). They also showed the solution that was obtained by the weighted-average method. Their recommendation is alternative 4.

REFERENCES

Altay, N. and W.G. Green III (2006), "OR/MS research in disaster operations management," *European Journal of Operational Research*, 175:475–493.

Armstrong, R.D. and W.D. Cook (1979), "Goal Programming Models for Assigning Search and Rescue Aircraft to Bases," *The Journal of Operations Research Society*, 30(6):555–561.

Captivo, M.E. and J.N. Climaco (2008), "On multicriteria mixed integer linear programming tools for location problems—an updated critical overview illustrated by the bicriteria DSS," *Computacion y Sistemas*, 12(2):216–231.

Chow, W.K. and C.H. Lui (2002), "Numerical studies on evacuation design in a karaoke," *Building and Environment*, 37:285–294.

Cohon, J.L. (1978), *Multiobjective Programming and Planning*, Academic Press, New York.

El-Anwar, O, K. El-Rayes, and A. Elnashai (2009a), "An automated system for optimizing post-disaster temporary housing allocation," *Automation in Construction*, 18:983–993.

El-Anwar. O, K. El-Rayes, and A. Elnashai (2009b), "Optimizing large-scale temporary housing arrangements after natural disasters," *ASCE Journal of Computing in Civil Engineering*, 23(2): 110–118.

Georgiadou, P.S., I.A. Papazoglou, C.T. Kiranoudis, and N.C. Markatos (2007), "Modeling emergency evacuation for major hazard industrial sites," *Reliability Engineering and System Safety*, 92:1388–1402.

Goicoechea, A., D.R. Hansen, and L. Duckstein (1982), *Multiobjective Decision Analysis with Engineering and Business Applications*, John Wiley & Sons, New York.

Hillier, F.S. and G.J. Lieberman (1990), *Introduction to Mathematical Programming*, McGraw Hill, New York.

Hobbs, B.F. (1980), "Multiobjective power plant siting methods," *Journal of the Energy Division*, 106:187–200.

Hultz, J.W., D.D. Klingman, G.T. Ross, and R.M. Soland (1981), "An interactive computer system for multicriteria facility location," *Computers and Operations Research*, 8(4): 249–261.

Inter Academy Panel (2009), *Natural Disaster Mitigation: A Scientific and Practical Approach*, Science Press, Beijing.

Keeney, R.L. and H. Raiffa (1993), *Decisions with Multiple Objectives: Preferences and Value Tradeoffs*. Cambridge University Press, Cambridge.

Lee, S.M., L.S. Franz, and A.J. Wynne (1979), "Optimizing State Patrol Manpower Allocation," *The Journal of Operations Research Society*, 30:885–896.

Mid-America Earthquake Center (2006), *MAEviz Software*, available online, http://mae.cee.uiuc.edu/software_and_tools/maeviz.html, last accessed March 2, 2010.

Mileti, D.S. (1999), *Disasters by Design*, Joseph Henry Press, Washington, DC.

Narzisi, G., V. Mysore, and B. Mishra (2006), In *Proceedings of the IASTED International Conference on Computational Intelligence (CI 2006), ACTA press*, pp. 224–230.

Novoa, J.I., and A.H. Halff (1977), "Management of flooding in fully-developed low-cost housing neighborhood," *Water Resources Bulletin*, 13(6):1237–1252.

Passy, U. and Y. Levanon (1980), "Manpower allocation with multiple objectives—The min-max approach," in Fandel, G. and T. Gal (eds) *Multiple Criteria Decision Making: Theory and Applications*, Springer-Verlag, New York, pp. 329–344.

Price, W.L. and W.G. Piskor (1972), "The application of goal programming to manpower planning," *INFOR*, 10(3):221–232.

Ritzman, L.P. and L.J. Krajewski (1973), "Multiple objectives in linear programming—an example in scheduling postal resources," *Decision Sciences*, 4(3):364–378.

Saadatseresht, M., A. Mansourian, and M. Taleai (2009), "Evacuation planning using multiobjective evolutionary optimization approach," *European Journal of Operational Research*, 198:305–314.

Simonović, S.P. (2009), *Managing Water Resources: Methods and Tools for a Systems Approach*, UNESCO, Paris and Earthscan, London.

Wagner, H.M. (1975), *Principles of Operations Research*, 2nd edn, Prentice-Hall, Englewood Cliffs, NJ.

Zeleny, M. (1973), "Compromise programming," in J. Cochrane and M. Zeleny (eds) *Multiple Criteria Decision Making*, University of South Carolina Press, Columbia, SC.

Zeleny, M. (1974), "A concept of compromise solutions and the method of the displaced ideal," *Computers and Operations Research*, 1(4):479–496.

EXERCISES

7.1 An emergency management team has evaluated proposed sites for the location of temporary "tent city" using two objectives: available space (AS) and closeness to water supply (WS). The optimal performance of each site is given by the following (AS, WS) pairs: A(20, 135), B(75, 135), C(90, 100), D(35, 1050), E(82, 250), F(60, −50), G(60, 550), H(75, 500), I(70, 620), J(10, 500), K(40, 350), L(30, 800), N(55, 250), O(40, −80), P(30, 500), Q(30, 900), R(60, 950), S(80, −150), T(45, 550), U(25, 1080), V(70, 800), W(63, 450).

(a) Solve the problem graphically. List the excluded, dominated, and nondominated alternative sites.

(b) Identify the nondominated solutions by maximizing AS subject to WS $\geq$ b, where b = 200, 400, 600, and 800.

(c) Identify a nondominated solution by the weighting method for a relative set of weights to (AS, WS) of (20, 8).

7.2 A firefighter has job offers from five different stations and is faced with the problem of selecting a station within a week. The individual lists all of the important information in the form of a payoff matrix.

	Fire Station				
Criterion/Objective	1	2	3	4	5
Salary ($)	64,000	60,000	66,000	67,500	62,000
Workload (hr/day)	9	7.5	10.5	12	6
Location*	2	3	5	2	4
Additional income	9,800	9,600	7,600	10,500	11,000
Cont. training*	4	4	3	1	5
Additional benefits*	3	2	4	5	1

Note: *Larger numbers represent more desirable outcomes.

Acting as the firefighter, apply the weighting method, make your choice, and explain your decision.

7.3 For Example 2 in Section 7.4.1:

(a) Identify a compromise set of solutions for $\alpha_1 = 0.1$, $\alpha_2 = 0.9$, and $p = 1$, 2, 3, 20, and 100.

(b) Identify a compromise set of solutions for $\alpha_1 = 0.9$, $\alpha_2 = 0.1$, and $p = 1$, 2, 3, 20, and 100.

(c) Compare the solutions to the solution in Table 7.7. Discuss the difference.

(d) Use COMPRO to check your solutions.

7.4 Evaluate a plan for disposing of waste from an industrial spill. For the given situation, you want to use the best combination of two methods: (i) a filter for treatment and (ii) direct disposal into nearby river. The costs are disposal \$2 per cubic meter per second (m^3/sec); treatment \$5 per m^3/sec. Another objective is to maximize the water quality in the nearby river measured by dissolved oxygen (DO). The DO will decrease by 4 mg/L for each m^3/sec of direct disposal and increase by 1 mg/L for each m^3/sec given filter treatment. The flow of water filtered must be less than 3 m^3/sec. The emergency cleaning crew can handle maximum 8 m^3/sec; moreover, physical constraints on the capacity of the available equipment limits the amount that can be directly disposed to 6 m^3/sec and the amount of that can be treated to 4 m^3/sec.

(a) Considering both the economic and the water quality objectives, conduct a multiobjective analysis using the weighting method.

(b) What is your recommendation?

(c) Why?

7.5 Evaluate the three alternatives for the recovery of the region from major flood disaster presented below using a multiobjective analysis method of your choice. In your evaluation use three objectives: economic (NED), environmental enhancement (EQ), and regional development (RD). Three alternatives may be used in various combinations: flood control, hydropower, and water quality control (mostly low-flow augmentation measure in m^3/sec). The flood control alternative, a levee that cannot exceed 2 m in height, will yield \$1000/m in annual benefits for both the NED and RD objectives and destroy 20 environmental units per meter of levee height. The hydropower alternative cannot exceed 2 MW of power and will yield \$1000/MW for the NED account but only \$500/MW for the RD. It will add 10 environmental units per MW. The water quality alternative will yield \$500 per m^3/sec of flow NED. It cannot exceed 2 m^3/sec and will yield \$1000 per m^3/sec for RD. It will yield 10 environmental units per m^3/sec. The sum of each meter of levee height plus each MW of power plus 1.25 m^3/sec of flow augmentation must not exceed 5.0. Your goal is to maximize NED, RD, and EQ.

7.6 For the example of Dallas floodplane management, assume the following scales have been determined:

Bad—1, Fair—2, Good—3, Very good—4, and Excellent—5

Very few—1, Few—2, Some—3, Many—4, and Very many—5.

(a) Using above scales and evaluation provided in Table 7.1 apply the compromise programming method to find the compromise set of alternative solutions;

(b) Using above scales and evaluation provided in Table 7.1 apply the compromise programming method to find the best compromise solution;

(c) Using above scales and evaluation provided in Table 7.1 apply the compromise programming method to find the most robust solution;

(d) Compare your solutions to those presented in Section 7.5 and discuss the difference.

PART IV
Be Prepared

8 A View Ahead

The writing of this book started in the spring of 2009 after lengthy research and preparatory work. Here is a brief and incomplete review of what has happened around the world between then and now (spring 2010).

- *April 6, Italy*: An earthquake of magnitude 6.3 struck central Italy, killing more than 200 people and injuring another 1000. The town of L'Aquila was the epicenter of the earthquake, but as many as 26 towns in the area were affected.
- *June 25, Czech Republic*: At least 10 people died in flooding in the eastern Czech Republic, and rising river levels prompted flood warnings across central Europe following heavy rains. The 10 Czechs died near the country's border with Poland and Slovakia, with most of the damage near the town of Novy Jicin, 260 km (160 miles) east of Prague. Rescuers evacuated hundreds of people from wrecked houses and buildings threatened by high water, and the government moved to deploy up to 1000 soldiers to help.
- *August 7, Philippines*: At least 22 tourists on Mount Pinatubo were trapped and killed when heavy rain caused flooding and landslides.
- *August 10, Taiwan*: Typhoon Morakot was the deadliest typhoon to impact Taiwan in recorded history. It formed early on August 2, 2009 as an unnamed tropical depression. Because of the size of the typhoon, the barometric pressure steadily decreased; however, maximum winds only increased slightly. Early on August 7, the storm attained its peak intensity with winds of 150 km/hr (90 mph 10-minute sustained). Morakot weakened slightly before making landfall in central Taiwan later that day. Roughly 24 hours later, the storm emerged back over water into the Taiwan Strait and weakened to a severe tropical storm before making landfall in China on August 9. The storm gradually weakened as it continued to slowly track inland. The remnants of the typhoon eventually dissipated on August 11. Typhoon Morakot wrought catastrophic damage in Taiwan, leaving 461 people dead and 192 others missing, most of whom are feared dead and roughly US$3.3 billion in damages. The storm produced copious amounts of rainfall, peaking at 2777 mm (109.3 in.), surpassing the previous record of 1736 mm (68.35 in.) set by Typhoon Herb in 1996. The extreme amount of rain triggered enormous mudslides and severe flooding throughout southern Taiwan. One mudslide buried the entire town of Xiaolin, killing an estimated 150 people

Systems Approach to Management of Disasters: Methods and Applications, By Slobodan P. Simonović

in the village alone. In the wake of the storm, Taiwan's president Ma Ying-jeou faced extreme criticism for the slow response to the disaster having only initially deployed roughly 2100 soldiers to the affected regions. Later, additions of troops increased the number of soldiers to 46,000 working to recover trapped residents. Rescue crews were able to retrieve thousands of trapped residents from buried villages and isolated towns across the island. Days later, the president publicly apologized about the government's slow response to the storm.

- *September 2, Indonesia*: About 60 people died when a 7.1-magnitude earthquake hit the island of Java, the most populous area of the country.
- *September 9, Turkey*: More than 30 people were killed when fast-moving floods caused by heavy rain swept through Istanbul.
- *September 28, Philippines*: Almost 90 people died in and around Manila in flooding caused by Tropical Storm Ketsana, which resulted in about 17 in. of rain in 12 hours. The floods were Manila's worst in about 50 years.
- *September 29, Samoa, and American Samoa*: An underwater 8.0-magnitude earthquake caused a tsunami that killed more than 115 people.
- *September 30, Indonesia*: A 7.6-magnitude earthquake hit the island of Sumatra, leaving more than 1000 people dead and thousands trapped under the rubble of collapsed buildings in the city of Padang.
- *November 9, El Salvador*: A small, low-pressure storm originating in the western part of the country brought an enormous amount of rainfall that caused flooding and mudslides. About 140 people were killed and some 1500 homes were destroyed. The unnamed storm coincided with Hurricane Ida. Initial reports blamed the devastation on Ida, but officials later said Ida was not responsible.
- *January 12, Haiti*: The 2010 Haiti earthquake was a catastrophic, magnitude 7.0 M_w, earthquake, with an epicenter approximately 25 km (16 miles) west of Port-au-Prince, Haiti's capital. By January 24, at least 52 aftershocks measuring 4.5 or greater had been recorded. As of February 12, an estimated 3 million people were affected by the earthquake; the Haitian Government reported that between 217,000 and 230,000 people had been identified as dead, an estimated 300,000 injured, and an estimated 1,000,000 homeless. The death toll was expected to rise. They also estimated that 250,000 residences and 30,000 commercial buildings had collapsed or were severely damaged.
- *February 20, Portugal*: The Madeira floods and mudslides were the result of an extreme weather event that affected Madeira Island in Portugal's autonomous Madeira archipelago. At least 43 people died and at least 100 have been injured. The final death toll was uncertain. Damage was confined to the south of the island. The rainfall was associated with an active cold front associated with an Atlantic low-pressure area that on February19, 2010 was over the Azores and moved northeastward. Between 6 a.m. and 11 a.m. local time, 108 mm (4 1/4 in.) of rain was recorded at Funchal weather station and 165 mm (6 1/2 in.) of rain at the weather station on Pico do Arreiro. This storm was one in a series of such storms that had affected Spain, Portugal, Morocco, and the Canary Islands with flooding rain and high winds. These storms had been bolstered by

an unusually strong temperature contrast of the sea surface across the Atlantic Ocean. Abnormally warm waters had been widespread off West Africa, whereas relatively cold surface waters had stretched between western Europe and the southeastern United States.

- *February 27, Chile*: An earthquake occurred off the coast of the Maule Region of Chile, rating a magnitude of 8.8 on the moment magnitude scale and lasting about 3 minutes. Eight cities experienced the strongest shaking—VIII (destructive) on the Mercalli intensity scale. The earthquake was felt in the capital Santiago at Mercalli intensity scale VII (very strong). Tsunami warnings were issued in 53 countries, and a tsunami was recorded, with an amplitude of up to 2.6 m (8 ft 6 in.) high, at Valparaíso. Government confirmed the deaths of at least 723 people, although later reports reduced the estimated death toll to 279. Many more had been reported missing. Seismologists estimated that the earthquake was so powerful that it might have shortened the length of the day by 1.26 microseconds and moved the Earth's rotation axis by 8 cm or 2.7 milliarcseconds.
- *and counting ...*

Intensity and frequency of hazardous events is on the rise. Larger numbers of hazardous events together with general growth of population is causing an increase in (a) the number of disasters, (b) their devastating impact on loss of life, and (c) material damage. In this ending section, I would like to bring to the attention of the reader two issues that are already shaping the world as we know it—climate change and population growth and migration. The book will end with the further strengthening of the message of the need for a systems approach in managing the disasters of the future—what new knowledge is required and how to put this knowledge into practice.

8.1 ISSUES IN FUTURE DISASTER MANAGEMENT

I defined a disaster as a result of complex dynamics involving interaction of innumerable systems parts within three major systems: the physical environment; the social and demographic characteristics of the communities that experience them; and the buildings, roads, bridges, and other components of the constructed environment (see more details in Chapter 1). Change in each element of the three systems of a disaster affects the way how systems respond. The systems' response is a measurable description of the disaster—a number of displaced people, total material damage, recovery cost, area flooded, and similar.

Future disasters, in my opinion, are to be dominated by the change in two main system components: (i) climate and (ii) population growth and migrations.

8.1.1 Climate Change

Many of the impacts of climate variations and climate change on society, the environment, and ecosystems are caused by (a) changes in the frequency or intensity of extreme weather and climate events, and (b) sea-level rise. The IPCC Fourth

Assessment Report (IPCC, 2007) concluded that many changes in extremes had been observed since the 1970s as part of the warming of the climate system. These included more frequent hot days, hot nights, and heat waves; fewer cold days, cold nights and frosts; more frequent heavy precipitation events; more intense and longer droughts over wider areas; an increase in intense tropical cyclone activity in the North Atlantic; and sea-level rise.

Temperature Extremes Recent studies have confirmed the observed trends of more hot extremes and fewer cold extremes and shown that these are consistent with the expected response to increasing greenhouse gases and anthropogenic aerosols at large spatial scales (Allison et al., 2009). At smaller scales, the effects of land-use change and variations of precipitation may be more important for changes in temperature extremes. Continued marked increases in hot extremes and decreases in cold extremes are expected in most areas across the globe because of further anthropogenic climate change.

Precipitation Extremes and Drought Recent climate research has found that rain is more intense in already-rainy areas, as atmospheric water vapor content increases (Allison et al., 2009). Recent changes have occurred faster than predicted by some climate models, emphasizing that future changes will be more severe than predicted. In addition to the increases in heavy precipitation, there have also been observed increases in drought since the 1970s. This is consistent with the decrease in mean precipitation over land in some latitude bands. The intensification of the global hydrological cycle with climate change is expected to lead to further increases in precipitation extremes, both increases in very heavy precipitation in wet areas and increases in drought in dry areas. While precise prediction cannot yet be given, current studies suggest that heavy precipitation rates may increase by 5%–10% per °C of warming, similar to the rate of increase of atmospheric water vapor.

Tropical Cyclones The IPCC Fourth Assessment found a substantial upward trend in the severity of tropical cyclones (hurricanes and typhoons) since the mid-1970s. A trend is toward longer storm duration and greater storm intensity that are strongly correlated with the rise in tropical sea surface temperatures. It concluded that a further increase in storm intensity is likely (IPCC, 2007). Several studies since the IPCC report have found more evidence for an increase in hurricane activity over the past decades. However, the scientific debate about data quality continues, especially on the question of how many tropical cyclones may have gone undetected before satellites provided a global coverage of observations. On the basis of the complete analysis of satellite data since 1980, a global increase of the number of category 4 and 5 tropical cyclones is estimated to be 30% per 1°C of temperature increase.

Other Severe Weather Events Allison et al. (2009) concluded that there are insufficient studies available to make an assessment of observed changes in small-scale severe weather events. However, recent research has shown an increased frequency of severe thunderstorms in some regions (particularly the tropics and southeastern

United States). In addition, there have been recent increases in the frequency and intensity of wildfires in many regions with Mediterranean climates (Spain, Greece, southern California, southeast Australia) and further increases are expected.

Sea-Level Rise Population density in coastal regions and islands is about three times higher than the global average. Currently, 160 million people live less than 1 m above sea level. This allows even a small sea level rise to have disastrous consequences. They may be caused by coastal erosion, increased susceptibility to storm surges and resulting flooding, groundwater contamination by salt intrusion, loss of coastal wetlands, and other issues. Sea level rise is an inevitable consequence of global warming for two main reasons: ocean water expands as it heats up, and additional water flows into the oceans from the ice that melts on land. Since 1870, global sea level has risen by about 20 cm (IPCC, 2007). The average rate of rise for 1993–2008 as measured from satellite is 3.4 mm/year, while the IPCC projected a best estimate of 1.9 mm/year for the same period. Actual rise has thus been 80% faster than projected by models (Rahmstorf et al., 2007). Future sea level rise is highly uncertain. The main reason for the uncertainty is in the response of the big ice sheets of Greenland and Antarctica. Sea level will continue to rise for many centuries after global temperature is stabilized, since it takes that much time for the oceans and ice sheets to fully respond to a warmer climate. The future estimates highlight the fact that unchecked global warming is likely to raise sea level by several meters in coming centuries, leading to the loss of many major coastal cities and entire island states.

Summary In summary, the climate change, as the most significant challenge for humanity, will play a major role in future disaster management:

- Increases in hot extremes and decreases in cold extremes are continuing and are expected to amplify further.
- Climate change is leading to further increases in precipitation extremes, both increases in heavy precipitation and increases in drought.
- New analyses of observational data show that the intensity of tropical cyclones has increased in the past three decades.
- Satellite measurements show sea level is rising at 3.4 mm/year since these records began in 1993.

8.1.2 Population Growth and Migrations

Population Division of the Department of Economic and Social Affairs of the United Nations Secretariat (2009) reports that the world population at 6.8 billion in 2009, is projected to reach 7 billion in late 2011 and 9 billion in 2050. Most of the additional 2.3 billion people expected by 2050 will be concentrated in developing countries, whose population is projected to rise from 5.6 billion in 2009 to 7.9 billion in 2050. In contrast, the population of the more developed regions is expected to change minimally, from 1.23 billion in 2009 to 1.28 billion in 2050.

The population of the less developed regions is still young, with children and young people under 24 accounting for 49%. The most important factor in controlling the population growth is the access to family planning, particularly in the least developed countries. If fertility were to remain constant at the levels of 2005–2010, the population of the less developed regions would increase to 9.8 billion in 2050, instead of the 7.9 billion projected by assuming that fertility declines. Without further reductions in fertility, the world population could increase by nearly twice as much as currently expected.

Negative impacts of disasters are directly related to the population trends and changes. Larger number of people is directly translated in larger exposure and potentially higher risk from hazards. Population and climate change are connected. Most of the climate change is attributed to anthropogenic impacts. One of many active feedbacks includes: population growth → increase in food production → land use change → emissions of greenhouse gasses → global warming.

Disasters, climate change, and population interactions will have one more dimension of complexity in the future—climate migrations. One of the observations of the Intergovernmental Panel on Climate Change (IPCC, 2007) is that the greatest single impact of climate change could be on human migration—with millions of people displaced by shoreline erosion, coastal flooding, agricultural disruption, etc. Since then various analysts have tried to put numbers on future flows of climate migrants (sometimes called "climate refugees"). The most widely repeated prediction is 200 million by 2050 (IOM, 2008). But repetition does not make the number any more accurate. The scientific argument for climate change is increasingly confident. The consequences of climate change for human population distribution are unclear and unpredictable. The available science translates into a simple fact—the *carrying capacity* of large parts of the world is compromised by climate change.

The disasters that will move people have two distinct drivers: (i) *climate processes* such as sea-level rise, salinization of agricultural land, desertification, and growing water scarcity; and (ii) *climate hazard events* such as flooding, storms, and glacial lake outburst floods.

It is necessary to note that nonclimate drivers, such as government policy, general population growth and community-level resilience to natural disaster, are also important. All contribute to the degree of vulnerability people experience. The problem is one of time (the speed of change) and scale (the number of people it will affect).

8.2 A SYSTEMS VIEW

Successful management of disasters requires integration of systems view into the considerations of the daily activities of everyone who has an influence on future losses. This, in turn, represents a major shift in cultural approach to disasters and their management. This book's main goal was to introduce an array of tools to those who are ready to accept that shift. Some of the tools are simple, some are more difficult, but they all provide strong support for the implementation of a systems view of disaster management.

I am aware that time is necessary in order for these tools to be accepted and used on a regular basis. My expectation is that the use of the tools will lead to a higher knowledge and understanding of how to deal with complex issues involved in management of disasters. Some may be critical of taking this path to developing a different view of the disasters management. However, my point is that implementing the tools presented in the book will train the mind of those who are using them. That training will lead in time to the switch in thinking—move from linear to systems thinking. Development of systems thinking will form a new culture of sustainable integrated disaster management.

This book is just the beginning of the lengthy path. One of the significant issues is not dealt with within this text—the issue of uncertainty. All the tools presented are for deterministic problems. Their extension to problems under uncertainty is available. I hope that the continuation of this work will present the systems tools for disaster management under uncertainty.

To bridge the gap between theory and practice, I suggest that the material from the book be used with various forms of education and training—short courses, workshops, formal training, development of case studies, and similar. In that way, the tools will be easier to understand and master. However, the systems view needed for sustainable integrated disaster management cannot be introduced through the study of just any one of the specialized fields available today. The next generation of disaster managers must be educated across the disciplines that have knowledge related to the management of disasters under the umbrella of the systems view.

The final words are for my two children and my only grandchild—I hope that some of this work will help you weather the storms of the future (Hansen, 2009).

REFERENCES

Allison, I., N.L. Bindoff, R.A. Bindschadler, P.M. Cox, N. de Noblet, M.H. England, J.E. Francis, N. Gruber, A.M. Haywood, D.J. Karoly, G. Kaser, C. Le Quéré, T.M. Lenton, M.E. Mann, B.I. McNeil, A.J. Pitman, S. Rahmstorf, E. Rignot, H.J. Schellnhuber, S.H. Schneider, S.C. Sherwood, R.C.J. Somerville, K. Steffen, E.J. Steig, M. Visbeck, and A.J. Weaver (2009), Updating the World on the Latest Climate Science, *The Copenhagen Diagnosis*, The University of New South Wales Climate Change Research Centre (CCRC), Sydney, Australia, 60 pp.

Hansen, J. (2009), *Storms of My Grandchildren*, Bloomsbury, New York.

International Organization for Migration (IOM) (2008), *Migration and Climate Change*, IOM Report 37, Geneva, 64 pp.

IPCC (2007), "Summary for policymakers,"in Pachauri, R.K. and A. Reisinger (eds) *Climate Change 2007: The Physical Science Basis. Contribution of Working Group I to the Fourth Assessment Report of the Intergovernmental Panel on Climate Change (IPCC AR4)*, IPCC, Geneva, Switzerland, p. 104.

Population Division of the Department of Economic and Social Affairs of the United Nations Secretariat (2009), *World Population Prospects: The 2008 Revision, Highlights*, United Nations, New York.

Rahmstorf, S., A. Cazenave, J.A. Church, J.E. Hanen, R.F. Keeling, D.E. Parker, and R.C.J. Somerwille (2007), "Recent climate observations compared to projections." *Science*, 316: 709.

Index

Systems Approach to Management of Disasters: Methods and Applications, By Slobodan P. Simonović

PLATE 1 Many residents had to shovel their way out of their homes (courtesy of Winnipeg Free Press).

PLATE 2 Soldiers and many volunteers shoring up protection for homes in Winnipeg (courtesy of Winnipeg Free Press).

PLATE 3 Flooding in the Valley (courtesy of Winnipeg Free Press).

PLATE 4 The floodway pushed near its record flow of 65,100 cubic feet of water per second (courtesy of Winnipeg Free Press).

PLATE 5 Ste. Agathe fell prey to floodwaters within an hour when the dike failed early on April 29 (courtesy of Winnipeg Free Press).

PLATE 6 The 15-mile Brunkild Z-dike was constructed out of mud, sand, and limestone in 72 hours (courtesy of Winnipeg Free Press).

PLATE 7 Looking southward from floodway gates on April 30 (courtesy of Winnipeg Free Press).

PLATE 8 Ed Hallama's 100-year-old photographs were among the valuables damaged when the river swept into his Grande Pointe home (courtesy of Winnipeg Free Press).

PLATE 9 No man was an island at Rosenort (courtesy of Winnipeg Free Press).

Printed in the United States
By Bookmasters